→ Aral Sea (p. 82)

→ Queensland's NCC payment penalty (p. 193-94)

→ SA's challenge of NSW irrigator license (p. 195)

→ criticism of the Commission as weak (p. 201)

→ native title (p. 208-211)

→ water harvesting in Queensland (p. 229-230)

→ Cubbie Station, WAMPs, & dam moratorium in Queensland (p. 226-241)

→ naming of the Murray (p. 71)

→ storage capacity on the MDBC (p. 67)

→ angry irrigator letter (p. 141)

→ description of licenses, allocations & reliability (p. 146)

→ ~~irrigator~~ irrigator acceptance of cap (p. 149)

→ threats to farm survival under cap (p. 168-173)

→ return of water to Snowy (chapter 8)

Ticky Fullerton is a reporter with *Four Corners*, ABC TV's flagship current affairs program. In her seven years with the ABC, she has also been the national political and business reporter for *Lateline*. For two of them Ticky presented the ABC's rural program, *Landline*, from Brisbane, travelling throughout Australia to cover country issues where she developed a passion about water. In researching *Watershed*, she has travelled to every State talking to those who use, care for, manage and regulate Australia's great resource.

Prior to this, Ticky spent ten years working in corporate advisory for global investment bank, C S First Boston, both in London and Sydney. It is a rare understanding of business, political and farming interests, combined with a strong belief in conservation, that has driven Ticky to write about one of Australia's most contentious issues.

Ticky Fullerton has a BA in Law from Oxford University.

TICKY
FULLERTON

watershed

DECIDING OUR WATER FUTURE

ABC
BOOKS

Published by ABC Books for the
AUSTRALIAN BROADCASTING CORPORATION
GPO Box 9994 Sydney NSW 2001

First published October 2001

National Library of Australia
Cataloguing-in-Publication entry
Fullerton, Ticky.
    Watershed: water use in Australia: juggling the needs of
    farmers, politicians, conservationists, big business,
    ordinary people – and nature.
    Bibliography.
    Includes index.
    ISBN 0 7333 0999 2.
    1. Water use – Australia. 2. Water-supply – Australia.
    3. Water – Pollution – Australia. 4. Water conservation –
    Australia. I. Australian Broadcasting Corporation.
    II. Title.
333.9100994

Cover designed by Reno Design Group
Maps and diagrams by Ian Faulkner
Set in 10/11 pt Veljovic Medium by
Midland Typesetters, Maryborough, Victoria
Colour separations by Colorwize, Adelaide
Printed and bound in Australia by
Griffin Press, Adelaide

5 4 3 2 1

Acknowledgments

Most of all, my thanks to the many people I have met and spoken with on the *Watershed* trail, who agreed to be part of this. Without you, there would be no book. There are many more who I spoke to for background.

Next to Mum and Dad, whose Christmas holiday went to water. Dad, your hours spent reading, prodding and generally chewing the cud with me were brilliant.

Special thanks too, to Rob Carroll for being there, Peter Cameron for quiet words, Kerry Lonergan for his faith, Tanya Atkins for late night checking beyond the call, Julie Rosenberg and all my other long suffering mates.

To the publishing team at ABC Enterprises: Matthew Kelly, Stuart Neal, Mary Rennie and Jane Finemore for getting me across the line.

Last and far from least, thank you to Murray my magnificent 1983 Mazda who has been to more river catchments than any four-wheel drive in Australia.

Mum and Dad
Lux in Tenebris

Contents

Introduction

This book is for all Australians. Whether you drink strong lattes on upturned buckets in The Cross or Fitzroy, read and digest James Joyce, watch *Big Brother*, grow wheat or do all of the above, the point is, *Watershed* is not just for water buffs. The story is too important.

During some of the time I wrote this book, there were reports that a third of New South Wales was under water. With the exception of Tasmania, Melbourne and the Western Wheatbelt, the last few years have been lush in this country.

But all this is deceiving. As sure as the sun rising, El Nino's droughts will return. The truth is that we are at a watershed. The decisions made in the next few years will determine whether Australia remains the safe, beautiful, plentiful and remarkably cheap land that we all enjoy, or whether we join a downward spiral of degradation that will cost too much to fix.

This book is deliberately not sponsored by anyone. Water is far too controversial. It is also unashamedly about people and places, from the Kimberley to the Coorong, from Sydney to Windorah. The aim has been to walk a mile in the shoes of every group that should have a say: the Greens, farmers, Indigenous people, miners and businessmen. *Watershed* is these folk telling their stories, telling it like it is, in interviews I did along the trail between October 2000 and May 2001. At times it is humorous, at times sad, but always thought provoking.

I have an ornithologist cousin who spent most of his life in Canada. On hearing about this book, he said to me, 'I'll know whether you've done your job properly if I see the freckled duck in the index.' The duck made it but one of the first things to understand about Australia's water is that each river catchment, town and country, has its own story. All deserve telling but if they were, *Watershed* would never be finished.

Being researcher and writer on such an emotional topic as water is a huge responsibility in terms of the balance the book achieves.

Having done it, I can honestly say that I do not believe it is possible to journey through Australia's water story without becoming a shade greener. But as you will see, there are many people, especially farmers, with worthy grievances especially about water reform.

There will be many who disagree with *Watershed*. Good. My aim is to haul the debate up both the community and political agenda.

Watershed is designed to be read from cover to cover, but equally most chapters are quite distinct. I do encourage you to use the maps; particularly in the rural parts of *Watershed* they help make great sense of what is going on.

My first book and it's been a true baptism.

TICKY FULLERTON
Sydney
August 2001

1

The Primary Resource

*Pure water! Cool clear water. Hey! Step up schmuck, let's
get some dealin' done. H_2O, that's my go. Hey! Don't you
understand? This is water! You can't live without it!*

Mad Max III

Oasis Living

Bottled water is a great Australian fashion accessory. Thousands of mini
plastic bottles adorn the desks of bright young personal assistants in the
city and double as Valentine vases in February. These bottles say, 'I am
healthy', and Australians pay more for water than they do for petrol or
milk, much to the delight of Coca-Cola Amatil and Schweppes, who are
pumping water into new bottles as fast as they can manage.

Not all the water is local. Plane loads and shiploads of water arrive
from the apparently special hills of France as a matter of course. It is
one of the lunacies of free trade that lies somewhere on Seattle Man's
agenda.

The truth is that unless you live in Adelaide, $1.30 a litre for water
is a fashion fad. The naked emperor certainly would have carried a
bottle. Gucci has made little water bottle holders. While many other
parts of the world have bottled water for health reasons, Australia's tap
water is bloody good. We enjoy oasis living—witness the billboards of
small children laughing over bubblers as water squirts up to their
faces. Safe water. Never failing water.

Now what if this was not quite true?

When a young shop assistant in David Jones says, 'Oh, water,
yeah, they say all the next wars are going to be about water,' as if it
might be the central theme of one of the missing *Star Wars* movies,
one wonders if at least a subconscious message is slowly drifting
through the population.

The shop assistant is right, of course. The top Commonwealth think
tank, Commonwealth Knowledge Network, predicts that within 25 years,

at least 65 nations will be seriously short of water. 'Water is likely to figure as a national security issue for many countries in the 21st century. Lack of water can have profound economic and military consequences.'[1] There will be eight billion thirsty human throats on the planet by 2025.[2] 'I am convinced that resource conflicts, conflicts over water are going to be a major issue in the next century,'[3] Maurice Strong, Chair of the United Nations Earth Council, has said.

Water wars have started already: the battle over the Golan Heights between Israel and its Arab neighbours; Turkey's dam scheme on the Tigris and the Euphrates, which threatens Syria and Iraq downstream; Egypt's fears of Ethiopia and Sudan taking its Nile waters; and India and Bangladesh fighting over rights to take water from the Ganges are just a few.

During the Gulf War, Canadian businessman Dan Colson and his wife observed extraordinary behaviour from one of their neighbours. At 9.30 one morning a large delivery truck arrived and men started carrying cases of bottled water into the neighbours' house. This went on for the best part of the day. A few days later the husband knocked on the Colson door for some relief. 'It's my crazy wife,' he exclaimed, 'she ordered 500 cases of Evian and I can't move in the house. She's convinced Saddam is going to poison the water supply.' It took them four years to drink the water.

Thankfully this madness is still a long way from us. But it would be very wrong to think that Australia's water issues are nothing but a bit of surface tension. There is an old saying that if you turn off your neighbour's water you have war, and that is exactly what is happening. Australia is already at war over water. We have our own civil war all right, and like all civil wars, the casualties—people's livelihoods— are going to be heavy. It's not simply between feral hairy armpitted greenies and red-neck farmers, although that is one front. It is much more inclusive. Ask yourself why people like Kerry Packer, Peter Garrett, swimmer Tammy van Wisse, Dick Pratt, Jeff Kennett and Alan Jones have got strong views on the subject. Anyone who has a stake in water today, be it a businessman, politician, water skier, cotton grower or dam buster, is having to face up to the fact that there's only so much to go around.

Actually, we've stuffed it up. Over the last two centuries, we have completely misjudged what the land could handle and have a salinity problem beyond comprehension. We can blame the policy makers of the last century, as year after year, water authorities dished out more water than there was to go around, mainly through ignorance but also in response to greed and cronyism. And as we'll see throughout *Watershed*, every year decisions on water get harder, not easier. According to the Commonwealth's Water Resources Assessment 2000, water use has shot up from 14,600 gigalitres in 1983/84 to 24,000 gigalitres in 1996/97.[4] Perhaps we should not be surprised that the policy makers of the new century are leading from behind.

Of course water is no good unless it is safe, a fact that both Sydney and Adelaide woke up to in 1998, when they learned what cryptosporidium meant. It was enough to keep Sydneysiders away from tap water for six weeks. Ask a schoolboy in Adelaide about water and he'll tell you how bad people have poisoned the Murray River where he gets his drinking water. If we can't fix the Murray up, word is that by 2020, on two days in five the water for Adelaide, a city of over a million people, will be undrinkable.

At the top levels, farmer and conservation bodies have got together and said the country needs $65 billion over ten years to fix up Australia's water problems, because 'she won't be right, mate'. Sixty-five billion dollars. That's about $325 a year for every man, woman and child, and we need to plant 40 billion trees. But there is much, much more to the water story than salinity and the Murray–Darling Basin.

Water: yawn. Salinity: double yawn. How many of us have turned the pages of those weekend magazine specials 'Sold down the river' or 'A river used to run through it'? As if a mere mortal could do anything about it. It's the same at water conferences, which have suddenly become all the vogue. If there was an Olympic swimming pool worth of water for every time a speech or paper started with the words 'Australia, our country, is the driest inhabited continent on earth', we wouldn't have a water problem. It's true that we are the only country with a dry boat race, the Henley on Todd in Alice Springs. The race was cancelled in 1997 because there was water in the Todd River, no good at all if your running feet are sticking through the boat's bottom. But ask the people in Tully about the driest continent and prepare for laughs. The Queensland town gets about 11 metres of rain a year.

We are a land of extremes, where most of the rain falling on the tropical north creates huge flooding rivers which still run free to the sea. Only about 6 per cent of our rain lands in the Murray–Darling Basin on which we rely to produce 40 per cent of our agriculture and 90 per cent of our irrigated agriculture. These are extraordinary figures if thought about for a moment. Aside from the southeast and a couple of pockets in the west, we are living in Dorothea McKellar's land of drought and flooding rain, with the highest variability of rainfall on earth. And most of the lakes and rivers of inland Australia exist only on paper.

Yet for all the public and government apathy over Australia's water, there are amazing forces at work trying to push a new vision for Australia. There is a call for a whole new era of 'green engineering', which may end up being more powerful than anything in software or genetic engineering. As it stands now, science can't answer the problems we have. In the meantime, to save our stressed rivers, reformers, scientists, environmentalists and many politicians are busy putting into place a regime which is pushing the price of water up, up, up and clawing back water from users. But higher costs won't solve Australia's looming environmental crisis on their own. We're going to have to think of whole new ways of producing the fabulous irrigated crops that bring in $12 billion to $16 billion a year for Australia and help to keep this economy going. Such new ways mean drastic changes in farming and rural society, with much pain down the track.

Such is the fear in the southern States about sick rivers that it is now politically incorrect to talk of dams and dam building. Yet the truth is that the weirs, dams and channels that regulate our rivers today are crumbling and leaking, many of them built by the State in the 1920s and 1930s. They desperately need rebuilding in ways that are more 'Green' friendly. But no one wants to pay for it, least of all government.

Behind the scenes, some of Australia's most powerful people are thinking very differently and much bigger than $65 billion, with a plan to fight the droughts and floods (which if you believe the greenhouse warnings are only going to get more extreme). In a land smattered with open earth channels and uncovered storages where evaporation can be 3 metres a year there are calls to re-plumb Australia in a project that would dwarf the Snowy Scheme: water for all, regulated dams and pipes that would bring an end to the misery of extreme weather, capturing flood water and delivering it to dry country. Some of these people also want to make something of the massive water resource in the Australian tropics. Dick Pratt is calling for $500 billion to be spent over 50 years.

There are rows about whether our land can cope with any more milking of its resources, whether we are blindly setting off to exploit the remaining pristine parts of the country, having learned nothing from the past. But it is not just Pauline Hanson warning that if we do nothing, we face others from the increasingly desperate north, short of land and water, moving in to make the most of these untapped vital resources.

It's a potent brew, with yet another ingredient: personal investment. Water is becoming a very big game in this country. Bruce Gunning runs his own real estate agency in the New South Wales town of Moree. 'A water licence, you could get off the government in 1979 for $24. Now I can sell you that same one before the sun goes down for $1.4 million.' Ten years ago, it was worth $400,000. Not bad for a product that comes without the dot.com risk of evaporating overnight. And it's a market that could take many years to level out.

Magnates are appearing, although their names are unfamiliar to the average Aussie. Already, water is worth more than the land it runs through. Should anyone be surprised that Kerry Packer has spent more than a nanosecond thinking about water? It has brought money, but it will also bring power. If I had money, I'd buy water.

Choices

The unforgivable sin of Australian society is to be unaware of the hard choices we have to make about our water future. The most important knowledge we need is about the environment. We are running up a national debt in salinity, destroying the gene pools of precious plants and animals and messing up water quality. The way we are going will compromise both the lifestyle of our children and grandchildren and, in the longer term, the health of humankind. If Australians are to relegate the country's wetlands to the Discovery Channel, at least we should do it knowingly.

We are the country that bore the first Green movement, and are the home of the Green Games, yet this is not reflected in practice. As you'll see in *Watershed*, some of the brightest and most passionate Australians barrack on the Green side, battling to make people understand why this country is different, why its frail, unique make-up means that it cannot take the abuse that Europe and the United States endure. In 1947, *Australian Surveyor* remarked that 'even the most reliable streams in Australia are, by world standards, very unreliable'.

There can be few tougher lives than those of the pioneer and taming the bush was a magnificent achievement. Sadly there is a down side: as land use has intensified, the arm wrestle between man and nature in this country has been tipping nature towards the table. 'Eighteen million people have done as much damage as 200 million in the US in a shorter period of time,' according to Monash University ecologist, Professor Sam Lake.

One of the problems with oasis living is getting the message through that water is worth worrying about at all. For most city Australians, tap water has never been recycled. Compare that with water drunk from a London tap which has been through an estimated seven other people first! Recycled effluent? Tell 'im he's dreaming.

Yet the bigger challenge comes not from city water. It is to convince the chardonnay set and the *Groundforce* watchers that the dramas of water in rural Australia have any relevance to their busy lives. This is notwithstanding the fact that the Hunter, Barossa and Riverlands vineyards perform annual miracles for them changing water into wine. Why *should* city people care? The longer answer runs through every chapter of *Watershed*, but the short one is that fresh foods at Woolworths wouldn't exist without water for the inland. In fact almost everything we consume and much of what we wear is watered in rural Australia.

In boffin terms, it's all to do with the 'eco footprint'. Each of us consumes a certain amount of food, produces a certain amount of garbage, and consumes a certain amount of energy. Cities are the end point of a linear process that begins in the countryside. Turn on a light and we burn coal; eat and we're consuming a portion of the land on which the food was grown. The chart opposite gives an idea of how much water food needs to be produced. One slice of bread demands 50 litres. A cup of plain yoghurt uses 400 litres. And a serving of steak uses 5,600 litres. For each of us, this translates to 'our ecological footprint', a term first coined by Bill Rees, Professor at the University of British Columbia.

To sustain a person in an Australian city, the eco footprint for every person is 4.5 hectares of *arable* land. And your fair share of arable earth, if everyone was to have an equal portion, is 1.5 hectares. Basically we need another two earths.

Building the bridge between city and country is fundamental. We have lost connection and lost empathy with rural Australia and supermarkets have a lot to answer for. We now shop in an environment where meat doesn't smell and shoppers buy 99 per cent fat-free mince, cook it in olive oil and then wonder why Australian beef has gone off. There is a big and unclaimed corporate responsibility here for placing a wedge between the city and rural Australia.

Most of our water is in the hands of farmers, and today's policies are making very big changes there in the interests of saving our rivers. Dairy farmer Max Fehring will tell you, 'I'm proud of what I do as an irrigator. I've been able to give Melbourne and Victoria total security in food.' When the talk turns to clawing back water use, he is blunt: 'If they want to change it, by taking away half my water, they need to take some responsibility for it.' The cities—and city people—must get involved. Indeed, our cities are totally dependent on rivers often

The Water our Food Uses	
	Litres
Lettuce (1 cup)	14
White sugar (1tb)	32
White bread (1 slice)	50
Whole white bread (1 slice)	32
Tomatoes (120g)	36
Fresh broccoli (80g)	50
Brown rice (30g)	73
White rice (30g)	114
Pasta (60g)	164
Egg (1)	287
Plain yoghurt (1 cup)	400
Chicken (230g)	1,500
Hamburger (115g)	2,800
Steak (230g)	5,600

Source: Max Fehring

kilometres away. Massive diversions shift water from the Shoalhaven and Thomson catchments to Sydney and Melbourne, and water from the Murray River to Adelaide.

The parts of the country where tensions in water are greatest are foreign to most Australians. You enter a world where the talk is of catchments and rivers, not suburbs and towns. People live up on the Lachlan, the Clarence, or come from the Namoi or Goulburn Murray Valleys. And in the country what matters more than anything is water.

The Murray–Darling river system has a big hold on *Watershed*. Look at the map on page 11. The river runs through four States. It's an amazing river. Quite apart from the food it produces, it supplies drinking water to three million people, including those in Adelaide and the nation's capital.

The Murray–Darling story is that of the pioneer settler, the soldier settler, the engineer and the cultural melting pot. But it is also a tale of how we have managed to screw up our best river system and of our desperate attempts to right the wrongs of the past. It involves ignorance, opportunism, greed and betrayal—of many different people.

Charged with sorting out the mess on our behalf are the politicians of no less than four States, a Territory and the Federal Government. And there's no love lost between them. Today, 1.7 million tonnes of salt flow down the Murray River and out its mouth. That's 1.7 million one-tonne trucks of salt. On a recent visit from Canada, David Suzuki, the Green soothsayer, exclaimed, 'For heaven's sake, you're primarily-desert. Water is a great gift. Something like the Murray–Darling water basin, I mean that's a national treasure, and from what I've heard since I've been in Australia this time, you're really trashing that system.'

Every river in Australia has its own great yarns, whether it be the Snowy River that finally knocked Jeff Kennett off the perch, or the Cooper Creek that still runs free and will continue to do so as long as the colourful blokes up in the Channel Country draw breath. From water magnates who store water in dams the size of Sydney Harbour, to the privatising of rain, or a crisis that scared four million Australians away from their taps for weeks, but gave no one as much as a runny tummy; or the race to beat water cut backs by trapping water before it gets to the river or drawing it from underground.

Of all the primary resources of this sunburnt country, water comes first. 'There are references to water being compared to love, as a source of trouble' noted the same *Australian Surveyor* just over 50 years ago. In 50 years' time, we may or may not have a prime minister who is half Greek, a quarter Irish and a quarter Indonesian and we may have a population of 30 million or 50 million, but we'll all be talking about water.

There is a growing insecurity about water. And time and again throughout *Watershed,* people are shown to be cottoning on to its value. What is missing at almost every stage is the leadership to make decisions that will take Australia forward.

Lingo

Like every industry, water has its own jargon and newspeak. There has been a determined effort in this book to get rid of most of it. You shall not see the phrase 'ecologically sustainable development' (or ESD as the hip protagonists refer to it) anywhere else in this book. For one thing, many environmentalists, for example Mary E. White, see it as 'an oxymoron when the economy is the yardstick and the paramount driving force'.[5] For another, it is a huge turn-off for normal people. Economics and ecology are both derived from the same Greek word *oikos*, 'home', and yet they are strange bedfellows. We are now trying to marry them up by putting a value on the ecosystem, an effort

THE MURRAY - DARLING BASIN

Source: Murray–Darling Basin Commission

which CSIRO Land and Water chief Graham Harris describes as a cynical response that ignores the values which we know society holds dear.

Another gem is the word used for the harm done to our rivers from towns, farming and industry. In the water industry it is referred to as 'externalities', an appalling euphemism much like 'collateral damage' used to detach Western viewers during the Gulf War from the smart but devastating aerial attacks which on the ground were far from surgical, far from clean. Unfortunately, it has been hard to excise this one from *Watershed*.

One cannot go far in Waterworld without coming across 'Syd-harbs', a favourite measurement of journalists and even some experts in the industry. A body of water is so many 'Sydney Harbours' full. The fact that most of these people do not actually know how much water there is in Sydney Harbour doesn't seem to put them off. The point is it's a lot of water. Sydney Harbour, with an area of 55 square kilometres, is actually just over 500,000 megalitres or about 250,000 Olympic swimming pools. (A megalitre is one million litres or a thousand cubic metres and is equivalent to about half the water in an Olympic swimming pool). Everyone talks megalitres in the water business or, in this book, 'megs'.

To save the fingers of the author and eye of the reader, 'environmentalists' are generally referred to as 'Greens' and 'environmental flow', the water we have come to realise is so important for the health of rivers and the fish, ducks and other friendlies that live in and on it, as 'duck water' or 'fish water'.

Finally, the matter of 'free water' should be cleared up. To the Greens, water is still 'free' in this country and should not be. It is true that where most of the water is used, in rural Australia, users are not actually paying for the water itself. They pay in other ways. First a licence is needed, entitling the holder to 972 megs (500 Olympic pools). We've seen earlier that this is a one-off cost which could range from $400,000 to $1.4 million. Secondly, users are charged for the *delivery* of the water. In town, water pricing reflects delivery charges, but there is also a volumetric element which came in with recent water reforms. The more you use, the more it costs. Of course, buying and selling water in the market is a different thing again, which we will come to in chapter 7.

Overall, water costs much less in the country (say $20 to $60 per meg) where users are charged bulk prices compared with the city where it costs around 70 to 90 cents a kilolitre, or $700 to $900 a meg.

At times, bureaucratic systems have done their best to make the subject of water unintelligible, a great irritation given that users are mostly ordinary folk. As a former head of South Australia's Department of Environment and Natural Resources, John Scanlon told State MPs recently, the push to reform the Murray–Darling was 'really complicated and, in my view, those who have expertise I could count on one hand'.[6] *Watershed* is the DIY guide to Australian water, and yes, the issues are as complex and fascinating as water itself. We should expect nothing less of a substance that the Karajarri people of the Kimberley call 'living water', a powerful spirit that can be solid, liquid and gas, dissolve minerals, make the desert bloom and fish breathe, supply cities, generate electricity and store energy, and that makes up over 95 per cent of our bodies.

2

Cryptic Water

'Truth is never pure and rarely simple.'
Oscar Wilde — The Importance of Being Earnest

Crisis, what crisis?
It seems just a distant memory. That long six weeks between 29 July and 19 September 1998, when the comfortable life of Sydneysiders was thrown into turmoil. The whole of the city was held to ransom, residents warned to boil water for weeks on end or face swallowing a potentially lethal cocktail of bugs, cryptosporidium and giardia.

Finger pointing between the parties involved, hyped by a media frenzy, took the crisis to levels of hysteria as the chase began to find out who was responsible for putting Australia's raciest city in such distress. First the outbreak was blamed on two dead foxes, or possibly dogs, which had fallen into the system, but the dead are not accountable. Media focus soon turned to the water treatment plant. The owners of the plant cried foul and claimed the source was really bat poo in the distribution pipes. Then the rain was blamed for causing human sewage to spill over into the water, and the list of culprits went on.

What must seem very odd for Sydneysiders is that for all the fuss no one became ill. Hospitals reported no more than the usual admissions for the runs and, anecdotally, no one seemed to notice their dog or cat (clearly ignoring the boil water instructions) suddenly suffering from the squitters.

Nevertheless, the crypto crisis cost the taxpayer $75 million, according to the State Auditor General's report of December 1999, and claimed the scalps of both the chairman of Sydney Water (nipping in the bud a promising political career) and his managing director, who stood down after the inquiry. Coca-Cola Amatil, Schweppes and Pepsi did much better out of the panic. Bottled water sales went through the roof, the glue on the Mount Franklin labels hardly dry before leaving the factory. Such was the demand that bottled water had to be trucked in from Melbourne, Brisbane and Adelaide, sales jumping a whopping 30 per cent for the year.[1]

There are many reasons why this extraordinary story is worthy of mention. The whole saga showed how desperately vulnerable urban populations can be, if anything actually *does* go wrong with the water supply. The events may indeed have led to a phony crisis, quite overblown. Yet in one sense Sydney is very lucky to have experienced it, because the crisis opened the book on water management to find extraordinary deficiencies and weaknesses in the structure of our industry. It is the ill wind that has blown Sydney a great deal of good.

There is now an urgent choice to be made: we can either not give a toss about the quality of our rivers and dam water and spend millions cleaning up dirty water to drink, or we can take the much tougher political route and stop some of the rot in our big river catchments at a time of growing urban sprawl and intensive agricultural development. The tale of Sydney's crypto crisis also raises serious questions about whether, and how, our water systems should be privatised.

Giardia and cryptosporidium are both very unpleasant bugs. The less dangerous of the two, giardia, has a shorter lifespan and can be killed by chlorine. Crypto at its worst is much harder to eliminate. Symptoms include the runs, stomach cramps, vomiting and fever, and can last several days. That's for healthy people. For those with low immunity, however, such as AIDS sufferers or patients undergoing chemo and operations, crypto can be fatal, turning into something closer to cholera.

While crypto exists in water, its favourite place to reproduce is in the guts of animals, including humans. The bugs produce eggs (or oocytes) which are passed through the faeces and into water systems. It might come as no surprise to hear that one of the most common sources of crypto infection is the local swimming pool.

David Casemore from the British Public Health Laboratory paints an almost too vivid picture. Babies' bottoms, it seems, behave like bacterial machine guns. 'When the first episode of diarrhoea occurs— I'm sorry if this is just before you're going to have your lunch—you'll get a little squirt of material which is totally fluid and doesn't even look like faeces, and that disperses very rapidly in the pool. And you can have 100 millilitres containing 10^5, 10^6 oocytes per millilitre dispersed in the swimming pool very rapidly.'[2]

As a water-borne disease, crypto is a slippery little sucker. Until 1984, it had not even been detected in water, and while there are many varieties of the bug, only *Cryptosporidium parvum* can kill you. Even then, some strains of parvum are benign. Not only is identification of crypto in the lab very difficult, it is also pretty hard to tell whether the bugs are dead or alive, fairly important in terms of the damage they do.

If this sounds a bit academic, crypto has certainly done enough practical damage to scare authorities around the world. The worst incident was in the United States in Milwaukee in the early 1990s when 403,000 people got sick, nearly 8,000 of whom were hospitalised and 100 of those, who were HIV positive, died.[3]

The drama unfolds

The way the Sydney crypto crisis unfolded is all the more intriguing if the responsibilities of the city's water company, Sydney Water, are explained. Sydney Water was corporatised in 1995, a process which meant it was still government-owned, but was restructured to run at a profit. So Sydney Water's responsibilities back then were to supply safe drinking water, look after the environment and make profits.

For his part, the NSW Minister for Health had emergency powers to protect the public from unsafe drinking water, but he relied heavily on Sydney Water for information. The fallout between Sydney Water and the Health Department was one of the most revealing things about the crisis.

The first time crypto and giardia were noticed in the system was on Tuesday 21 July, when low readings from the treatment plants at both Prospect and Potts Hill were taken from routine samples. The levels were not high enough to raise any alarm bells and by Wednesday the situation was all clear. It didn't last. On Thursday, there was contamination near Sydney Hospital, but the view was taken that the bugs were likely to be a local problem relating to cross contamination within the grounds.

The 'local' problem persisted into the Friday, and the Health Minister, Andrew Refshauge, was advised of the positive readings. Sydney Water didn't feel the need to advise Craig Knowles, Minister for Urban Planning, responsible for Sydney Water's operating licence.

It was on Saturday 25 July that the first high readings were gathered, still in the eastern CBD. Sunday's results were again high in this area, yet still nothing was done about a public health response. The following Monday, the monthly meeting between Sydney Water's managing director, Chris Pollett, chairman David Hill, and Minister Knowles, Chris Pollett advised neither the chairman nor the minister about the bug readings.

It was not until Monday afternoon that it was agreed to release a 'boil water alert' just for the eastern CBD, with the issue still being treated as a localised incident. Minister Knowles and David Hill had finally been informed, the minister later saying he was told the problem could have been the result of earthworks on the Eastern Distributor then under construction.

A row then developed over just what should go into the boil water alert, the Health Department suspicious that Sydney Water was playing down the seriousness of the situation.

On Wednesday 29 July, at around 5.30 p.m., Sydney's water problems took on a new level of seriousness. Sydney Water received word that bugs had been found in the city's main treatment plant at Prospect, the plant which filters water for 85 per cent of Sydney. Sediment in Prospect's number one tank, taken off line for routine maintenance the week before, was found to be contaminated. The implication was that the number two tank, which was keeping the system running, was almost certainly contaminated as well.

By 7.30 p.m., there were also contaminated results from the Potts Hill reservoir, in Sydney's Inner West, servicing a much smaller part of Sydney. Amazingly, it was decided only to put the Potts Hill supply on a boil water alert. But when high readings turned up to the north of the city at Palm Beach the following day, the authorities finally gave in and a Sydney-wide alert was declared. Minister Craig Knowles took charge of keeping the public informed and water was deemed safe by 4 August.

That was the first event. While there were a couple of smaller readings downstream of the Prospect filtration plant on 13 and 14 August, the next big contamination (the second event) didn't arrive until 24 August, causing another frenetic bout of water boiling and bottle buying. Finally, just when the public thought it was safe to go back to Sydney water, a third scare occurred on 5 September, which ran another two weeks.

The big question

So what went wrong? As the media dug deeper, it seemed more and more likely they would never get to the bottom of the crisis. Was it the privately owned plant, faults at the research laboratory, a dirty water catchment—or was it all a lot of fuss over nothing?

Phillip Street is Sydney's shark alley, home to most of the top barristers in the State, including the man given the task of answering the big question, Peter McClellan, QC, now a Supreme Court judge. He has been involved in legal water issues for 25 years but little could have prepared him for this case of Murphy's Law. For, indeed, what could go wrong duly did. With a small team under immense pressure, he produced his five weighty reports for an anxious and impatient State Government. He met with 130 people, and dealt with 200 submissions. In this day and age, the results are all on the Net, and highly recommended reading, even if voluminous, at www.premiers.nsw.gov.au.

The first report was commissioned on the run. Peter McClellan had ten days to come up with some answers on the cause of contamination. What he did deduce was that there didn't seem to be any big outbreak of disease, and that the crypto and giardia detected might already have been dead or decaying. He still believes this today. In fact, one of the reasons the Milwaukee incident killed so many could have been the short time between the release of bugs from an abattoir into the water system and people drinking that water. Nevertheless, Peter McClellan's inquiry supported the decision to boil water.

Covered behinds

There turned out to be not just one, but many, Achilles heels in the water system. But what is inexcusable is that the people handling the city's biggest water crisis were out of their depth.

What emerged from Peter McClennan's reports was a picture of a managing director unable to stand up to his chairman's pressure and a chair more concerned about Sydney Water's corporate performance than Sydney water drinkers. It all boiled down to the handling of the crisis on the evening of Wednesday 29 July.

Remember by 5.30 p.m. the Prospect treatment plant that supplied 85 per cent of Sydney was found to be contaminated. Amazingly, Managing Director Chris Pollett was not aware of these developments and left the office at 6.35 p.m. that evening. Author and former political adviser Christopher Sheil, in his book on the crypto crisis, wrote of Chris Pollett's action: 'What frees up a man paid hundreds of thousands of public dollars to go home for the evening without bothering to inquire about the status of a potentially deadly infection in the product for which he has official responsibility; a product which is essential for the daily lives of some four million people; a product which he knew had been under threat for over a week?'[4]

By 7.30 p.m., the contaminated results from the Potts Hill reservoir had come in and during the early evening Sydney Water staff set about trying to work out what the readings meant. According to the inquiry's report, there was no doubt that the evidence now available pointed to the Prospect plant as one of the sources of contamination.

It followed that the whole system could be contaminated.[5] At 8.45 p.m., Managing Director Chris Pollett was called back to the office and at 9.31 p.m., he rang Chairman David Hill, who came in shortly thereafter. Before David Hill arrived, his managing director was preparing a boil water alert for the whole of the Prospect system. By the time the press release got out, it was just for Sydney's Eastern Suburbs. In Peter McClellan's view, Hill's role in events is clear.

> There was significant evidence that the Prospect plant may have been the cause of the problem and there was no evidence or suggestion that any other cause was more likely. Nevertheless, a decision was taken to limit the area to the Potts Hill system. This decision was significantly influenced by Hill. In my opinion, the decision did not reflect appropriate concern for public health.[6]

Peter McClellan's inquiry includes interviews with the people involved in the Wednesday night decision. Very little interpretation is provided in the report because very little is needed. The interviews were revealing enough. And entertaining enough to play on Richard Glover's ABC radio program when the report was released. In his office, two years later, Peter looked amused. 'I've never spoken to Glover, but it did make me laugh as I was driving along in the car.' When David Hill arrived shortly after 10 p.m., he found a room of ten people on the 23rd floor in a high state of excitement. 'Not a situation that I would say, in management terms, is under control,' according to the chairman, and Peter McClellan had no doubt his description was correct.[7]

In a tense atmosphere, with those assembled about to make the difficult choice between a Sydney-wide alert and a much narrower one, David Hill then allegedly issued some very challenging statements, including something to the effect: 'I hope you know what you blokes are doing. Do you realise that what you're doing here will affect the organisation for the next ten years, and probably longer than that?'[8] When interviewed by Peter McClellan later, the chairman agreed he had said something like that. 'I said, "This will do irreparable damage to the company for a number of years, if it ever recovers." But that was, I think, prophetic, wasn't it?'[9]

Chris Pollett's recollection of David Hill when he entered the room was also telling. According to Pollett, Hill said 'very clearly, very strongly' to him, 'We need to consider this carefully—we don't want to cause undue harm.' Once again, the chairman's memory of events is very similar. 'Bearing in mind the bedlam I had observed on the 23rd floor,' he said, he remembered going on to say to Chris Pollett,

'We should calm everybody down, tell them to have a cup of tea and send most of them home.' A piece of advice, Peter McClellan remarked, that his managing director wisely did not take.[10]

Yet, in contrast to his determined chairman, Chris Pollett under questioning appeared to be less clear. 'My recollection, my recollection is that we were still talking about, I mean it was early days. I means it's not very long. I mean . . .', he stated later in the inquiry witness stand.

Peter McClellan's interview with David Hill left him in no doubt that Hill had reservations about Chris Pollett's ability to deal with a crisis.[11] It can only be speculated as to why David Hill was worried and, in particular, whether his concerns were motivated more by the commercial damage that a Sydney-wide alert would do to the company in the Olympic city than by a rudderless management in time of crisis.

That night, when the manager of the water distribution network, clearly not happy to discount the Prospect risk, pushed for an area-wide alert to be issued, he claims David Hill whitewashed his concerns as speculation. 'I don't want to know about your theory. I want to work on actual data you've got, where there is a problem.'[12]

In the chairman's view, it was not responsible to force people to boil water unnecessarily. Yet when it came to the crunch, he denied that he took the final decision to limit the alert. 'I didn't then and I don't now understand where all the pipes run. I couldn't make that calculation and I couldn't even make a contribution to the calculation, so I accepted their advice.' Asked whether he thought he influenced the situation, Hill replied, 'I don't think in any way it would have influenced them in any improper way, no.' He also had no recollection of anybody mentioning Prospect.[13]

When Chris Pollett was later grilled by Peter McClellan on the influence of his chairman, he claimed David Hill had warned, 'The precautionary notice should only cover affected areas which can be supported by facts and data. To go wider would be reckless and cause unnecessary alarm.'

The inquiry's conclusion? 'Although Pollett may have made the actual decision, Hill defined its parameters. There would be few managing directors who would make a decision which the chair described as reckless.'[14]

The drama continues. Enter the Health Department, which was not told of the Potts Hill or Prospect bug results until 9.30 p.m. on that Wednesday night. The Department was in an invidious position. Under a rather weak memorandum of understanding, which was not legally enforceable, Sydney Water was committed to supply 'safe' drinking water and to notify the Health Department the moment a potential public health hazard emerged. It meant that the Health Department was dependent on Sydney Water to define the health risk, and national guidelines did not even specify levels for crypto and giardia. Unfortunately, the Health Department did not have any clear authority to issue public health alerts on drinking water. New laws have thankfully now changed this situation.[15]

That Wednesday evening, the Director General of Health, Mike Reid, had rung Chris Pollett for an update. The inquiry found that Mike Reid was entitled to believe from what Pollett had said that the alert would be for the whole Prospect system. The Health Department duly issued its own Sydney-wide alert. Imagine David Hill's fury when he found out. This is his own version of what he said to the department's media officer, Shari Armistead: 'I said on the telephone, "This is irresponsible. It's unauthorised and inaccurate." Armistead said, "You people should have put the release out earlier. That's why we put it out." I said, "You're in enough shit already. Don't argue. Just retract the bloody thing and get Mike Reid to ring me." Then I hung up.'[16]

Sydney Water then contacted the media and tried to kill the Health Department release and replace it with its 'watered down' version. This release was restricted in area, did not mention crypto at all, removed any reference to 'severe diarrhoea' and caution about boiling ice which may have been frozen during the big scare. Inevitable confusion followed. When Mike Reid later rang Sydney Water, David Hill said to him, 'The Department has behaved with the ultimate irresponsibility. Your people, without authority, have released information on a serious issue that is inaccurate.'[17]

One suspects you have to read between the lines of the inquiry report for what Reid said next. Peter McClellan leaves it as, 'I understand that Reid was not aware that he was speaking to the chairman of Sydney Water, made a curt reply and terminated the call.'[18]

The discovery of contaminated readings at Palm Beach the next day justified the Department's action. Minister Craig Knowles and Premier Bob Carr held a press conference, blaming the Prospect plant or somewhere above it. Finally, at 7 p.m. on Thursday, the first city-wide boil alert was issued.

David Hill and Chris Pollett both stood down after the inquiry. Hill stood as a Labor candidate for the seat of Hughes in an election in which the Carr Government enjóyed a thumping victory. For David Hill, it was not to be.

The imperfect plant

As deplorable and comic as the management of the crypto crisis may have been, it also overshadows the basic technical problems that may have contributed to the problem in the first place.

The water catchment for Sydney extends an impressive 16,000 square kilometres. The big collecting tank is Warragamba Dam, built in 1960, the largest city-supply concrete dam in the nation, supplying about 80 per cent of Sydney's water. That's about four times the volume of Sydney Harbour. A staggering 20,000 kilometres of pipes deliver water from 11 treatment plants to the house tap.

The Prospect plant, a mammoth filter for most of Sydney's water, was seen as a weak link in the chain of supply and it is quite possible that the bugs passed through the plant or were stored in it. Unfortunately, Prospect was never designed for total removal of bugs, but it was believed to be able to filter more than were measured during the crisis, particularly the second time the bugs hit.

In 1993, a confidential Sydney Water Managing Director's Report, later tabled in NSW Parliament, stated, 'The most difficult of the pathogenic pollutants to treat are cryptosporidium and giardia. These are best treated by keeping them out of the system at all catchment levels.'[19] There were, however, factors other than its original design which made the plant vulnerable. Prospect works by a process of dilution, during which time chemicals are added to coagulate particles that are trapped through sand filters. It means regular backwashing of filters is essential. About every 48 hours, the coagulate was back-washed and stored in lagoons near the plant.

In July 1998, when the crisis began, the two water tanks that receive water once it's filtered were being given their first mainte-nance ever. To conserve energy, Prospect was using the lowest doses of coagulants necessary and the filters were only being backwashed up to 72 hours apart. At least four times during July, dilution flow from pumps was interrupted. It is also possible that the operator reduced water levels in the tanks too much then rushed water back in too fast and stirred up contamination in the sediment, flushing out the bugs.

The plant was not the only weak link. A problem occurred up-pipe at Warragamba Dam, Sydney Water's responsibility. A thermocline in the dam creates layers of water at different temperatures, which change with the season.[20] Crypto likes to live low down in the cool, and unfor-tunately the dam operators were removing water at just this level.

It is still not clear whether the plant was at fault, or whether there were just too many bugs going through even for a plant operating efficiently to catch.

Peter McClellan did not think upgrading of Prospect was justified. A new form of membrane, for example, would cost \$600 million. Instead, he recommended new operating targets for the plant.

Managing Director of Sydney Water, Chris Pollett, clearly felt the plant was at fault. At the height of the crisis, on Thursday 30 July, he actually misinformed both his minister, Craig Knowles, and the public by declaring that the whole plant was being by-passed, a story that he continued with until Friday. This was never the case and just added to the confusion.

Overall, Peter McClellan found that 'Sydney Water seriously failed to discharge the obligations it owed Minister Knowles'.[21] What few people are aware of is that during the crypto crisis, Craig Knowles' wife was very ill and undergoing treatments that suppressed her immune system.

Interestingly, Sydney Water does not own Prospect. The plant was privately built and is now operated by Australian Water Services (AWS), a joint venture between French company Lyonnaise des Eaux and property developer Lend Lease. When the accusations started flying, AWS was quick to blame scientists at the Sydney Water laboratory for wrongly identifying some of the bugs which were in fact just algae.

Back in the lab

If there were faults at the plant, there were also problems at the Sydney Water-owned laboratory, Australian Water Technologies. Scientists had real difficulties in testing for the bugs. Those at the lab said they did find crypto and giardia, and while the inquiry accepted this, it had big questions about the levels of contamination and whether the bugs were dead or alive.[22] In his final report, Peter McClellan says, 'The reported levels caused the international scientific community to doubt the accuracy.'[23] An independent audit also found serious quality-control failures in a lab placed under huge stress. It concluded that the data from the lab should not be used to make public health decisions until all deficiencies were corrected. Lab staff had reportedly explained to Sydney Water management that the extra work created by the crypto crisis would affect quality and Peter McClellan noted, 'It is clear that Sydney Water was prepared to accept data for which the quality control was inadequate.'[24]

Unsurprisingly, eyebrows were raised about the lab being run by Sydney Water, prompting Peter McClellan to comment, 'I do not believe that a laboratory owned by Sydney Water is appropriate to provide testing for public health purposes. I recommend there be established an independent testing laboratory which provides testing services for all regulatory agencies.'[25]

Since the inquiry, there has been much debate over whether the bugs were crypto or just algae posing as crypto. Dr Jennifer Clancy, who was one of the international scientists consulted during the inquiry, published her views in a US journal that crypto and giardia were not present and that any that showed up were due to inadequate controls in the lab, which led to cross contamination.[26] She also believes that monitoring the bugs is not a practical option for predicting water quality and the sooner regulatory bodies realise this, the better. Apart from anything else, by the time the mice are dead in the lab, it can be a little late. Perhaps more would have been made of the piece if it had been published in Australia. Most in the scientific community would rather let sleeping dogs lie.

Even now, there is a race on in the world scientific community to understand crypto. The boffins still can't agree on just how good water filtration is in eliminating crypto, nor on how it should be tested, nor even on what the health consequences of crypto in water are. Crypto, however, is not going away.

Sins of the catchment

It's an oldie but a goodie—'Prevention is better than cure.' During the chaos and finger pointing of six weeks of crisis, there was not nearly enough focus on the other possible cause of the outbreak: the 16,000 square kilometre catchment which supplies the city, running from the Coxs River near Lithgow in the Blue Mountains, to Goulburn and the Mulwaree and down to the headwaters of the Shoalhaven River near Cooma.

Just before the second outbreak of bugs, heavy rainfall caused Warragamba Dam to fill from 60 per cent capacity to overflowing in about ten days. Worse, the heavy rains had been falling on the back of a drought. It had been a long time since the land had been cleansed by a good downpour. The sheer volume of water seeping into the sewerage system caused sewage treatment plants to overload. To put no finer point on it, a great deal of shit was washed into the rivers.

Jeff Angel from the Total Environment Centre was adamant: 'During a US presidential election, I think a couple of terms ago, we had a big banner hanging from some major part of the landscape which said, "It's the economy stupid," and I felt, in the last couple of months, that we should have hung from the office of Sydney Water Corporation, "It's the catchment, stupid".'[27]

Peter McClellan was also interested in the catchment. His inquiry's findings were shocking. The floods overloaded the system. 'The Goulburn plant was unable to irrigate its treated effluent and this was released into the Wollondilly River. Sludge ponds at Bowral were flushed into the river, and the Bowral, Mittagong, Bundanoon and Berrima plants were all required to operate at extraordinary levels. The consequence is that significant volumes of poorly treated sewage were released into the catchment and may have found their way to Warragamba Dam.'[28]

Top of Peter McClellan's list of recommendations was the creation of a separate catchment management authority, one independent from Sydney Water, and one with teeth. Until *Watershed* he had done no formal interviews on the crypto crisis since the inquiry, presumably to keep its conclusions finite. But this is an area he clearly feels passionately about—we must find a long-term solution that will protect the Sydney catchment. 'What I had in mind during the inquiry, I was driven by concerns that unless we address the issue, it would forever erode before our eyes. I was also conscious of trying to drive a long-term perspective, 2001 and also 3001. It's why I said about the Sydney Catchment Authority, "someone should wake up in the morning owning the issue".'

Peter McClellan had wanted many more changes to clean up the catchment and by the time his final report was complete, he was able to state that as a result of earlier recommendations,[29] there was better legal protection with a new Sydney Catchment Authority in charge of managing Sydney's drinking water catchments set up under the *Sydney Water Catchment Management Act 1998*.

The 1998 Act also required a full audit of the Sydney Catchment Authority. Completed in December 1999, the audit committee included some of the top names in CSIRO, the auditor himself being hydrologist Dr John Williams. At his base on the side of Black Mountain in Canberra, John Williams was interested in Peter McClellan's reaction to the audit—'Was he disappointed in the progress that has been made? You see, some people who read my report say, "John, you didn't push them hard enough".' In spite of these doubts, for a layman at least, the verdict of the self-described 'snapshot of the catchment environment, land-uses and human activities' is shocking. Indeed, the reader of John Williams' audit report has only to flick through to see photos captioned 'Meat processing adjacent to Mulwaree River', 'Dairy cattle adjacent to Wingecarribee Reservoir', 'Severe erosion in Shoalhaven catchment', 'Extractive mine site adjacent to Shoalhaven River' and 'Garbage disposal in gully near Wollondilly'.[30]

Conventional wisdom has been to divide the catchment into two, with an inner catchment of special areas kept pristine to act as filters. The audit found this whole strategy flawed. The special areas are small, for instance, less than one-third (33 per cent) of the Warragamba catchment,[31] and less than 1 per cent of the Shoalhaven catchment. More importantly, what they do not cover is the headwaters in the outer catchment, which is where much of the damage is being done. As pressure continues, special areas simply won't be able to cope.

The Sydney region, including satellite areas, is developing rapidly. Whether it's millionaires in Bowral or Kangaroo Valley, young families in the new suburbia of Bong Bong, or on hobby farms, poo in the upper catchments is a big problem. John Williams' audit estimated that in Sydney's upper catchments, 30,000 people are living in unsewered homes, meaning up to 10,000 septic tanks, or the like, many of which are not up to scratch. There are also 63 licensed sewage treatment plants that release effluent or dispose of biosolids within the Sydney water catchment area. And thus far, the Sydney Catchment Authority has had little or no ability to monitor failure of treatment plants and shoddy tanks. 'The pattern of development appears to be that reticulated water is provided to encourage urban and peri-urban development in these highly desirable living areas, but practically no planning or attention is given to the disposal of wastewater, effluent or, in some instances, solid garbage.'[32]

Information is sketchy on the number of intensive animal operations in the catchment: pig and poultry, meat and wool processing. But as at December 1999, the audit contrasted the Catchment Authority's estimate of 13 poultry farms in Mulwaree Shire with the audit's independent review panel reckoning of around 130!

Twenty-two coal sites were also identified within the special areas, 11 still operational with attendant risks of subsidence and mine water discharges, including of chemicals. Add to that vast areas of cleared land, unsealed roads and stock access to rivers, increasing extractions from rivers and, of course, sewage effluent and biosolids from treatment plants and it is bizarre crypto hasn't 'come on down' earlier. The audit was not optimistic. 'If not appropriately managed and controlled, a range of human activities within the Warragamba and Shoalhaven catchments can be expected to provide increasingly significant loadings of contaminant to the Sydney Water storages.'[33]

The biggest concern the audit raises is whether the Catchment Authority has the teeth that Peter McClellan had ordered for it. 'Failure to support the Authority with adequate legislative powers and effective institutional arrangements is the paramount hazard facing the hydro-logical catchments that supply Sydney's water.'[34] On paper the chair of the Sydney Catchment Authority has been given a regional environment plan for the outer catchment legislated through the Department of Urban Affairs and Planning. But as John Williams point out: 'To implement it, he hasn't got staff on the ground over a catchment from Lithgow to Cooma just about, to say nothing of Kangaroo Valley and the Shoalhaven, so he's got to have a relationship with the National Parks and Wildlife Service, the State Conservation and Land Management Department and other State instrumentalities, and all he can say is that it's got to be in accordance with the regional environmental plan. But the REP is implemented by local government. You can just work your way through that if you want.'

Pressure from developers is enormous, according to John Williams, who says he got all sorts of irate people thinking that he was trying to stop development. There is also a suspicion in various quarters that the Prospect plant has been directly responsible for the gung ho development, because of the security it gave: any unhealthy water would be fixed well before it reached the drinking tap. CSIRO's urban water expert, Andrew Speers, also a former Sydney Water employee, agrees. 'I think the big problem with Prospect is that it led to a downgrading of catchment protection because the feeling was "We don't have to worry about the catchment any more".'

Andrew Speers is hesitant about the idea of banning all further development in the outer catchment area but he too has said the teeth of the Catchment Authority need to be sharp. 'What did it cost Sydney Water, $40 million, $50 million for the crypto in terms of compensation and cost of inquiry and so on? You can buy a lot of protection equipment and regulation for that sort of money. Whether that means no further development I think is a debatable point.'

The good news is that, finally, the concerns of John Williams and his team are being taken seriously. On 28 June 2001, the Carr Government announced that $132 million would be spent by the Sydney Catchment Authority over the next five years on ageing sewerage systems, stormwater, industry waste, old mines and erosion in the catchment. The question is, how far will that money go and can it keep pace with the ongoing sins of the catchment?

For peat's sake
One of the biggest environmental tragedies in the catchment coincidentally occurred during the crypto crisis, not far from Robertson in the Southern Highlands, where a rare peat bog, formed over 12,000

years and covering over 400 hectares, had been acting as a perfect natural filter of water which then collected in the Wingecarribee Reservoir. But the peat was also the source of 95 per cent of Australia's potting mix, extracted over 30 years by a Queensland mining company, Emerald Peat. The operation had been shut down temporarily five months before the disaster. Despite a licence being issued to mine peat, no less than seven separate organisations, including Sydney Water and the Environment Protection Authority, had called for the mine to be closed after the buffer zone separating the mining and the reservoir had reportedly been breached.[35]

Nevertheless the damage had been done. In August 1998, severe rain flushed a mining dredge and a reported 1 square kilometre island of dirt and vegetation into the reservoir. Pictures published in newspapers showed a valley ripped to bits. There were fears of outbreaks of blue-green algae, crypto and giardia in the reservoir. A year later, John Williams' audit found that water flowing into the reservoir was turbid, with a high level of biosolids.[36]

Emerald Peat was taken to court. Ironically the QC representing the peat mine in court was none other than Peter McClellan! When I asked him how the head of Sydney's crypto inquiry could have ended up representing the perpetrator of such catchment damage, there was a half smile. 'Well, as you know, at the bar we have to take whatever job comes along. If I don't, I get disciplined . . . it always makes for interesting dinner party conversation, though, and of course where you often get the least understanding is at home, from your wife!'

What now?
Peter McClennan's Sydney Water inquiry did give Sydney Water a kick along and improvements in monitoring, operating and communication have been made. The Potts Hill reservoir has been covered and new sewerage systems put in around the catchment. But Sydneysiders should know that the system is far from foolproof.

All immuno-suppressed people are advised by NSW Health to boil water at all times. In New South Wales, laws are now in place for Sydney Water to report on the quality of water to consumers every three months via their water bills.[37] The company also publishes test results for parasites in treated water, over the Net, with the accompanying statement, 'NSW Health continues to closely monitor all water quality results 24 hours a day. Sydneysiders will be notified immediately should there be any risk to public health.'

However, despite Peter McClellan's recommendation, independent testing of crypto and giardia only started in July 2000, two years after the crisis. For the first time in September 2000, specific guidelines on crypto and giardia were recommended by the National Health and Medical Research Council,[38] which also provides up-to-date information on the bugs and guidelines on acceptable levels in drinking water and on their detection. They are only guidelines, however, and the extent to which the States will pick them up is open to question.

In contrast, Britain has now made it a criminal offence for drinking water to contain more than one crypto oocyte in 10 litres. The rules are expected to cost British water companies (and inevitably the consumer) £8 million (about $A23 million), and that is just for monitoring![39]

It is worth noting that the demand for bottled water during the crisis turned out to be no blip. In additon to a 30 per cent increase during the year of the crisis, sales soared a further 15 per cent the following year.[40]

Public or private? – a water quandary

One of the more challenging and emotive issues that the Sydney crypto crisis raises is privatisation (in all its guises). But before we look at Sydney, it's worth reflecting upon the broader privatisation experience.

Australians can justly feel that they have had a good dose of privatisation. For some, the sale of our national bank, airline and telephone companies has meant a rather nice nest egg from the stock market. Yet for others, privatisation is like seeing governments flog off the family silver. In later chapters, we will come across some farmers who now believe government is out to privatise the rain that falls on their own land.

Despite the heated debates, enough sell-offs have occurred around the world to make some fairly solid generalisations. In a utility industry like water, the two main options are to be run by a public authority or a private company. A third is to corporatise.

Public ownership has traditionally suffered from a 'cost-plus' mentality, where frequently decisions are not properly costed. Bureaucracies can often be sleepy and over-staffed, with little incentive for innovation among employees. Public ownership can be expensive and inefficient *in economic terms*. Public authorities are also Treasury-dominated, and suffer from the myopic nature of this department, which is always reticent to commit long-term investment capital (an attitude otherwise known as NIMTOO – not in my term of office).

On the plus side, however, these organisations tend to be safety conscious, consumer conscious and environmentally conscious. They do, after all, represent the public interest, and the public generally likes the idea and feels safe. Corners are less likely to be cut and the public authority is accountable to the Minister and Parliament and to the Treasury.

Private firms offer something very different. Efficiency and cost consciousness to the consumer are the clear benefits, but a major attraction is also capital investment. Provided the investment makes a positive return, the market takes care of that. Inevitably, however, with a cheaper service and focus on the bottom line comes less focus on safety and the environment.

It is interesting that during the Thatcher years in Britain, which critics saw as a time when the Tory rule was 'If it moves, privatise it and if it doesn't, shut it down', a line was drawn at water catchments. Nicholas Ridley, the then Secretary of State for the Environment and reportedly as brilliant as he was right wing, put up the controversial plan of privatising the entire water cycle, from the catchments to the sewer. Despite his zeal, his Green Paper never even got to a Parliamentary bill. This was privatisation gone too far, even for the Tories. What were sold off were the water supply and treatment companies.

One of the real irritants about privatisation for many Aussies is that it involves foreign take overs of our companies. Whether you believe that this takes money and control out of the country or destroys the culture that the Arnotts biscuit is a part of, foreign investment is one of the most emotive topics on talkback radio.

If there is any action happening in the water industry, it is likely to be by the Frogs and the Poms: names like Vivendi and Lyonnaise des Eaux, North-West Water and Thames Water. The reason is that the privatisation wave rolled into Europe about ten years before it hit Australia. In that time these companies have had the capital and freedom to pretty much do what they want, which includes taking big chunks of Australian companies when they are up for grabs, like British Airways having a munch at Qantas.

When privatisation first arrived in Australia, the ideology of economic rationalism (cuts = efficiency = productivity = profits) was all-powerful.

Peter McClellan remembers, 'I sat next to Bob McMullen when Keating lost, back in 1996, and asked him when he thought Labor would get back in. He said we won't do it until we can generate a new political philosophy for the 21st century. He said Labor are the generators of ideas and conservatives merely the caretakers of them, I think that's quite a nice way of looking at it. This was four to five years ago, when privatisation was driving the agenda.'

Today's nationalist talk and fears of privatising, however, overlooks an important fact. Privatisation in the form of whole-hog sell-offs came to an abrupt stop when it got to water. Even Jeff Kennett couldn't do it.

Very little is mentioned on water in Australia's seminal book on privatisation,[41] partly because of how little privatisation of water has been done, but a book published more recently goes straight for the jugular. In *Water's Fall* the author, Christopher Sheil, mounts a frontal attack on economic rationalism, arguing that making money and running companies for the benefit of the community are two incompatible objectives. To its core, a commercialised company is a profit maker at almost any cost.

Water's Fall takes two classic examples involving what could be called 'bad' privatisation. The first is 'The Big Pong' in Adelaide, which happened in 1997, when the sewerage system failed and it was not long before reports were about of 'the city of churches stinking to high heaven'. The second example is the contracting process for the Prospect plant.

Peter McClellan gave the book a ringing endorsement on the back cover, and back in Phillip Street, he explained why. 'It was a good book, written from a particular perspective, and I think we do need to challenge the thinking. As you will have gathered from the inquiry, I have concerns about the culture we've created by expectations that water management be a cash-generating exercise.'

Water's Fall is eloquent and convincing, but should carry a health warning for any investment banker in the CBD. Such people would soon find themselves frothing uncontrollably at the mouth with outrage at a Quigginesque[42] thesis which blames almost all incompetence on greed and a general bias in favour of the shareholder over the consumer.

There have been many passionate critics of privatisation but to be fair, privatisation can be deeply misunderstood. There is good and bad and very necessary private involvement in our utility industries, particularly when it comes to big spending where the public purse is firmly closed. Where there should be debate about privatising water (and there is) is in the contracting out and selling off of vital infrastructure, such as water treatment plants. You need look no further than Sydney.

Privatisation, Sydney style

The privatising of part of Sydney's water system was driven largely by a government unwilling to pay the huge sums involved in repair and development of big water infrastructure. Until the Prospect treatment plant was built in 1993, Sydney's water source was the Prospect Reservoir, constructed in 1888. The original plan for this reservoir was for water to be stored for up to six months, which would allow for natural settlement and filtration. But a century later, water was only being stored for one or two days and authorities had resorted to throwing massive doses of alum into the reservoir, to drag the sediments to the bottom.[43] By the 1980s the situation was becoming desperate, with water standards falling below national guidelines. State Treasury, according to Christopher Sheil, turned a blind eye to the water problems. When asked about the water authority, its response was 'Wait for it to crash, you only get money in a crisis.'[44]

With the dominance of economic rationalist thinking in the corridors of power and talk of a new efficiency and effectiveness to be delivered to the 'shareholders' of New South Wales, the public, time was ripe for privatisation.

In 1990, the NSW Premier, Nick Greiner, a great 'ecorat' enthusiast, agreed to a program that would give Sydneysiders water that met 1980 national guidelines by 1991 and the health aspects of 1987 guidelines by 1994. At the centre of this was a new treatment plant at Prospect, a 'BOO' scheme, in which a private company would own the plant outright. In November 1992, Australian Water Services (AWS) was selected to build, own and operate the plant.

What then followed was astonishing. It would be hard to dream up a scenario that could condemn privatisation more soundly. In summary, government lost control of the tender process and ended up wearing many of the risks it should have passed on to the private company—one of the real plusses of privatising.

Attracted by the cheapest bid, government locked itself into a contract with AWS *before* it had confirmed that the plant technology could actually achieve the 1987 water standards. Further, if government did not sign the formal contract within eight months, by July 1993, it began paying a financial penalty. And the State also committed itself to paying for any upgrades if the plant's design needed changing.[45]

The Milwaukee disaster of March 1993 which killed 103 people seemed to have no impact on the process. Those making the environmental recommendations were kept out of the loop on a key report[46] and, while environmental sign-off was eventually given, there seemed to be no formal process within the Water Board at that time to confirm that AWS had proved its technology. Surely a fundamental requirement.

Peter McClellan was asked in the course of his inquiry to look into the legal contracts which set up the responsibility and accountability of Sydney Water. He had to use the powers of a Royal Commission to get some of the documents from AWS handed over and then only after squeals of 'commercial in confidence'. It turned out that there was no record of sign-off by Sydney Water on AWS's plant technology. The inquiry found that to this day the 'final reporting of the AWS test results to the Operating Executive remains a mystery'.

A final embarrassment to government happened when AWS agreed in 1993 that it would be responsible for removing 99.9 per cent of the parasites in water. When it came to legal enforcement, however, the inquiry found the commitment held no water at all. 'A court would not require AWS to accept such an obligation under the present contract.'[47]

Interestingly, in its June 2001 announcement the Carr Government pledged to spend $18.8 million to complete the construction of an additional water filtration plant at Prospect Reservoir and almost $54 million over the next four years to upgrade other water filtration plants around Sydney.

The Prospect plant contract was undoubtedly a 'how not to do' example of privatising infrastructure development. But it does not rule out BOOT (build, own, operate and transfer back to government) schemes. Just ask Don Richardson who works for British United Utilities (formerly North West Water and not to be confused with United Water) and is quite proud of the company's MacArthur treatment plant in Sydney, commissioned at around the same time as Prospect, about which there are few complaints.

A little bit pregnant

Where grey heads are beginning to ache among policy makers is in the business of corporatisation. Corporatised companies are effectively a halfway house, still owned by government, but trading as a corporate entity, with decision making by managers, who are independent of government and all in the expectation of profits that any private company might have. Unfortunately, high profits (just like gambling) usually involve risks.

It seems as well that Australia's State-owned water companies are getting a taste for the entrepreneurial life, attracted by potential profits. 'SA Water is in Western Java, Western Australian Water Corporation has pulled out of Johannesburg, it did try to get into Libya . . . most of them are doing something,' commented Don Richardson. 'I have no problem where governments are guaranteeing the payment and it's an overseas aid type thing. But when you're going over doing an entrepreneurial business, that's a totally different ball game.'

He continued, 'You've got the big public authorities who are trying to do the privatised thing. They're under pressure, they also want the ego thing. I was told that SA Water and Sydney Water are both benchmarking with the object of proving that they're just as good as the private companies.'

Don Richardson is all too aware of the situation. United Utilities is based in Adelaide, where the company has picked up various pieces of private business. Like the other big international water companies, it was wrong-footed by public antagonism that has seen the pendulum swing away from privatisation, before it really got going in water. Aside from a couple of contracts to build, own and operate water treatment plants in Sydney and Melbourne, business has all but dried up.

Don Richardson is not happy. 'We came to Australia under a wave of UK privatisation thinking that there would be ten years. Australian utilities will see privatisation, full flotation. That has not occurred. The public bodies didn't push any further, and they started to do their own design and build and so the projects after 1995–96 started becoming smaller and we realised that there was nothing in it.'

Don Richardson would say this, of course, wouldn't he? But the argument is not new. Had the NSW insurer GIO been owned by government and not shareholders when it collapsed after a spate of world disasters, it would have been the taxpayer picking up the bill. In high-risk industries like communications there may be good reason to shift risk away from the taxpayer onto those who are happy to take on risk for a higher return.

In water, however, there is a strong argument that key parts of the industry remain in government ownership and control: the Sydney Catchment Authority being a critical example. Perhaps the most significant development following the crypto debacle is that the NSW Government introduced a law that changed Sydney Water from a corporatised company independent of government back to a 'statutory State-owned corporation' so the minister responsible for Sydney Water would have greater power to access information. In June 2001, Sydney Water also announced it was abandoning plans to privatise its research unit, AWT.[48]

Getting real

Privatisation is not all good, but nor is it all bad. The inquiry was not able to conclude that AWS and the Prospect plant contributed to the crypto crisis. Had it been more serious, however, it is likely that the privatisation process would have come under greater scrutiny. This said, there is a place for private involvement.

The reality is that governments are all unenthusiastic about infrastructure spending, unless it is a flash in the pan that the public can see, like the Olympic site, or a political statement, like the Alice-to-Darwin railway.

For the anti-privatisation lobby, this is simply not good enough. Governments should be spending more money on infrastructure, and to accept that privatisation is necessary because governments aren't spending is to cop out and further entrench the politics of fiscal belt tightening. Implicit in this view seems to be the assumption that if government were owning and running the joint, everything would be all right. Sadly, the history of bureaucracy tells us this is not so. Poor service to the public was an instigator of privatisation. Are we to imagine that the government would have been spending the money the private sector would not? As Christopher Sheil himself admitted, NSW Treasury's view was wait until there is a crisis. And support for privatisation is not limited to the Liberal Party. Just talk to those in NSW Labor who have spent years trying to get electricity privatised.

In Sydney's crypto case, it could be there was an inadequate plant, unflushed pipes, poor dam management, woeful catchment management, and a laboratory that could not cope. Yet how much of it is the fault of privatisation? And would the situation have been much different under public ownership? Civil servants, after all, have been known to cover their behinds as well.

Significantly, one of the more popular theories is that there was crypto in the water, but it was dead. Had it been alive, and caused deaths, the bedlam of 1998 would have magnified to terrifying levels.

Government, on behalf of all of us, needs to carve out crystal clear lines of responsibility for any risks, and by now most States have had enough lessons in negotiating. In the case of contracting out, it means coming to a very clear understanding of liability. More generally, it demands a very sophisticated regulatory regime, something the Brits, through trial and error, have turned into an art form with regulators like OftWat (water), OfTel (telcos) and OfRail (railways), but an area with which Australia is still grappling.

In an emergency, we all need to know who is in charge and who takes the rap. That's all. If this is not possible, then there are strong arguments to leave assets and industries government-owned. Theoretically, Sydney Water would be a strong candidate for privatisation, but it may have blotted its copy book too much for the public to trust the private sector.

Is Peter McClellan ideologically opposed to privatisation? 'I don't even have a problem privatising Prospect. No.'

3

City Limits

Yesterday it rained, and now the garden will go crazy. It is early summer and for weeks the ground has been parched. I have feared for the roses, tended them with an irresponsible hose pipe. But yesterday the skies growled, the lightning flashed, and the heavens opened. In my garden, precious lives are no longer at stake; water has returned.

Philip Ball—H_2O: A Biography of Water

The ugly consumer

Without guidance and a conscience, human beings can be ignorant, wasteful and selfish creatures. When it comes to water, most of us stand guilty of some or all of these charges.

On average, each Australian household uses around 250 kilolitres a year.[1] This is probably not quite as much as the professional consumers in the United States (as the Australian Bureau of Statistics and the Victorian Government would have us believe[2]), but nevertheless, on this dry island continent, it is impressive enough. While we may go on and on about this dry continent, for most Australians nothing could be further than the truth. Our cities are the urban oases which have bred a very complacent attitude among their inhabitants.

We all know what we get up to: the indulgence of the power shower (an efficient shower head and shorter showers could save 28,000 litres of water each year); leaving the tap running; leaving the tap dripping (try putting a plug in the bottom of the basin with a tap left dripping and see how fast a few litres of water arrive). We still use single flush toilets when dual flush saves an estimated 36,180 litres per household per year (or 5 litres a flush). Then a personal favourite: 'hose sweeping'. Using the hose as a broom wastes an estimated 1,000 litres per hour![3] The fact that this practice has been partially replaced by air- and noise-polluting blower machines should not be applauded.

Oh, for the days when those strapping young men used something called a broom.

Lawns in our urban centres are treated as mini-irrigation areas, some city councils being the worst offenders. For a typical household, watering the garden accounts for half of its total water use. Over-watering is rife, sadly locking plants into a dependency on the garden hose, as roots have no need to run deep looking for moisture.

And what washes into our drains! Oil leaks from cars and mowers, detritus of all sorts washes faithfully down stormwater drains. Despite graphic television ads of stormwater pouring into the rivers we swim in, every Saturday people are back on the streets lathering up their vehicles with detergent which joins all the other flotsam and jetsam, including the murderous plastic shopping bag, down the drain. Per hectare, no one is a higher polluter than the city dweller. Incidentally, car washing with a hose once a fortnight uses an estimated 13,000 litres over the year.

Yet more unpleasant stuff finds its way into stormwater drains. Each year for example, 34,000 *tonnes* of dog poo is washed into Melbourne's Port Phillip Bay![4] Then one day a year, Ian Kiernan gets us all revved up and thousands go off and pick up rubbish to Clean up Australia, like a 'Poohsticks' outing.

These few thoughts already touch on a good part of the water cycle in cities. What is interesting about urban water is that each city deals with the problems in its own way. The reasons vary from differing circumstances to mismanagement. To unravel what is really going on, to understand what the demands are, what damage we're doing and what can be done about it, it is worth starting at the beginning—in the catchment.

Straight out of the tap—a human right
It is not given much thought by consumers: every day, without fail, clean water appears, from high in the hills or deep under the ground and through thousands of kilometres of pipes. A population of 13 million is supplied with water—in 1999–2000 1.2 million megs to residential customers.[5] It has become a basic human right, really, something those in the Third World have not got, but we have.

Before we go on, there is one very good question about urban water which should be knocked on the head early in the piece: why should city people feel guilty about water usage when the millions of us, taken together, account for just 12 per cent of all the water used in Australia and a massive 72 per cent is sucked up by agriculture? As we saw in chapter 1, city consumption drives agriculture and city folk should take some responsibility here. But, in one of the many Wyuna Roads in Australia ('wyuna' being Aboriginal for water) lives the CSIRO expert who has another point. 'Water availability is one thing and water storage is another,' said Andrew Speers, an expert on city water issues. 'You can have all the water in the world but if you can't store it somewhere until you need it, you have real supply problems. And cities as they grow consume the storage volumes; there's less and less per capita stored for people and the cost in environmental and capital terms of building new dams and storages is immense.'

In almost all our cities, the storage space for water is running out. Those who should know, like the National Competition Council's Graeme Samuel,[6] believe we have perhaps 25 to 30 years before new dams are demanded, unless alternative solutions are set in place.

In water resource terms, Australia's capital cities have been blessed in very different ways.

The bulk of Melbourne's water (90 per cent) is so pristine that it needs only the lowest levels of treatment.[7] Water is collected from 140,000 hectares of natural forest in the hills behind Melbourne, some of which have been closed to the public for over 100 years. The water is then left to stand for up to five years to enjoy a natural settling process. In the year to June 2000, Melbourne Water apprehended 111 people over security breaches and recommended the bulk of them for prosecution.[8]

The Thomson River in Victoria's Gippsland district, dammed 20 years ago, is Melbourne's largest water storage, providing 60 per cent of the city's water as well as water for irrigating rural areas. Melbourne's 2000–01 summer was the hottest for 20 years, prompting the Victorian Government, in March 2001, to announce a major new resources strategy to look for more water. City thirst comes at a cost in Victoria. The loss of flow in the Thomson and intensive dairy operations in the Gippsland area have combined to create nutrient-rich and oxygen-deficient water which has spoiled Lake Wellington, a massive body of water in East Gippsland. Fish kills have been the inevitable consequence.

In stark contrast to Melbourne, Adelaide folk have every right to complain about water quality. The city's water is infamous for being the most unpleasant tasting of all Australia's cities, and it has been that way for some time. South Australians are in the invidious position of relying on the bottom end of the Murray–Darling system for 40 per cent of their drinking water in a normal year and up to 90 per cent in a drought. Already there are some claims that salination of Murray River water is beginning to eat away the city's copper pipes.

Ian Wilson, a former Federal Minister for Home Affairs and the Environment from Adelaide, keeps a bottle of tap water which he first slammed on the dispatch box in Canberra in 1977 in protest at the quality. The now aged brown water in the plastic bottle looks even more dubious.

In the west, Perth is becoming increasingly reliant on its precious ground water from the Nangarra mound, now supplying 52 per cent of Perth's demand compared with 40 per cent three years ago.[9] While a new dam has recently been commissioned, salinity is an increasing problem and the city is running out of options for surface water supply, particularly if a 20-year dry cycle continues. Last year (2000), the former West Australian Liberal Government announced it was looking seriously at desalination as an option for the city, and the new Labor Government has not ruled this out.

Growth is left to Brisbane and Darwin. Brisbane gets its water from three dams: North Pine, Wivenhoe and Somerset. In purity terms, the dam water is closer to Warragamba in Sydney than the Thomson in Melbourne. The dams are surrounded by a mix of farming and some development with a buffer zone of only 400 to 500 metres of reserve. According to the city council's manager of water and sewerage, Mark Pascoe, south-east Queensland will be short of water by 2015 unless something is done.

Darwin seems to be thinking further ahead and more ambitiously than the other capitals, and for good reason. Still a small town of less than 90,000, some forecasts put Darwin at a million people by 2070. The relatively pristine city catchments also enjoy 2.5 metres of rain a year and there are major proposals for dams nearby to accommodate the many thousands of new citizens that are expected. According to head of the Territory's Resources Development Office Dr Howard Dengate, it's possible these dams will be built sooner rather than later, 'perhaps with private-sector money in return for access to water for irrigation in the interim, until the city gets big enough.' Howard acknowledged that that might lead to tensions between irrigation and city users down the track, however. 'Now that could lead to political conflicts down the track so there are political decisions that need to be made about that.'

Shake up in the city

Now that most Australians are suffering from reform fatigue, industry change conjures up anything but excitement. In fact, city water reform has been a remarkable success, and much better received by all involved than reform in the country.

Reform began in 1993, at the height of the 'competition policy' era pioneered by Fred Hilmer, now chief executive of John Fairfax. Each of the States and the Territories, through the Council of Australian Governments (COAG), agreed it was high time for a shake-up in the water sector. It certainly was a shake-up. Policy and regulation were separated from water supply and wastewater operations, benchmarking of performance was introduced, pricing of water became largely consumption based, real rates of return were demanded for service delivery businesses and cross subsidies to domestic consumers were eliminated.

At about the same time the entire water sector was undergoing its own reorganisation, driven by the new thirst for corporatisation. Sydney Water was corporatised in January 1995 and on the same day Melbourne Water was disaggregated into a head works corporation and three retail businesses to encourage competition by comparison. SA Water and the body running Canberra water were corporatised six months later in July 1995. In January 1996, in the largest deal of its type in Australian history, Adelaide's water and wastewater services were outsourced to a private consortium. Institutional reforms were also introduced in Western Australia.'[10]

For consumers, the most important change was in the pricing, in the move to 'user pays'. Water metering is now in place in all capitals except Hobart, where it was argued that the cost of metering was not worth the gains that would be made. On average, about half the consumer's bill reflects actual use and half a fixed charge.

Water prices across Australia declined an average 9 per cent over the five years from the 1991–92 financial year, a phenomenon driven by New South Wales and Victoria. And while the fall is open to interpretation, what is clear is that the move to consumption-based charging brought a major bonus—consumption went down. In total, water supply per property across Australia in major urban areas fell by 19 per cent in the six years to 1996, a saving sufficient to supply the urban centres of Adelaide and Perth, with urban populations of 1.3 million and 1.1 million respectively.[11] The El Nino-driven drought kicked the next two years back up 6 per cent or so, but the numbers are now steady with an average 256 kilolitres consumed per property last year (2000).[12] According to Andrew Speers, Sydney has expanded by its last 250,000 people without any major increase in total water demand, mainly because of pricing. People are using less water.

For industry too the change in pricing has had huge ramifications. Payments based on consumption rather than land area mean that water is now a controllable cost instead of a tax. The extreme example, according to Andrew Speers, was of a shoe repair shop with a single hand basin at Wynyard Station in Sydney which was paying $5,000 a year in water rates because of the land value! Heavy industry also has to be focused on using less water. The changes in pricing have made it worthwhile for several plants to put in their own treatment facilities so that they can re-use their wastewater. On the consumption side, there is little doubt that reform in urban water is achieving what it set out to do.

You might ask, who is losing from this process? Costs to consumers are dropping, water supply companies are making money and less water is being consumed. One answer is the 9,000 workers in the water business who don't have a job, after a 40 per cent cut in positions.[13] Yet, aside from the job losses, it turns out that the push for efficiency and profit has had other critics.

At the end of autumn 2000, Victoria was suffering from a severe drought. The massive Thomson Dam was down to 38 per cent capacity and water restrictions seemed inevitable. It was then that suspicions were raised about water distribution companies which were now basing their earnings on consumption and had little incentive to encourage customers to cut back. As Dr Peter Fisher wrote in August 2000's *Financial Review*:

> There was still a reluctance to impose water restrictions, which by the way, can change mindsets towards more sparing usage. Some 30 megs is used annually in Melbourne just hosing down paths and aprons. The letter to the editor columns of the *Age* were awash: could it be that water authorities were unwilling to impose restrictions because of the impact on profit?[14]

It was an accusation the water authorities hotly denied.

Waste not, want not
It is now time to roll up the sleeves, for the supply of water is obviously the clean bit of the process. Supply and disposal require 180,000 kilometres of pipes in Australia, that's roughly four and a half times the girth of the earth. Once water leaves the home, little more thought is given to it by Joe Public. As recycling expert Jenifer Simpson put it, 'Disposing of our personal waste is not one of our greatest achievements. We have difficulty in admitting that we have bowels, a rectum and an anus—and that which emerges from them is quite unmentionable. We don't like to spend thought or money on this rather unsavoury aspect of our human activity until we are absolutely forced to.' We'll meet Jenifer Simpson shortly.

Happily, one of the benefits of reform has been that by using less water, there is a little less to clean up. Nevertheless, whether we drink it, wash in it or just waste it, we create the problem of dirty water.

Before going into the smellier side of the business, it is as well to get the meaning of a few important words quite clear. 'Sewage' is what everybody thinks it is — whatever disappears down the loo, along with all other wastewater in the household, like that from the shower and washing machine. 'Sewerage', on the other hand, is the piping and infrastructure that takes the unmentionable contents wherever they need to go. Once treated, sewage becomes 'effluent', an important distinction from stormwater, which is normally much cleaner stuff.

The water set-up in Australia is wildly extravagant. For most towns and cities, water supply and discharge operates in a straight line. Water is extracted from rivers far away, is piped into the home, then piped out again to treatment plants, before most of it makes its way either into rivers again or the sea. This country has spent billions of dollars over the years, ensuring that all water delivered to the home is treated to a level of pure drinking water, yet we actually drink just 1 per cent of it. Just one-hundredth of all that pure water. Even more staggering, all of the water that goes to urban industry is pure drinking water as well and while sectors such as the food industry might need such purity, the system seems most entravagant. Needless to say the capital costs of changing the entire ridiculous system would run to more than the cost of setting it up in the first place, but that is to give in too easily.

Australia is a newcomer to recycling. Today, we recycle just 11 per cent of our effluent nationally (although a large part of this, however, happens in rural areas, along rivers, and not in the cities) but just four years ago this figure was 7 per cent.

So what happens to all this unrecycled waste? As with the supply of water, cities have fared differently in the scale and nature of the problem of waste, but how they have managed the challenges goes a long way to explaining their situations.

How to upset a surfie

It should be no surprise that one of the most passionate groups trying to draw attention to the gaffs in wastewater management are the surfers. After all, it is these latter day national heroes who have had to dodge the turds and tampons over the years when things went wrong.

Nowhere is this situation more tense than in Sydney. The most populated city in the country, unlike Melbourne and Adelaide, has not had the luxury of space, which is so important for waste treatment.

Sydneysiders have been dumping their raw sewage into the ocean for decades. Back in the 1930s, this was through shallow outfalls close to shore. Untreated sewage spewed out of the North Head outfall for 50 years and then for the last 20 years at only primary treatment level.[15]

The differences in treatment levels can be confusing, but essentially primary treatment removes the solids through a mechanical process, secondary treatment involves a biological process with bugs that digest the sewage, and tertiary treatment actually filters the effluent. That said, some processes of secondary treatment are almost up to tertiary standards without the need for filters. 'To be blunt, what really counts is the amount of shit you can get out,' said Andrew Speers.

As at 1998 there were around 140 public sewage outfalls discharging a total three billion litres a day of human effluent and industrial waste into the sea around Australia, land of 7,000 beautiful beaches. By far the largest outfalls are three in Sydney, which pump out around 840 million litres a day.[16]

It was the public outcry urged by the surfies about smells and illnesses picked up at Sydney beaches—and some damage to Manly pine trees—during the 1970s and 1980s that goaded the State Government to look for alternatives to the 50-year-old system. Manly, slap next to North Head, is one of Australia's most famous beach resorts. As the mayor of Manly, Jean Hays, recalled, 'In those days you could really be swimming with raw sewage, not that you would of course.' Big spending on a major secondary treatment plant was ruled out in the mid-1970s, the government's engineers pointing out that Sydney's major sewerage systems are already established, and 'it would not be economically or physically feasible to consider significant changes to the basic system layouts.'[17]

The solution to Sydney's beach smells eventually agreed on was the construction of three huge deepwater outfalls, opened at the North Head, Bondi and Malabar treatment plants in 1990 and 1991. Together these plants now account for around 80 per cent of Sydney's sewage treatment. They work on the basis that 'dilution is the solution to pollution'.[18]

There is no doubt that the old outfalls were sub-par. Effluent at North Head was pumped out just 100 metres from the cliff face from a single pipe near the surface and diluted little more than one-to-one with sea water. The new outfalls run out 4 kilometres and are 60 to 80 metres deep, which is the critical factor. Then there are a series of 'diffuser heads' which dilute the primary-treated effluent between 200 and 1,000 times.

An extensive survey by the NSW Environmental Protection Authority (EPA),[19] completed between 1989 and 1993, found that effluent from the deepwater outfalls was trapped below the ocean surface for most of the time (75 per cent in winter and 97 per cent in summer) and headed to the Southern Ocean. The marine environment also seemed remarkably robust, with the exception of the rocky sea floor near North Head. Finally beach swimming was free from visible sewage 90 per cent of the time. It was a good rap, helped by new laws which prevented treatment plants from discharging sludge (solid effluent) into the ocean. Instead, sludge is taken back inland where it is used in agriculture. Pressure on the system was also eased considerably by stricter trade waste policies introduced in the 1990s.

Not everyone is happy about making the ocean the secondary treatment plant. Melbourne, Brisbane and Adelaide all treat their sewage to at least secondary level but in Sydney 77 per cent is only primary treated.[20] Critics have long accused the government of simply aiming to get the stuff out of sight and out of mind rather than worrying about what diseases might be unleashed. They have branded deepwater outfalls as white elephants, simply postponing the day the State is forced to upgrade coastal plants to treat sewage to secondary treatment levels.[21]

Dr Rex Campbell from the Surfrider Foundation accepts that beach health is better, but still worries about unknown factors like viruses which can live for long periods in ocean water. 'The last two governments have both promised to work towards re-use of treated effluents, but this is not happening at the big plants. The trouble is that it's only when effluent is treated to high secondary level that industry and gardens want the water. Secondary and tertiary treatments need slower rates of throughput but the big plants are being put under more and more pressure to munch through sewage faster. It's all a short-term solution but I guess that's Sydney Water's mandate.'

Andrew Speers accepts that the North Head treatment plant is not exactly optimal. 'They call it a high-rate primary treatment plant. Sounds nice doesn't it? "High-rate" just means it goes through faster, so there's less treatment.' But he also believes that overwhelming public concern about beach pollution has distorted the agenda. 'I don't want the next comment to be taken [to mean] ocean outfalls are necessarily good, or necessarily bad, but the five-year ocean monitoring study on the impact of ocean outfalls showed virtually no environmental impact. I know a lot of people don't want to hear that, but it did.

'Secondly, it's a political issue. I mean North Head outlet is right here where people live on the north side of Sydney. This is where they swim and therefore it's like the argument whales versus salinity. I'm not for a moment saying that whales shouldn't be protected, but they're a focus because they are "charismatic mega-fauna".'

To the charismatic surfers, though, you have to be out there on a board. For almost ten years now, the Surfrider Foundation has been calling for the closure of all ocean outfalls around Australia. Surfrider says that in the last five years effluent management has become worse, not better. In 1998, the holiday seaside town of Lorne was hit by two sewage spills 'with raw sewage running down the streets and onto the beaches'.[22]

There is another town where sewage management is in trouble. Attempts to use evaporation ponds as a settling process in Alice Springs have so far been a complete failure according to the Northern Territory's head of Resource Development Howard Dengate, the ponds having become a haven for mosquitoes. 'They haven't found any way to use that sewage water in any productive sense. As you fly into Alice Springs you'll see ponds and they're leaving it to evaporate. There [have] been several proposals to use it for citrus and dates, but it's all proven too difficult for Native title reasons. There's a real issue there, the impact on human health with Ross River and Murray Valley encephalitis, because they're not handling the water.'

The one thing everyone can agree on is that there are better ways to deal with wastewater. Whether by fortune or good planning, other cities seem to be on the right track.

The sewage munchers

Melbourne's Western Treatment Plant at Werribee, 35 kilometres from the CBD, consists of 10,000 hectares of land and vast lagoons alongside the ocean, and is where 52 per cent of the city's sewage is broken down naturally to create a very high-quality effluent. The process works by moving the sewage flow through a series of lagoons, allowing solids to settle and micro-organisms to digest the organic material. Mechanical aerators speed up the process. The area was first developed as a sewage farm in 1892 to counter diphtheria and typhoid epidemics. By 1930, grass filtration systems had allowed Werribee to become a permanent winter treatment with pastures supporting 20,000 to 30,000 sheep. In the 1980s there were major upgrades to use lagoons as stabilising ponds and methane from lagoons is now trapped and used to help run the plant.

By 2003, the Werribee plant will be producing around 300,000 megs of effluent a year, enough to irrigate the entire 65,000 hectares of the Mornington Peninsula on the other side of the bay. At present, most of the treated effluent runs into the ocean, but to reduce the impact of ocean outfalls, the Victorian Government has given Melbourne Water a target of 20 per cent effluent re-use by 2010. Whether the company makes it is another question, but already the corporate newspeak is to see waste as an opportunity, not a problem.

Treated effluent is now taking on a real value, which can only be a good thing. What happened at Toowoomba, just inland from Brisbane, gives an insight into just how important effluent may become. After the 1991 scare that saw a 1,100 kilometre long blue-green algae trail down the Murray–Darling, Toowoomba, at the top of the river system, was on notice to clean up its town sewage treatment plant. 'We fixed up some of it, but there [was] still 60 per cent of it going into Garry Creek and on to the Darling,' said Toowoomba Mayor Dianne Thorley.

A solution presented itself when a new power station was commissioned not far away at Millmerran on the Darling Downs. The power station offered to take all the crude effluent and pay handsomely for it. The Downs has seen very dry times recently and water in this prime agricultural area is at a premium. But the solution created another problem. Disgruntled irrigators on Garry Creek had enjoyed the shitty, nutrient-rich water for many years. Suddenly, shit had value, which brought up the interesting question: who owns the effluent? The answer is that the council does until it hits the river. From there, it becomes the property of the State, which must deal with the irrigators. For the council, pipes were the answer.

The resulting compromise was that Millmerran would take 1,000 megs a year, for which it handed over $3 million up front to Toowoomba Council and would pay $350,000 annually for 40 years. Irrigators would get the rest. Not bad for shitty water! The Millmerran pipeline is almost complete and the pipe to irrigators is expected to be a BOOT (build, own, operate and transfer back to government) scheme. Irrigators will have to pay for the effluent. According to Diane Thorley, 'They're not very happy about it, but it's better than no water,' and pointed out that it is also a win for the Murray–Darling system.

Perhaps the greatest expertise in re-use comes from Adelaide and the largest of its four treatment plants, Bolivar. Initiated by Premier Thomas Playford and built in the late 1960s, Bolivar can cope with the waste of 1.3 million people (90 per cent of Adelaide's population today). Like Werribee, after initial treatment, the wastewater moves to lagoons and collected methane drives almost half the plant. Most of the tertiary treated effluent is pumped up the Virginia pipeline to some of the 1,000 market gardens on the Northern Adelaide Plains, with the remainder heading for the ocean.

It is a project that SA Premier, John Olsen, is particularly chuffed about. As he said to me: 'These market gardeners previously had unlimited access to underground water and were effectively just drawing it down and down and down until the sea water was coming into the underground aquifer. We went through two and a half years of negotiating with the range of market gardeners, which was like the United Nations, in terms of the groups that were there.' Now over 20 billion litres of irrigation water a year is produced for growers, that's about the same as their bore water consumption and is half the total outflow from the Bolivar plant.[23]

Brisbane has tackled its sewage problem a little differently. All of the city's waste is treated at secondary treatment plants, with effluent flowing into the Brisbane River and on into Moreton Bay. The Luggage Point treatment plant, dealing with about half the waste, is being upgraded, but recycling of effluent is limited to about 11 megs a day, which is re-used industrially by BP and local golf courses.

The city's water and sewerage manager, Mark Pascoe, explained: 'A joint State and federal study was undertaken about five years ago to find out what the community expected and decide on the best technologies to deliver on that expectation. For instance, if we'd needed to make the Brisbane River clean enough for kids to swim in, we would have made different decisions.' But he pointed out that other contributors to river pollution such as industry would have had to change as well.

In Brisbane there are dreams of replicating a Virginia pipeline, which would take treated effluent inland to the Lochear Valley around Grafton and perhaps even up the hill to Toowoomba and the Darling Downs. The hill is quite a hill, making pumping costs considerable, and there are questions about the salinity extra water might bring to the Lochear. Most of all, money is the issue: Mark's estimate is $300 million to $400 million. The plan is still in its infancy.

The Big Pong

One cannot sing the praises of South Australia's Bolivar plant without mentioning the massive stink bomb that hung over Adelaide for three months in 1997. Bolivar was found to be the main source of the 'Big Pong', as the great absorber of smells began to work in reverse, letting off putrid, sulphurous odours that daily made their way on the north wind 20 kilometres down to the city.

It is a disgraceful story, one which fuels the argument of the critics about privatising or outsourcing public assets. The Bolivar treatment plant was sold off by the South Australian Government in 1996 to the private company United Water (owned by Vivendi and Thames). But while the lust for profits may have contributed to the Big Pong (through badly maintained equipment), in fact the rot started in early 1993, before consumption-based pricing was introduced and almost three years before the wastewater management services were contracted out to United Water.

The problem began with the filters through which the effluent needed to pass before it reached the stabilisation ponds. Unlike the ponds, filters do give off some odour. SA Water Corporation had been required to fix the problem, but rather than spend money, filters were progressively turned off or down. Inevitably, the lagoons became overloaded. When United Water took over in January 1996, the situation deteriorated rapidly, the main gate in the plant failing and a replacement also breaking down.

By the end of that year, the design load limit for the lagoons was being exceeded most of the time, and by April 1997 they were smelling badly. Bacterial treatment in lagoons had simply broken down under the load. As the auditor brought in to get to the bottom of the drama observed, 'This heavy loading has come about not because of plant overloading but largely because of action to reduce biofilter odour production and operating and maintenance costs.'[24]

It was a disaster waiting to happen, and indeed the Government's handling of United Water's winning bid for the 15-year contract to service Adelaide in late 1995 was surrounded by its own stench of scandal. This included the bid being lodged hours after a final deadline (when details of other bids had already leaked out) and the inability of government to deliver on promises to the electorate that the private company contracted would be at least half Australian owned.

Stormwater

Dealing with the sewage problem is not optional. If the system goes wrong, the public quickly knows about it. But there is another underground problem in the suburbs that Australians seem perfectly oblivious to—and that is stormwater. While effluent is the 'treated everything' from inside the house and the garden drain, stormwater is all the rest: rain and all the nasties that get washed down with it into the mouths of stormwater drains that line the roads.

It's hard to get the stormwater message through to the public. Surfrider's stencils of 'Drains to the beach—do not litter' on storm drains help make the link between suburb and beach, but surveys done in Perth by CSIRO show just how short-lived such a message is. Australians are at first enthusiastic about doing the right thing but quickly become jaded after the novelty wears off and CSIRO survey visits stop.[25]

Run-off is a bigger issue than ever in cities and much of it can be blamed on the fashionable changes to the good old quarter-acre block. After all, neighbours do need to keep up with neighbours. The typical result is that a perfectly good house with a proportionate garden is rolled in favour of a massive concrete limpet with double garage and a handkerchief lawn—which is more often sacrificed for a patio. Leaving the eyesore aside, the effect of this change, multiplied across suburbs, has been to reduce dramatically the amount of ground available to absorb water.

Public opinion has not been helped by the plethora of lowest common denominator lifestyle programs like *Groundforce* and *Backyard Blitz*. These shows with gleaming, beaming and squeaky clean gardening 'experts' promise to change the perfectly good backyard of lawn and Hills hoist into a sort of pastiche UHT nightmare, a wall-to-wall showroom of sapphire blue tiles, pagodas, numerous over-fertilised pot plants, and fountains for good measure. Is it any wonder that the first reaction of those who have been kept in the dark about the 'blitz' burst into tears when their new 'little Italy' is revealed to them?

These are precisely the changes which increase the amount of water whooshed out to sea, or worse, into our sewers, which have enough to deal with already. In crude terms 90 per cent of the rain that falls in urban areas runs off the land. Compare that to 10 per cent in rural areas.

The nightmare for the wastewater people is of stormwater invading the sewerage pipes, overloading them with extra water and rubbish that is forced to go through already overstretched treatment plants. It also makes it impossible to separate the problems of effluent and stormwater. This is something of a hobbyhorse for Andrew Speers. 'It's not just cracks, it's illegal connections, people connecting their stormwater system to the sewerage system, and it's manhole covers that are poorly designed so they are too low.' The common garden tap can be just as bad. Properly designed, the drain under the tap should be raised enough above ground level to stop rainwater running off the patio and garden and into the sewerage system. Unfortunately, the chances are that yours is not.

Then there are the cracks in home pipes. Amazingly, half of the stormwater infiltration into the sewerage system comes when water seeps into the house service line, which the water authority is not responsible for and no one else looks after. 'I just replaced the sewer line in my backyard because of tree roots,' said Andrew Speers. 'Now, I have no idea how long it's been in that condition. And people just don't check. So long as the toilet continues to flush, nobody bothers to ask whether the pipe is leaking or not.'

The time that is really unpleasant is when sewerage systems overflow, often into rivers or harbours. Pipes are designed with overflow points in them, because if effluent can't get out because of a blockage somewhere it will come back up through the toilets. Attractive thought.

According to United Water in Adelaide, around 80 per cent of all sewer overflows are due to blockages caused by tree roots clogging up the system. The other reasons are illegal dumping of solid materials, cooking fat being disposed of in sinks, and lastly power or equipment failure. Is it any wonder that during storms, these people wince. Studies on many of the stormwater outlets at beaches show that bacterial density increases are so high that it is not wise to swim after a good downpour.[26]

Yet again Sydney draws the short straw. The city lies on impervious sandstone, so big sewerage tunnels end up sitting in carved-out rock filled with stormwater that leaks into any crack in the piping. Until recently, pipes were overflowing at discharge points, albeit with diluted effluent, straight into the harbour about 20 times a year.[27]

To fix the problem, a new tunnel has been built to take 90 per cent of the wet weather sewerage overflows from four major points around the harbour and feed it by gravity to (guess where?) North Head treatment plant, the idea being that the contents would be pumped up from a holding area and treated when the storm passes.

Unfortunately the new tunnel does not contain all overflows from the old tunnels. A furious Manly mayor could not be appeased by the situation.

> Even with around $500 million having been spent on the North-side Storage Tunnel (originally estimated to cost $375 million), we are still faced with the prospect of having sewerage overflows into Sydney Harbour, and further investment in entrenching 'end of pipe' solutions will do nothing to alleviate this problem, particularly as on Sydney Water's own figures, biosolids (sludge) collected at sewerage treatment plants will increase by about 100 per cent over the next 15 years due to population growth and treatment plant upgrades.[28]

'North Head now takes 42 per cent of Sydney's sewage, all the way from Blacktown,' said the mayor, explaining that despite the deep ocean outfalls, up to 60 sludge trucks a day have been rattling through the residential areas of Manly for nine years now and locals are sick of it. The mayor might well be angry: Premier Bob Carr's $3.01 billion Waterways Package announced in 1997 promised to avoid a 'quick fix' solution and decrease the reliance on ocean discharge.[29]

A State Parliamentary inquiry into the sludge problem was announced in August 2000 but, due to a backlog of another kind, is still under way. Chair of the inquiry, Richard Jones MP, commented early this year that using the sea for more and more low-level treated effluent as Sydney expands westwards was becoming ridiculous. According to Richard Jones, land has now been put aside in Sydney's west for treatment facilities and in his view Sydney needs to look seriously at keeping sludge inland where it is close to farming and treating the effluent to a drinkable level where it could be put back into dams or rivers below dams to increase flow. And how much is this likely to cost? 'Oh, several billion,' he said. Que sera, sera.

As if fixing these leaks and overflows is not made hard enough by nature and the home owners, control of the system is completely fragmented. '[The] drain that I inspected out near the Olympic site on one occasion,' explained Andrew Speers, 'rose in private land, flowed to local council, which was a natural creek, then flowed to Sydney Water, which was a channelised stormwater drain, then flowed back into a natural creek owned by a different council, then flowed into a tidal mudflat controlled by the MSB Waterways. Now how can you manage that asset? Sydney Water obeys its operating licence, and it cleans out the silt once a year and it stops at its boundaries. Next rainfall, it all washes down again. And the flow accelerates massively through the Sydney Water area because it's channelised and then it runs through the creek downstream which is not channelised, and because there's so much energy in that water, it just scours it.'

In terms of the big picture, the economics are stacked against sensible sewerage management. While water authorities may charge for stormwater collection in their areas, it is usually a flat fee, regardless of the size of house or business, so there is no financial incentive to do anything innovative about run-off (beyond what a council might require). Much of the money raised in stormwater charges is not even spent on maintenance of the system.

Australia might do well to look at a few ideas from across the Pacific. In the United States, there are now regional stormwater businesses, public and private, which charge people according to land use or zoning, and discounts are given where developments make an effort to create areas where seepage of rainwater back into the ground can take place. This is not blinding science, just commonsense.

You might wonder with all these problems why on earth the good old rainwater tank isn't brought back to help solve both consumption and run-off. Water would certainly be better in the whisky. Alan Jones, Sydney's most powerful morning voice, has pondered it:

> What is wrong with a tank beside your house, for God's sake? I mean that's the way we were brought up. We were utterly self-sufficient in relation to water. Is someone trying to use water as a means to control our lives? They're going to now start charging us for it. And you know it seems to me that these are the things that are properly the province of the individual . . . The whole of Sydney stood still when someone looked like threatening one frog on the Homebush Olympic site and yet here is this city of three million people where it is now no longer acceptable, and hasn't been for years, to put up a tank. So we whinge every time the Warragamba Dam drops below a certain level, but every time it rains, all this stuff just carts up, gathers up, all the slush and muck and oil and waste, and we pour that into the harbour and so we ruin the harbour as well and we ruin the beaches. Incidentally, it is legal to put up a hot-water tank in Sydney; it's just not done.

It is staggering that even in the largest inland river town, Canberra, with its low-lying, spacious and expanding suburbs, that rainwater tanks are not promoted. The redeeming news is that storage tanks are exactly the way our top scientists are thinking.

Weaning Adelaide off the Murray

Recycling is a hot topic in scientific circles, and some of the hottest thinking is coming from South Australia. The reliance of Adelaide on the Murray puts it in a uniquely vulnerable position; 40 per cent of its water in a normal year is sourced from the river. But what if Adelaide were able to do without Murray water?

'Enough water falls on Adelaide in an average year to supply all its water needs. Just falls from the sky,' said CSIRO Land and Water's Tom Hatton. 'Now, we don't ever consider stormwater as a source of water. And Adelaide might even be quite fortunate because it has aquifers below it that can be used for artificial storage and recovery, which means you could store a proportion of their water.'

So far, we have discussed re-using what should be the dirtiest water, sewage water. What the boffins are working on is not effluent, but stormwater. A system where stormwater would cease to be a flood problem. It would be a major resource, perhaps 55,000 megs a year or one-third of Adelaide's water needs, stored in aquifers underground and then harvested for both agricultural and domestic use.

As early as 1952 in South Australia there was a proposal to use aquifers close to Adelaide to store rural run-off from the catchment through winter when the dams spilled over. In the end, the pipeline from the Murray was favoured. 'It was water they could see, I guess,' said Peter Dillon, part of CSIRO's Adelaide groundwater team. But Adelaide's dependence on the Murray is also limiting growth in parts of the State like the Yorke Peninsula, and this has put aquifer storage back on the drawing board.

Working properly, the process is remarkably cheap and efficient. Instead of huge outlays for dams of 10,000 or 50,000 megs, these local storages could be 10 or 100 megs, involving a well, some pipes, and pumps to inject the water into the aquifer and then extract it. And like much of Australia, Adelaide is blessed with numerous underground water storages, typically about 100 metres deep.

Better still, aquifers not only store water, they also play a part in cleaning it since many bugs die in the dark environment. The first work on aquifers started at Andrew's Farm in 1993, with the CSIRO working with the idea of a developer who wanted a model village in a rural setting where any excess irrigation water would be collected and trapped in wetlands, recharged to an aquifer and then pumped out again to water vines.

Peter Dillon set about testing the limits of aquifers as water clean-ers. 'We were expecting major problems, injecting water with high levels of suspended solids, you know, it looked brown.' To everyone's surprise, results were very positive. Most of the solids, including heavy metals and phosphorus were removed through the natural filtering process of the water's trip to the aquifer. Bacteria were not at all happy—90 per cent die off every 10 to 20 days. The process filtered out pesticides and even killed viruses, including some of the crypto sporid-ium and giardia. Work is far from complete, the team now getting to grips with the biotic effects in the aquifer—it might be, for example, that some of the bacteria secrete enzymes which are anti-viral.

Already urban developers have caught on. Floodwater manage-ment is a major problem in cities. Some councils are now demanding that land is set aside as open space to keep flow levels down. The good news is that developers too have worked out that parkland or wetland can be symbiotic with urban sprawl. Houses opposite parks command higher prices and houses opposite perennially green parks command even more. According to Peter Dillon these values more than offset the cost of irrigation systems and an aquifer pump to water the park in summer. Projects like Northfield and Kaurna Park in Adelaide are marketed as 'green parks'.

The other water reclamation work is at Bolivar, where some of the water for irrigators to be sent up the Virginia pipeline is being stored in test aquifers. And remember, this is treated effluent, not storm-water, which had proved problematic in some European trials. 'The Europeans didn't give us a snowball's chance in hell,' said Peter Dillon, 'because using organic-rich material can stimulate microbial growth and create a slime which clogs up the well.' Once again, though, the project looks very promising.

The potential is enormous. In the summer months, the Virginia pipeline is already pumping at near capacity, but in the wet winter most of the treated effluent is not needed and goes out to sea. If this could be stored, it could be a boon for agriculture. The same is true of many of the pipelines running across the State. There is even talk of recharging brackish aquifers with treated water that could shandy the storage enough to be useful. But that's down the track.

Head of the CSIRO's Policy and Economic Research Unit, Dr Mike Young, believes he will get the thumbs up from the State Government to do a feasibility study. 'It wouldn't be a total weaning off the Murray but perhaps in normal years we could get to the stage where we weren't taking any water from the river. The added attraction is that with growing salinity around Adelaide, we could leave up to 70 gigs of water in the river for environmental flow.'

Not all parties are thrilled with the news, however. It is precisely Adelaide's dependence on the Murray that has driven the campaign which has so successfully increased awareness of the plight of our most important river. Adelaide may not need the Murray, but the Murray needs Adelaide. 'You've been talking to some of the private water companies—Graham Dooley advocating this, is he?' was the reaction from the South Australian Premier, referring to the head of United Utilities. John Olsen seems well aware of the political implications, and in particular the fear that States upstream from South Australia might be less careful with their share of the Murray–Darling system. 'That is just simply one part of the equation. That question might be idealistic and romantic and something that we ought to seriously look at. But it doesn't address the question of food and beverage exports, which from our State's perspective, we're going gang busters at the moment and have the potential to continue to do so. So you cannot allow the States upstream to relax.'

The real Watter Quandary

High in the lush hills around Keel Mountain, on Queensland's Sunshine Coast, lives a rather extraordinary woman. For almost ten years, Jenifer Simpson has been on a crusade to make the Great Unwashed understand and appreciate the value of water—she is one of the few who is actually trying to get the message through to the average Aussie. An industrial chemist and donkey breeder, Jenifer Simpson is revered in bureaucratic circles as an ogre, more of that in chapter 6. It is very unwise to mention the 'D' word around Jenifer. She has been campaigning against dams for years. She also has an intoxicating giggle. As a producer of numerous publications for the Australian Water Association, from academic to cartoons, education is her main focus, and it starts with language. Take the time Jenifer was asked to participate in Queensland's water recycling strategy, run by the Department of Natural Resources (DNR). 'I arrived late, of course, because you can't get to Brisbane from here at 9 o'clock, because you start earlier and it just takes longer, so I was ten minutes late and it became extremely apparent after about five minutes that the 20 people there were from different government departments, you know, education and business, tourism, health, all there, and I would suggest something like 15 out of the 20 didn't know the difference between wastewater and effluent.'

Jenifer gets quite exasperated sometimes. 'Even our editor calls it a sewerage treatment plant instead of a sewage treatment plant. "Sewage" is the raw stuff we all know about. "Sewerage" is the piping that takes it away.' But her biggest battle has been with people in the recycling business who talk about 're-using wastewater'. 'We don't re-use wastewater, we re-use water. It's not a waste bottle, it's a bottle.' She is pedantic, yes. Almost a touch of Joyce Grenfell, for those who can remember. But such pedantry is not without good reason. As we'll see shortly, the mindset of Australians about re-using water can be remarkably closed. If I say I'm re-using and drinking wastewater, that's not nearly as pleasant as me saying I'm re-using and drinking water. And Jenifer Simpson is passionate about recycling.

By recycling, Jenifer Simpson means not just stormwater, she means recycling effluent, to drink. Recycling is not new—some of the earliest work on recycling came from outer space, where NASA's innovation was pretty important, given the obvious restrictions. So why not look at recycling effluent? After all, effluent is the only source of water that increases with population.

There's one teeny problem with recycling effluent. The knee-jerk reaction of most Australians to recycling—repulsion—is understandable perhaps, but nevertheless utterly ridiculous! It is an outrageous luxury to use water only once. Londoners on average drink water from the tap that has been through about seven people before it gets to them, and most of them seem perfectly all right. In fact, in a recent report, Essex and Suffolk Water have won approval to take water from the Chelmer River, downstream from where they discharge tertiary level effluent.[30] 'If you get into conversation as I do with taxi drivers or people on trains', enthused Jenifer, 'and always talk about water for some reason, and everybody I've met in the UK knows perfectly well that their water, particularly coming down the Thames, goes through all these people, and they agree it's a bit funny, but you know, it doesn't make any difference.'

In scientific terms, it's hard to imagine anything purer than H_2O when fully treated. Yet perhaps readers will not be surprised to hear that a seminal community survey in Perth demonstrated that the majority of Australian city dwellers are not prepared to drink recycled water. 'Only 9 per cent said they would be prepared to drink recycled sewage,' said Andrew Speers. 'Fifty per cent were prepared to accept recycled water for use in the laundry, but not for more personal use such as showering or cooking.' Perhaps these people should have been asked what type of water they believe the folk up and down the Murray–Darling system drink. It's recycled water, every day of the week. If it wasn't recycled, there would be even less flow than there is now.

In her fight to educate people on what she calls 'the dilemma', Jenifer Simpson some years ago created a cartoon character, Watter Quandary.

> Watter is a drop, and he's a fairly posh sort of drop and he went to a good school and he's got expensive parents and he wants to be drinking water. And he goes out one day and he's caught up in some run-off. It's been raining, and he rescues a rather sort of scruffy-looking drop that's very distressed and takes her home, and she's so upset that they get stuck into the gin. And as a result, the gin, the alcohol, overcomes their surface tension and they coalesce, and so he gets very polluted and he's so disgusting, and he has to go to the doctor.
>
> And the doctor says the only place for you, my boy, is at the sewage treatment plant. So he has to get into the toilet system and he gets flushed into the sewage treatment plant. And then he becomes effluent.

As would be expected with the pen of Jenifer Simpson behind Watter, the end is far from tragic. 'No, no, he gets cleaned up,' explained the author. 'He meets a water quality consultant called N. Vera Mantle, and she gets him beautifully clean through reverse osmosis and so forth.' In fact, Watter becomes purer than drinking water, but the drama isn't over, because when he checks in for an interview at the water treatment plant they refuse to take him. 'Clean or unclean, you can't possibly become drinking water. You're effluent and we never make drinking water out of effluent!'[31]

The solution was to head for the river, where treated effluent water is returned to the system and runs freely and is taken up by country towns as drinking water. In the end, said Jenifer, very succinctly, 'Watter and Vera end up in the town of None Better, where the local weir has got a blue-green algae bloom and they absolutely can't cope and they're out of water, so they go and advise them about water recycling and then they get married.'

It is not fair to single out Perth residents without mentioning protests elsewhere. The 'ugh' factor, as it is known, also emerged in California, when the city of Los Angeles announced in 1999 that it would be trialling a scheme to recycle sewage water for home use, a 'Water to Toilet' program. Treated effluent from sewerage plants would travel about 6,000 feet (2,000 metres) over a period of several years before it even hit the first production well. The debate was heated, with some experts questioning the safety of the recycling scheme and the testing methods. Periodically, there are also scares in Britain about whether the pill is putting hormones in the water system which accounts for the lowering fertility rate.

However, another revolt occurred much closer to Keel Mountain, in the Caloundra and Maroochy shires. As Jenifer Simpson tells the story, the Queensland town of Landsborough had piped water, but no main's sewage drainage. With septic tanks overloading, all the septic muck was going virtually untreated into Ewen Maddock Dam, which was rapidly growing in nutrients and becoming disgusting. This was a recreational dam, a place where kids swim.

The community agreed the area needed sewering and the decision was made to put the sewage plant outside the catchment of the dam. Importantly, the community group involved decided that rather than have effluent irrigating a forest, the water might be cleaned up to a higher quality and put back in the dam.

Consultants were brought in to educate the community about recycling, but this backfired. 'They think they're doing really well if they contact 10 per cent of the community, and then they extrapolate that to the rest of the community,' said Jenifer Simpson, pointing out that the people who are actually contacted end up with a much greater understanding of water quality and waste management than the people who aren't, so they're no longer typical of the rest of the community. The Queensland Environmental Protection Authority (EPA) gave the council a draft licence to put the effluent into the dam and on this basis a 'you beaut' plant with ozone, activated carbon and ultraviolet was built.

Unfortunately, at the same time the Citizens Against Drinking Sewage was launched by a local, who had for some time been harbouring political ambitions in the area. The feeling whipped up in the community by CADS was enough to give the EPA cold feet. No formal licence was issued and to add insult to injury, according to Jenifer, 'now they're building a pipeline to take this virtually drinking quality water down to Kawana and putting it out the outfall.'

The whole debacle, said Jenifer, was a lesson in community relations. 'When you give people some information and then take them to a treatment plant, show them around and look at the water and what happens, they actually convert, quite cheerfully. There'll always be some that don't, but you can't keep everyone happy, if they don't want to drink recycled water they can have a rainwater tank. They can buy a bottle of mineral water.'

Again, though, the economics work against the Greens. Not surprisingly, councils would like to be paid for improving the quality of water. Take the mayor from Wagga, who claims, 'we put potable water back into the Murrumbidgee that's used downstream, and we're saying that we should be given some compensation for that. We've got 24,000 megs of water as an allocation from Riverina Water for Wagga Wagga and out of that we only use 17,000 megs.' But, said Jenifer Simpson, 'Governments don't want a bar of that because it means they are losing revenue. They are extremely reluctant to admit that if you discharge effluent in a river and it's extracted downstream for irrigation or for urban water supply or whatever, it is in actual fact recycling.'

It will be a long crusade to convert the bulk of Australians to effluent recycling, but in pockets of the country, projects like the failed Landsborough plant may actually come off. Even if people could never be converted to drinking recycled water, we actually drink so little of the pure water that is supplied to us (1 per cent) we could go on almost indefinitely supplying this amount from fresh resources. It's the other 99 per cent that could be recycled water.

Water quality and wobbly consumers
An estimated $400 million a year is spent on water treatment in Australia. Despite this, and despite there being no fall in drinking

water standards in Australia, there appears to be a significant drop in consumer confidence. This is according to the nation's Productivity Commission, based on feedback from industry leaders. Crypto might be partly to blame, but the Commission believes the system is the major culprit.

The country's quality control is managed through 'guidelines' and 'standards'. 'Guidelines' are set by the National Health and Medical Research Council, but they are only guidelines. The actual 'standards', which have the force of law, are left to the States and Territories.

Overall water quality standards tend to be driven by levels set by the World Health Organisation, but as can be expected when the States get involved, implementation varies enormously in Australia. The NHMRC guidelines set out in 1996 were taken as long-term goals, they are still being phased in across the States. At the same time, mind you, these guidelines are being updated.

State regulation in Australia compared to other countries like the United States and Britain is a hotchpotch and overall pretty light on. This is good for flexibility within the industry, which undertakes a lot of self-reporting, but less good for compliance, certainty and accountability. Rather than legislation, tools like licences and 'memoranda of understanding' are used. (Remember from chapter 2 how successful the Sydney Water memorandum of understanding was for NSW Health.) The Productivity Commission has warned that 'differences in the quality of water across the country may prove to be contentious in the future unless the public understands the reasons.'[32]

The United States, with very high standards, has thrown money at water treatment. Australia, as the Commission has pointed out, has found itself in a more difficult position: 'In the absence of rigorous regulatory assessment, it is difficult for authorities to fully justify existing standards, which vary across and within jurisdictions. It is also difficult to make sound decisions about infrastructure investments, in the face of pressures to adopt new technology. Any further increase in standards is likely to require significant additional investment in water treatment infrastructure, with cost implications for consumers.'[33]

Consumers, however, are in the dark about water quality treatment. How many of us would know what level of bugs are 'safe' and what the risks are? Compared to both the Kiwis and the Americans, Australians get relatively little information on risks to public health. Without this knowledge, people are more susceptible to 'scare campaigns', and it makes it almost impossible for them to contribute to what goes into both the guidelines and standards.[34]

It was precisely this issue which got Melbourne Water worried. For some time, one of the tests being used worldwide for water quality was 'total coliforms'. Coliforms are micro-organisms that may cause disease. The worst of these coliforms are 'faecal coliforms', which, as the name suggests, come from the intestines of warm-blooded animals. Melbourne had next to none of the latter in its pristine catchments, but had a problem in that other coliforms occurred naturally and often grew in the system. The old tests only picked up faecal coliforms, but as testing was improved, even benign coliforms were being detected. What worried Melbourne Water was that, inevitably, these new test results would surface in the public arena and cause alarm. This would force drinking standards to be tightened and the options were not good: either a $500 million brand new treatment plant, or stuffing serious amounts of chlorine into the water to kill off the bugs to allay consumer fears.

What it did instead was some rather good research which has averted the concern. Dr Kit Fairley's team at Monash University's Cooperative Research Centre compared rates of gastro in 300 families which had a sterilising water filter with 300 which had a 'dummy filter'. The team found no difference in their health, showing that the micro-organisms in Melbourne's tap water were not causing any significant gastro problems, and were able to pre-empt a scare by releasing the results to the media. As it turns out, the World Health Organization is moving away from the total coliform test, just picking out the worst bugs like *E. coli* and enterococci.

There is a view that giving the public more information will simply lead to paranoia and a demand that water be much purer than is necessary or even healthy. Australians might then find themselves in the unfortunate position where *every* visit to India will be a case of Delhi Belly, not just the unlucky one. Perhaps we should give Aussies more credit than that, even the CADS and Sand Gropers. The whole process, however, just adds to the picture of muddled management in almost every part of the water cycle.

2050, a recycled odyssey

If one thought can be drawn from how urban water works in this country, it is that it's no longer possible to separate water supply from wastewater and stormwater. The three are critically interlinked. The amount we use impacts on the amount we have to clean up and discharge; stormwater invades the wastewater system every time it rains; and, if we can only get our minds around it, recycling could mean that both wastewater and stormwater can come again as drinking water. In this way, the thinking is no longer the traditional, linear transporting of water from a remote location to the city to another remote location.

The suburbs of the future are green. They employ a new 'third pipe', so called because the first is the pipe that brings your water in and the second takes waste away. The third pipe is the one that takes 'grey water' from the shower, sink and the washing machine and recycles them into the garden and the toilet.

These are not just ideas on the drawing board. The futuristic suburb of Mawson Lakes in Adelaide sits on two aquifers, one on top of the other. There are long-term plans are for one aquifer to be used as stormwater storage and for the other to store treated effluent which would be pumped back up for use in a separate 'lilac'-coloured third pipe, for use in the toilet and the garden. And yes, safe enough for kids to run around under the sprinkler. Similar developments are in progress at Fig Tree Place in Newcastle.

In Brisbane, Mark Pascoe's team is working on something even bigger. 'Icon projects, we call them,' he explained. 'We've got a development happening now of 800 houses with land on the market by 2002. There, we're looking at separating black water from grey water, treating the grey water a little locally, and then sending if off for use in a nearby industrial area. The idea is at least 50 per cent of the houses will have rain tanks.'

According to Mark Pascoe, the project is not quite Utopia. One of the biggest challenges is getting the urban bureaucracy all thinking along the same lines—health regulation, town planning—as well as the land owner and developer. He's also got to convince the community. 'There was a recent survey done in Brisbane and 59 per cent of those asked said they were concerned about water use issues. Even more said they'd be prepared to recycle grey water. But we weren't asking the question rhetorically. It's another thing altogether if you front up at their door with the plastic tank.'

Understandably, all this is right up Jenifer Simpson's street. 'I would like to see all new development with rainwater tanks, where people have got the choice. Then we have dual reticulation to them and they use their rainwater for all inside household jobs except for toilets—wash in it, shower in it and use it in the kitchen—and they use their recycled water for the garden and toilet.' Rainwater tanks would present a new architectural challenge in the city, but they could hardly be uglier than satellite dishes.

Andrew Speers, CSIRO's urban water expert, has more ideas for moving consumption from being demand driven to demand 'managed'. His computer modelling shows that a Melbourne household which typically caters for a flow of 300 litres per hour could reduce flow to just 40 litres through a combination of low-flow shower heads, dual flush toilets and recycling garden water to the toilet, although it should be mentioned that no allowance is made for fire fighting. But the dramatic change would be to put into each home a small water tank. The tank would be filled in off-peak periods and would be drawn upon by the household during peaks, but without affecting the main water supply. Reducing peak flow would not restrict water to the household. Water could at any time be drawn from the main system. However, it would mean pipes could be less robust and cheaper, so the cost of the third pipe option would be more affordable.

The cost of being green

Even if recycling of grey water and effluent does work in practice, the looming question is what all this might cost. As far as infrastructure installation goes, new suburbs are the most promising, but as with many other issues about water, the situation is confused because the true cost of delivering water is blurred. Just what the right price for water should be is hotly disputed. It is certainly more than we are paying now, if it is to reflect the real cost of being green, the cost of building and maintaining sustainable infrastructure. Andrew Speers is quite annoyed by the way that the whole concept of ecological sustainability, making things green-friendly, has been hijacked by business to mean *economic* sustainability. That's pragmatism. Jenifer Simpson gets infuriated when you-beaut technology is passed over as too expensive, because she thinks we don't pay the right price for water anyway. 'I think we could at least start by saying right we're not paying the right price and we don't expect people to pay the right price, but at least when we're doing the costing of options we have to use the right price, so that instead of saying water costs 50 cents a kilolitre we recognise in fact it costs $5 and use that costing. So when we're looking at alternative supplies, they start to look a bit more reasonable.'

It is also one of the unfortunate facts that when reformers look at cost–benefit analyses and 'user pays', they are very selective about the bits of the system they include. 'If you have to put a rainwater tank in a new development, plus a pump, plus some maintenance,' explained Simpson, 'the cost of that, which is quite expensive, is borne by the householder or developer. If you put in another dam, though, the council so far has always had the dam paid for by the State government.' The good news for recycling is that governments are themselves moving to full cost recovery on any new dam infrastructure.

Much more challenging is changing the old suburbs. 'Well it's always talked about as a nightmare, but that's because no one has worked out how to do it,' said Dr Mike Young of the Adelaide ground-water group. 'We're leaping from the systems of the 1800s and early 1900s to 21st-century technology. But we've got water utilities that are backed into contractual obligations with no incentives to think about change and there's a lot of institutional resistance.' Grey water recycling needs a third pipe, so for older suburbs the capital costs today are prohibitive. But futuristic thinking on effluent recycling may be more achievable. The latest ideas on the most economic scale for sewage treatment is just 5,000 houses and Mike Young talks of putting in small-scale plants in the Adelaide foothills. Treatment plants can even be put underground in future. And the cost? 'Sure, hundreds of millions, but so is the upgrade of existing treatment plants, and every time we spend huge amounts fixing old pipes *that* should be remembered,' he said.

According to Andrew Speers, where money is expected to have influence is on the size and position of our treatment plants. Around 80 per cent of the cost of a sewerage system is in the pipes, taking waste from the home to huge centralised treatment plants, 'an obsession with miles and miles and more miles of pipelines; and a lack of respect for the environment,' said Jenifer Simpson. 'And they assume that these things are going to last ad infinitum. But the life of an average pipe is about 76 years and a lot of our pipes that are in the ground at the moment are 75 and a half, and we have got no reserves at all to be able to refurbish or replenish or replace these pipes when they crack.' The plants themselves are so over-engineered that a former managing director of Sydney Water used to comment that they would be the safest place to be in a nuclear blast.

In a future world where the cost of treatment is likely to be driven by transport rather than connection costs, smaller local treatment plants could begin popping up around the suburbs.

This would also appease those fearful that water saving will bring constipation to the sewers! 'You think we've got a drainage problem?' asked water expert Don Rowe. We've got a sewerage problem. You've got these huge tunnels, all sorts of stuff going into them and the water that you 'waste' is there to flush that system out. Andrew Speers gives this theory short shrift. Reductions in water usage have got to fall a long way before we need to worry about constipation.

What futurists stress is that a dogged adherence to one solution is not going to help sustainability. Recycling, for example, which diminishes the flow in a river that depends on treated effluent, or for which pumping costs are ludicrous because of topography, will not be helpful. One of Jenifer Simpson's many cartoons is a bucket with a pin prick on one side and a gaping big 5 centimetre hole on the other. There's a fellow desperately plastering up the pin prick with million dollar notes and not noticing what is gushing out on the other side.

The technology is here today to close ocean outfalls, re-use water and perhaps never build another city dam. The companies, even if they are foreigners, are around to build the infrastructure: Vivendi, Thames Water and others. (Most in the industry agree that it is foreigners who have given Australia a kick-along in terms of innovation.) What stands in the way is cost. The cost of the infrastructure, but most important of all, the cost of the water itself.

The reformers have patted themselves on the back for reducing prices for consumers and yet the truth is that the value of water is much, much higher than this. At present, the cost to deliver water is between 70 cents and 90 cents a kilolitre, depending where you live in Australia. That's the equivalent of 2666 stubbies of water for 70 cents.

And the mysterious true cost? The true cost of water can only be estimated, but a relatively recent expert report put it at $2.50 a kilolitre.[35] Even if this true cost was not passed on to the average Aussie, at least if it was *recognised*, some of these futuristic schemes might not look so far fetched. There is something very Imelda-like about having 100 per cent of our water pure enough to drink and only actually using 1 per cent.

4

Pursed Lips

For his wild and beautiful river was in chains—fifteen hundred miles of it tamed to the yoke of civilisation, in a mastery he could not have dreamed. Never again, capricious as the clouds, could it appear and disappear at will. Its path was mapped out; its glory measured by the cubic foot, its water held back from the sea. Canutes of the twentieth century told the Murray to stay where it was— and the Murray stayed.

Ernestine Hill—Water into Gold

The swim

If you had been walking the banks of the Murray around the end of last year (2000), there's a chance you might have come across just one swimmer, doggedly crawling her way downstream—for three months.

Tammy van Wisse is not ordinary. The challenge the marathon swimmer set herself was to swim the entire length of the Murray River, about 2,440 kilometres from the Snowy Mountain waters at Corryong to its salty mouth near Adelaide.

Why? Well, partly to break the record set by Graham Middleton, the first person to swim the Murray in 1991. It took him 136 days and he finished 40 kilometres short of the mouth. Tammy Van Wisse did it 34 days more quickly. But there was another mission. She was to be a human piece of litmus paper for Murray water.

The swimmer was a fillip for the Greens, who are forever looking for ways to draw attention to the plight of our most important river system. Publicity shots had the 'River Nymph' posed in the water for the *Good Weekend* magazine, covered in a sort of grey-green mud pack, which apparently angered some of the locals, who maintained that Murray mud simply wasn't that colour.

Each day Tammy Van Wisse's fans could travel with her through a 'B Green' website that catalogued her daily progress, two million strokes, running into logs, dodging snakes and fighting a stomach virus as well as ear and eye infections that left her weak. She said, 'It gradually got wider and it gradually got dirtier, so much so that I couldn't see my hand in the water when I came into South Australia.' One hundred and six days and 2,440 kilometres after she first dived into cold Snowy waters, she hoisted the Australian flag in the muddy waters of the Murray mouth at Goolwa. Tammy Van Wisse had started in pristine waters less than 10 degrees in temperature, and by the time she finished in 27 degrees she could taste the salt.

The river

People hear about the importance of the Murray–Darling Basin, but it doesn't seem to sink in. The Basin covers one-seventh of Australia, 1,000 kilometres squared or about the size of France *and* Spain. The Murray–Darling river system provides drinking water for three million people (more than one-third of whom live outside the Basin) and supports 30,000 wetlands. Some of these wetlands have special status under a 1971 agreement signed in the Iranian city of Ramsar. Today there are 56 Ramsar listings in Australia.

When we're told that over 40 per cent of agriculture and 90 per cent of irrigated agriculture produced in Australia comes from the Basin, it means chances are that in Sydney, Melbourne and Adelaide, *almost all* our oranges, lemons, peaches, rice, bread, milk and wine come from there. Yet the Ugly Consumer wouldn't have a clue about these goodies being under threat.

So just how serious is the plight of the Murray–Darling? Doug Shears, arguably Australia's most powerful agri-businessman and head of Berri Ltd, says it couldn't be more so. 'If the deterioration over the next 20 years is anything like the last, the Murray will be unusable for anything, let alone agriculture.' The problem is twofold. There isn't enough water in the system and what is left is rapidly getting brinier and brinier.

The map of the Basin, covering huge areas of four States and the Australian Capital Territory, has become almost a separate country to water people, and it's worth taking a moment to have a look at it. Of the three major groups of rivers that form the system, the Darling, Australia's longest river, running 2,740 kilometres from north to south, contributes around 12 per cent of the river flow. The Murray itself, upstream of the Murrumbidgee junction, is by far the most important, providing about 75 per cent of the flow and carrying Snowy Mountains waters more than 2,500 kilometres to the mouth. Lying between the Murray and the Darling, the Murrumbidgee and

Lachlan rivers and Billabong Creek contribute around 13 per cent.[1]

From the source of the Darling in Queensland all the way to Adelaide, the river runs a gauntlet of suction pumps, diversions, dams, incoming polluting run-offs and salt infusions, built up over 150 years. Storage capacity of dams is a third higher than the average annual flow of the rivers.

If the Murray–Darling had a voice, this is what it might say:

I start near Toowoomba, where until recently my water sprang up near a local sewage outlet. I run first through the fertile Darling Downs but there are now so many big earth tanks trapping rainwater that I already feel low. But this is nothing to what I'm having to deal with at St George and Dirranbandi in southern Queensland. The storages are so big there that only the very biggest floods reach me these days. For miles and miles, my banks are pretty bare and most of the trees are long gone.

Cotton has done me no favours. The growers say they don't send chemicals my way but of course some get to me. From St George I run halfway into New South Wales, and they're still farming cotton.

My water from the east is just as punished. I don't know what drives the releases from the big dams—electricity? irrigation? Either way, river hygiene comes a poor third. And in the Riverina and all down the Murrumbidgee . . . rice. These fellows want blood, not water. With all the cow shit and superphosphate my Murray waters have to deal with, I'm a good soup by the time I'm through Victoria. And by this time, I'm salting up badly.

I get cleaned up, of course, by big salt interceptors. The thing is, I just can't flush like I used to. Some towns on my river banks are better than others. None, I can assure you, is as good as they say they are. And what makes me laugh is that where you'd think they'd want my best water, for their wine, is at the bottom of my system—you know, Mildura, Renmark and right down the bottom, the Barossa. I get to the end and this great barrage at Lake Alexandrina stops the sea rushing up to meet me, and all the fish that came with it. I'm pooped. In a good year, about 20 per cent of the river that I once was flows out of my mouth.

It's hard to know where to begin in talking damage to the Murray–Darling. Diane Thorley is the mayor of Toowoomba, the town that sits at the top of the Basin. She's the 'mayor with flair' and is well known for not mincing words. 'I basically see us as the tap and Lake Alexandrina or Adelaide as the plughole or the sewage.'

One of the earliest wake-up calls to the system's plight came in 1991, when a massive outbreak of the insidious blue-green algae appeared. The Darling River, which for generations had been a saviour to graziers, was now quite capable of killing stock. As author Mary E. White put it, 'It is hardly a matter of national pride that we have the world record for the longest stretch of river rendered poisonous to drink at any one time—the Darling in 1991 had 1,000 kilometres dangerously affected by cyanobacteria.'[2] To the residents' embarrassment, the town of Toowoomba was identified as a major culprit. It turned out that much of the poorly treated human waste from Toowoomba's population of 89,000 was going straight into Garry Creek, running into the Condamine–Balonne and on into the Darling.

It can even be dangerous to walk the dog along the Murray these days. 'Hazard' signs warn the public to keep dogs out of the water along some of the more picturesque stretches of the Murray–Darling. Even Tammy van Wisse's trip was delayed because of algal bloom and low water levels. Today, toxic blooms, virtually unheard of 20 years ago, cost us $200 to $300 million a year in damage and cleanup costs.

Dianne Thorley has had quite a job fixing Toowoomba's problems as we saw in chapter 3. But the problems run all the way down the system. It seems even the brumby is not free of blame. The *Australian* newspaper earlier this year kicked off its high-profile awareness campaign by identifying brumby excrement as a major source of crypto outbreaks in the pristine Snowy waters.

It should come as no surprise that the folk who care more than anyone about the Murray–Darling are those at the other end of Toowoomba and the Snowy, the bottom end, those who drink it. Adelaide is the only major city to depend on water from the Murray. As we noted in chapter 3, in a good year Murray water provides about 40 per cent of the city's water supply; in a dry year this can run to 90 per cent.

The mouth

There are water crises at many points down the Murray–Darling, but the most critical bit is the mouth. From the air, more often than not these days, the sand spits curl in from both sides like pursed lips. Records are not the best but since the whiteman arrived in Australia, the mouth has closed only three times, once around the turn of the century and twice in the last 20 years. We can now expect no flow out of our biggest river for four or five months of the year, according to Don Blackmore, Commissioner for the Murray–Darling Basin Commission.

Running away from the river mouth to the southeast is the Coorong, a stretch of water separated from the sea by a long, thin sand spit thousands of years old. There's a starkness about the Coorong. This is *Storm Boy* country where a small boy made friends with a pelican called Mr Percival; it is also the place of secret women's business on the now infamous Hindmarsh Island.

What has made the Coorong special has been the delicate balance of estuarine fresh and salt water, as sea water mixed with Murray water from the north and a wetland that drained in from the south-east. It has been home to a plethora of bird life, many species being migratory: ibis, egrets, stilts, spoonbills, gulls, terns, pelicans, and some very rare birds, like the freckled duck. Fish too have been plentiful in years gone by, such as the mullaway, which feeds off the carp and bony bream coasting down the river. Today, they struggle to survive.

The Coorong has been attacked from both ends. In the 1940s, huge barrages were put up to keep the salt water out of Lake Alexandrina, effectively cutting the estuarine area by 89 per cent and changing the balance in the fragile haven. At the other end, farmers on the wetland reclaimed land by digging channels that drained the fresher water straight out to the ocean.

Gone are the old floods of the 1950s that flushed the Murray out and triggered the great breeding seasons in the Coorong. As we shall see later (in chapter 6), there is now a war on between those who want to wind back the clock and others who want to leave things as they are.

Fish out of water

Fish problems are not confined to the mouth. The struggle of native fish up-river to keep up numbers is one of the most telling signs that all is not well in the river. Spare a thought for the Murray cod: a magnificent thick-lipped animal that can live perhaps for a century. The record size is 2 metres long, weighing 113 kilograms, but many smaller specimens hang preserved on pub walls throughout the Basin and tall tales abound. The classic story usually goes: there was a farmer who lost a large bait to a fish in the Murray and kept coming back with heavier and heavier tackle. Eventually he used a whole kangaroo on a meat hook as bait, connected to the steel cable of the winch on his tractor. The fish in the story then fought the farmer and his tractor to a standstill, resulting in the demise of the said tractor. Following this telling, you are usually invited to inspect the wreck of the machine under an old redgum near the pub, and are shown the stretch of water that was straightened by the fish in the struggle![3]

Nineteenth-century records of Murray cod and golden perch catches indicate that thousands of tonnes of fish were caught for several decades. But the Murray cod is up against it. Its natural habitat has been decimated. The deep holes, shade and natural debris it seeks out are rapidly disappearing. Snags in the river have been removed to create clear channels for irrigators, cattle and sheep have ruined the food supply and cover of the riverbanks. Worse, water extraction in some areas such as the Darling has left the river hot and low in oxygen, creating perfect conditions for blue-green algae.

The 3,500 dams, weirs, sluices and other barriers in the Murray–Darling system have turned out to be a highly effective piscatorial contraceptive. First, water is released when the irrigators need it, in the summer and autumn, precisely the wrong time of the year for fish breeding and everything in a fish's life is turned upside down. Many fish also need to travel long distances to breed, a problem if you hit a lock gate. Normally, our Murray cod covers about 90 kilometres before coming back to the same spot.

Many fish like a temperature of about 18 to 20 degrees Celsius to breed. The bigger reservoirs, like lakes Eildon in Victoria and Hume on the border near Albury, are very deep and, unfortunately, when they do release water, it tends to be from the bottom, where the water temperature can be a chilly 8 degrees. The effect on most fish is to start off the world's worst migraine. 'Within 20 years of Dartmouth Dam on the Mitta Mitta [river] going in, we lost all native fish in that system. And all the cold water does is favour introduced species like carp,' said Craig Ingram, the President of Native Fish Australia. The immigrant carp, with its well-earned name of 'rabbit of the river' has decimated the native population in both rivers and wetlands.

Oh, and fish really don't go for endosulphan. It sends them belly-up. 'Endo' is a nasty chemical used by the cotton and horticultural industries which in theory never gets into rivers. Except, say the Greens, for the time it spilled into the Ord in Western Australia or the time it ended up in beef in New South Wales, etc.

About one-sixth of Australia's 200 or so freshwater species live in the Murray–Darling, including the Murray cod, the trout cod, the Macquarie, golden and silver perches, the eel tail, catch fish and bony bream. Twenty-nine species are indigenous to the Basin.[4] Many are under threat, and there soon won't be any commercial fisheries on the Murray–Darling. Recreational fishing, one of Australia's most popular pastimes, is getting less and less recreational.

Fish fears run beyond the Murray–Darling. In Queensland, the Mary River trout is experiencing many of the same discomforts. Anglers and Greens have been fighting back on behalf of the fish, but impact is marginal. Migratory fish at various points on the

Murray–Darling now have access to fish ladders and fish lifts that create passages around man-made barriers. How do they work? The lifts are like elevators that continually carry up buckets of water from below the barrier and tip it out higher up. Fish ladders don't move; they are a system of chambers that fish can rest in as they move up against the fast water. They work better for trout than they do for the slower native fish, like the Murray cod and the golden perch.

There are also restocking programs in place, although hatcheries bring their own problems. 'Restocking is like trying to place a bandaid on a broken leg,' reflected Craig Ingram. 'Millions of dollars are being spent, but it's narrowing the gene pool and not addressing the major problem, which is that the river is too regulated.' A new industry in fertilisers made from carp has grown and even top TV chefs have risen to the challenge of cooking a tasty carp. But, as Craig Ingram has said, the long-term answer to river health is both simpler and much bigger: more fish and duck water is needed, back in the river. But before we get to the crusade to save the Murray, we should examine how we have almost destroyed our greatest river system.

White water

For the purposes of this chapter, the Murray's history starts about 200 years ago, with European settlement. While the river had a long and fascinating history before that, it remained essentially unchanged, the Indigenous people's demands being compatible with river health.

White folk started changing inland Australia and its waterways the moment they crossed the Great Dividing Range in 1813, but the sheer size of the Murray–Darling gave it a few years' grace before its length was fully charted. In 1824 the first big-river explorers, Hamilton Hume and Willian Hovell, made their way up the banks of a river they named the Hume. Almost six years later, in November 1829, Captain Charles Sturt began a boat journey down the Murrumbidgee and discovered 'a broad and noble river'. It was in fact the Hume, but thinking it a new river, he named it the Murray after the Colonial Secretary, Sir George Murray. Murray it remained. Sturt managed to make his way down as far as what is now the New South Wales town of Wentworth, near Mildura, where 'we were then roused to action by the boat suddenly striking upon a shoal which reached from one side of the river to the other,' he wrote. 'It was just as she floated again that our attention was withdrawn to a new and beautiful stream coming apparently from the north.'[5] Sturt later correctly identified these milky waters carrying fine alluvial soil from Queensland as the Darling, having discovered the upper portion of the river the previous

[handwritten margin note: naming of the Murray]

year. The same contrast in the waters of the two rivers can be seen today from the high wooden viewing platform at Wentworth. For the explorers, the atmosphere must have been every bit as exciting as the search for the source of the Nile in North Africa.

While Sturt did make it all the way down to Lake Alexandrina, the actual mouth of the Murray was not found until February 1831 by Captain Collet Barker, who was promptly speared by the locals.

Barker's demise and problems negotiating the waters around the mouth gave the river another 20 years before the first two paddle steamers, the *Mary Ann*, under William Randell, and the *Lady Augusta*, under Captain Francis Cadell, made the first trips from the mouth inland to the goldfields at Swan Hill in Victoria.[6] It was the beginning of an extraordinary trading era on free-flowing, navigable water (drought permitting).

By the late 1880s, several hundred paddle steamers and masses of barges operated on the river. Around the time of Federation, Wentworth, or the Darling Store as it was originally known, was even a candidate for the nation's capital. It would have been quite a sight—in-bound traffic carrying station supplies to isolated properties and on the return trip towing barges laden with wool bales, grain and fruit. Steamers went all the way up to Bourke but, according to locals, they had to rely on good water to get back. During one drought several took two and a half years to get back, two wool clips later.

The river might have been busy, but what was happening on its banks was to have a far greater and irreversible impact. By the mid-1840s, squatters were beginning to arrive along the Murray.

Every district in the Basin has its own stories of the settler days. The yarns from around Deniliquin in the Riverina of southern New South Wales, today one of our biggest irrigation areas, are some of the best, particularly with a spinner like Hunter Landale. 'In those days the forest was so thick that Urana was settled five years before Deniliquin was founded, just 60 miles away, in 1837.' Hunter loves talking history, indeed the handsome septuagenarian is history in the making. He lives in one of Australia's oldest homesteads near the Edward River in New South Wales. Mundiwa is a magnificent stone house with high ceilings, verandas and stables, and late in 2000 it was looking its best—rains had swelled the Edward River up between the gums. In the 1840s, Mundiwa was known as Deniliquin Run, a staggering 700,000 acre property north of the Murray. Claimed by Benjamin Boyd in 1845, the run was named after Denilakoon, the chief of one of the two tribes in the area, the Yabit Yabit people, said to be six foot four inches tall.[7] The other tribe, the Yorta Yorta, are today, 150 years later, fighting to claim back the river. More of that in chapter 9.

Hunter Landale's stories from the nineteenth century are all about water. The fertile grazing and growing country of the Riverina between the Murray and the Murrumbidgee was nothing without water, and the creeks that ran between the Billabong and the Yanco were far from secure. 'Forty-two was a drought period and Billabong didn't run for four or five years.'

When the *Lady Augusta*'s run reached to take in Deniliquin, land prices took on new values, but the problem of water remained. In 1855, Captain Cadell encouraged settlers on the upper Yanco to finance a cutting that would link the creek with the permanent waters of the Murrumbidgee. 'One hundred men were employed by October 1856 and a cutting of some miles was completed.'[8] It took another 40 years and two further attempts at deepening the trench before a permanent cutting was constructed, in 1896.

In the meantime, settlers were organising their own drought proofing. Dam building commenced up and down the creeks. It was then that the Billabong dam wars started. The tales of the Billabong and Yanco settlers blowing up each other's dams in the 1850s are folklore. They begin with George Desailly of Coree. According to author Michael Tolhurst, 'George had constructed a large dam completely blocking the creek and denying all water to settlers below him until such time as his dam overflowed. Those above him made no objection for it was to [their] benefit as the waters banked back for five miles. Those below had a different view.'[9]

On 25 October 1858, a group of angry settlers marched on Desailly's run, where they broke the dam. Ten days later three more dams were attacked. 'Desailly lost water for some hours, stopped the breach with sandbags and fortified the summit of the dam with a log house complete with loopholes for musketry. He let it be known that he would shoot anyone who might attempt again to open his dam.'[10] Desailly was not the only one to suffer. 'Brodribb at Wanganella, he was blown up,' said Landale. 'Fifty or sixty blokes from the bottom would just come and blow everything up.'

While the wars went on, sheep tied to within four or five kilometres of water by an invisible fence worked at denuding the riverbanks. They did a good job. In the late 1850s, bore wells were being sunk and the earliest of the canals built, taking the grazing areas out beyond the rivers where trees and shrubs were again attacked. It would take Australians several generations to accept the extent of devastation to our rivers from imported European farming methods. Early fortunes made on the sheep's back in the nineteenth century turned parts of Queensland and the Western Division of New South Wales into today's 'woody weed country' of brigalow and saltbush. In 1901, the Western Division was the subject of a royal commission—it was Australia's *Grapes of Wrath*, where the land was blowing away and farmers were going broke.

Yet at the end of the First World War, tree clearing had become a massive operation. When soldiers returned from Gallipoli and the Somme, blocks of land were handed out as a 'thank you' for services to the nation—and for coming back. Soldier settlements covered huge areas of mallee country, running just south of Mildura, right across the southern States. Here, the diggers found a new battlefront and many backbreaking years followed, pulling out tough mallee tree after mallee tree. Ernestine Hill's account from Mildura reads, 'Five thousand acres to clear! With axes and long-handled shovels, they "were over the top" and into it. The forest out on the rim of the red cliffs became a battle camp of bronzed heroes in war-worn khaki.'[11]

Over 10,000 soldiers took up blocks in Victoria alone, keen for an independent productive life. The reality was often much bleaker, and by 1939, 60 per cent of those Victorian soldier settlers had walked off their land.

After the Second World War, another scheme was introduced by the Federal Government, this time giving soldiers bigger properties and help with roads, houses and fences. But land *had* to be cleared; it was a requirement in all Crown leases. The frustration within today's farming community is obvious. One and two generations ago, their fathers and grandfathers were told to cut down the trees or lose the land. With the exception of Queensland, today's State laws make it almost impossible to clean up scrub, let alone trees, for development.

Yet the fact remains that 15 billion trees have been cleared since European settlement. Down some parts of the Murray, you can travel literally kilometres and not see a tree. With hindsight, this has proved to be one of the most dastardly acts ever committed by land users, even if it was done in complete ignorance and encouraged by ignorant policy makers hungry for development.

The rise and rise of irrigation

The late, great Alfred Deakin was one of Australia's better prime ministers. He was also the father of Australian irrigation.

From the threshold of his brilliant career, Alfred Deakin looked far into the future and saw the bare and blinding desert transmuted by industry and intelligence into orchards and fields of waving grain. The mountain would not come to Mahomet. The Murray River must be brought to the mallee. He set out immediately to master the subject and gave ten years of his life to the work.[12]

As a Cabinet minister in the 1880s, Deakin was appointed by the Victorian Parliament to go on a fact-finding mission to America. It was there, in Ontario, California, that he found the Chaffey brothers, George and William Benjamin, and lured them down under to start irrigating. Ernestine Hill's book *Water into Gold* published in 1937 recalled, 'There was a great deal of amazement and amusement, both on the Murray and in the capital cities. A fruit colony at Mildura! It was a jest. But George Chaffey was no fool. Here were the soils of California, the suns of Spain, and the third ingredient, water: 1,088,000,000,000 gallons of it—rolling past every year.'

The first vintage harvest from grapes planted by the Chaffey brothers was in 1881 and the brothers established a settlement at Mildura in 1887. According to Ernestine Hill, 'George Chaffey had a creative faculty that few men in the world's history can lay claim to even among the empire builders. He could stand on a ridge in a howling wilderness, and see a city there, and not only see that city, but build it himself, complete to the fountain in the square, like one of the Djinns of the Arabian Nights.'[13] Hopefuls arrived from as far away as Britain and the United States, attracted by promises of valuable produce and money.

But good times did not last. Early methods of clearing the deep-rooted mallee scrub created large-scale soil erosion, and settlers were unprepared for water seepage and evaporation from the irrigation channels. By the time the first fruit trees were in the ground in 1890, there were already early problems with salinity and poor plant varieties, the railway wasn't built and depression struck. Chaffey Bros went broke in 1895. George left for the United States, but William remained a leading member of Mildura community.

Alfred Deakin did not confine his interest in irrigation to the Chaffey brothers' work. In 1887 he reported on irrigation in Italy and Egypt, and in the years 1890 and 1891, the *Age* sponsored him to look at irrigation schemes in India. He would no doubt be pleased to know that after 100 years of irrigation, today in Victoria's Goulburn Valley farmers are getting 65 tonnes to the hectare off three-year-old peach trees.

While Victoria was powering ahead with irrigation, it was not until the early twentieth century that watering in New South Wales really got going. Intensive agriculture through 'closer settlement' had been encouraged from the 1860s onwards. The idea was to encourage smaller plots and more settlers, but development was patchy, not least because existing graziers were suspicious of land sharks and unwilling to sell off property. In 1902, Samuel McCaughey of the Riverina applied for a licence under the *Water Rights Act* of 1896 and by the following year, he had built 150 miles (240 kilometres) of channels at North Yanco, fed mainly by gravity, irrigating over 1,000 acres (400 hectares) of lucerne and sorghum. Despite a major drought he kept 16,000 sheep alive for three months.

To the north, around Leeton and Griffith, the Murrumbidgee Irrigation Area was developed between 1908 and 1912. At the time it was the largest capital works program of its kind in the world, built by men, horses and crude machinery. Irrigation spread slowly, hampered by post-war regulations that required settlers to prepare the land at some cost before they could get any superphosphate to grow the fodder crop lucerne. Felling and levelling property was very costly. 'In Parliament, Lawson, MLA, stated that one property with 1,500 water rights had not irrigated one acre for ten years, though the annual fee of £400 was paid for no return.'[14] Those rights would certainly be coveted today.

Farms on the Murrumbidgee in the early 1900s were also too small, some just a couple of acres. It became clear that 500 to 600 acres were needed for a farm to be viable. 'Farmers could then operate profitably, but only as a result of not paying interest on the capital investment to supply water.'[15] Subsidisation was to be a vital part of the history of irrigation. But what changed irrigation forever was a new era of nation building and the construction of the big dam.

After years of procrastinating, a water sharing agreement between New South Wales, Victoria and South Australia was finally signed in 1914, and the River Murray Commission created in 1917 to manage the river system. Between 1920 and 1980, 3,500 dams, locks, weirs and water storages transformed the free-flowing Murray–Darling into what Green groups would describe as a mere distribution pipe. For pioneering farmers, the pipe was a lifeline.

The building of the Burrinjuck Dam on the Murrumbidgee started in 1906 and finished in 1928, but irrigation from it started as early as 1912. The massive Hume Dam on the Murray above Albury, begun in 1919 and finished in 1936, was doubled in size in 1961. At the mouth of the Murray five barrages were built across the channels leading from Lake Alexandrina to turn the estuarine lake into a permanent freshwater supply for Adelaide. Incredible as it may seem, there's been almost no upkeep to these constructions. It's one of the major bones of contention for irrigators that we are still living off 80-year-old infrastructure that the States refuse to upgrade.

The watershed for irrigators right down the Murray was the commencement of the Snowy Mountains Hydro-Electric Scheme, ending a 60-year long settler struggle for secure water. This visionary scheme has over the years diverted millions of megs from the mighty Snowy River to drought-prone inland Australia. Sixty years on, the Snowy Scheme is now highly controversial. But sixty years or so is more than enough time to forget the social and economic challenges of post-war Australia and the mindsets of leaders of the time. Dams were hailed as miraculous engineering achievements, and they were. After wartime deprivation, the 'vision thing' was agriculture. Anyone who presented the Snowy Scheme's commissioner, Sir William Hudson, with an environmental impact statement would have been given pretty short shrift.

Deniliquin's Hunter Landale was taking no chances when Dartmouth Reservoir in Victoria (even bigger than the Hume) was mooted in the late 1960s. The proposal had been twice knocked back by Federal Cabinet, and an alternative at Chowilla in South Australia was gathering support. Firmly of the belief that politicians had no idea how important a dam would be to the region, he suggested to the Southern Riverina District Council that they should run a journalist tour to showcase the importance of irrigation. He then set about activating the schoolboy network and convinced two of Australia's top journalists, Keith Dunstan in Melbourne and Stuart Harris, the London *Times* rep in Canberra, to gather up some colleagues to be wined and dined in the Riverina.

Hunter Landale also convinced another contact, John Landy, the Samaritan former athlete and now Victorian governor, to chip in. In those days, said Hunter Landale, John Landy was working for ICI Agricultural and agreed to the company shouting all the food and accommodation for the media, a generous $6,000. As Hunter Landale tells the story, when John Landy asked when he would be speaking, Hunter Landale replied, 'You'll get more than enough of the benefit when the dam gets through.'

The expected rash of long articles followed and a short while later David Fairbairn, member for the federal seat of Farrer (and cousin of Hunter Landale), sent news that Cabinet had passed the development of the Dartmouth Dam by one vote.

Today irrigated agriculture makes an estimated $12 billion to $16 billion a year for the country, most of it in the Murray–Darling

Basin,[16] yet the irrigated area covers just 2 per cent of the Basin. Irrigation is one of the few bright spots in Australian farming, but it comes at a serious cost. Nothing illustrates this better than the two crops that have brought farmers so much success—rice and cotton.

The Green nemeses

There is little doubt that rice and cotton are the crops that have had the greatest impact on Australian rivers. Growers of these crops will point to the damage from grazing, but given that the damage is largely done, this is frankly unhelpful.

There are around 1,500 cotton growers in Australia farming over 452,000 hectares. Last year the industry produced over 3.2 million bales of cotton and exported 92 per cent of it, valued at a whopping $1.6 billion. Two thousand five hundred families growing rice on more than 150,000 hectares produce over 1.2 million tonnes a year. Together, these industries use close to four million megs a year. No one is measuring the hidden cost of paying for these industries.

It's rice

There is a strong view that Australia should not be growing rice at all. Businessman Doug Shears says, 'Growing rice south of the [Tropic of] Capricorn is mad. The dryness of the climate . . . I mean look at where most of the world's rice is grown. It's where rainfall is heaviest. I would say they'll have to shut the rice industry in 15 to 20 years.'

It is not well known that there's a long history of paddy fields in the outback and today the rice mill at Deniliquin is the largest in the southern hemisphere. There are 240 jobs to lose.

It seems the most obvious question in the world: how on earth did Australia, with our water problems, end up growing rice? Try almost ninety years of rice growing history. The Riverina can trace its rice growing origins back to a Japanese grower, Joe Takasuka, who actually started at Swan Hill in Victoria. By 1914, he was growing rice there commercially. Apparently, he sent his son on a motor bike laden with saddlebags full of seed from his first crop to the Murrumbidgee, another irrigation area just north of Deniliquin.

By 1926, 1,570 tonnes were harvested around Leeton on the Murrumbidgee, the Japanese already interested in exports from Australia. The rice was even said to be good enough for sake. Rice has been grown on the Murrumbidgee ever since.[17]

Today rice growing extends from the Riverina country of Leeton and Griffith (also famous for its hydroponic hash), on the Murrumbidgee, south to Deniliquin and Berrigan on the Murray, the main areas being Murrumbidgee Irrigation, Colleambally and Murray Irrigation. Miles of open channels and acres of emerald green paddocks bathing in water cover the landscape. Growers here prefer their paddocks described as being in 'permanent flooding' rather than 'waterlogged'. This is perfect irrigation country, with a dry, sunny climate. Providing the land is good, farmers can pretty much play God and control the weather.

Deniliquin, or Deni to the locals, is the heart of irrigated agriculture. From there, it's only a few hundred kilometres to the acres of vineyards and peach trees at Mildura in Victoria and about the same distance further on to Renmark in South Australia.

Most of these irrigated areas started as government-owned irrigation bodies and are now effectively privatised co-ops owned by the irrigators that are responsible for delivering water to the farms. The critical difference with rice is that the industry itself is a co-operative. In fact, the Ricegrowers Co-operative boasts that it is the most successful producer co-operative in Australia. Ricegrowers is the sole miller and marketer of rice in Australia, and exports 85 per cent of rice grown (including to Japan). The industry strongly defends its single selling desk, which gives farmers leverage on sale prices. Rice is a thriving industry.

Deni is the HQ for Murray Irrigation Area, Australia's largest private irrigation company, and its chief, general manager George Warne. Even if he is part of the 'wicked' irrigation business, George's frankness makes him hard not to like. He speaks fast: 'I think you'll find everywhere you go that people are blaming people upstream of them or they're saying they have no effect downstream of them. The truth is we all have an effect everywhere, and no one has a greater effect, I suggest, than we do. Our company takes 1.5 million megalitres out of the Murray, which from a point source is more than anyone else takes out and we take three-quarters of New South Wales' share of the Murray, and we've got a long history of irrigation and a huge dependence on it.'

Interestingly, it took until 1990 for rice to be introduced to all of the irrigators at Deniliquin, largely because the old Rice Marketing Board based on the Murrumbidgee did not fancy the competition. This is despite the building of a 100 mile (160 kilometre) irrigation channel, the Mulwala Canal, in 1935 to supply stock and domestic needs and irrigators. It still serves George Warne's irrigators today.

The area serviced by George Warne's Murray Irrigation Area, covering almost 800,000 hectares, now produces 50 per cent of Australia's rice crop and has overtaken the Murrumbidgee region. Water goes to

2,400 farms owned by 1,600 family businesses in the southern Riverina. Australia's yield, by the way, is double the world's average.[18] It's an immensely thirsty crop, but not at all labour-intensive compared to crops like grapes. In fact it's a perfect family farm crop with few additional employees needed. Australia also has the lowest use of chemicals for the rice industry worldwide and has improved both water efficiency and yield by 30 per cent over the last ten years.

But the industry certainly has an impact on the river. Rice uses a massive 12 to 13 megs per hectare and it can take much of the rap for the low flows in the Murray and has been blamed for the loss of native fish. And there are now major watertable problems, as land clearing and irrigation has inevitably built up underground aquifers. Even George Warne and Hunter Landale will tell you that there are large areas, such as much of Coleambally (developed since 1961), that simply should not be growing rice. In its defence, the industry points out that ricegrowers are the only irrigators who have industry-imposed limits on the amount of land they can put under rice. That limit may be too high, but at least, says George Warne, it is some restriction. Nevertheless there's no doubt that the pressure is on to drive rice out of business because of its massive water demands.

Cottoning on

About 800 kilometres to the northeast of the Riverina lies the other Green nemesis. The Namoi, the Gwydir, the Barwon–Darling, and up into southern Queensland and the Condamine–Balonne are all part of cotton country. You know you're there because of the tufts of cotton wool that fall off the trucks and fly up to meet you along the roads.

In the 1860s, when the American Civil War devastated cotton production there, Australia began to grow cotton to fill the gap. But as US production returned and world prices declined so did the fledgling industry and by the early 1950s, it was almost non-existent.

Australia's modern cotton industry began in the 1960s at Wee Waa in northern New South Wales. Americans Frank Hadley and Paul Kahl pioneered what was to become one of Australia's greatest agricultural export stories—irrigated cotton. The flat and fertile areas around Wee Waa were perfect for the crop, and the dry climate gave irrigators control over the weather. Cotton flourished.

Those not familiar with rural Australia will probably never have heard that cotton has a good side. The truth is that over the last 30 years, as inland abattoirs shut down, the sheep's back broke and beef prices collapsed, cotton has been an absolute saviour for many towns in northern New South Wales and southern Queensland: such as Bourke,

Narrabri, Moree, Gunnedah, Goondiwindi and St George. It is a great money spinner and a great employer, both on the farm and in the gins.

When the cotton price is strong, the dollars that can be earned in the industry are staggering. Cotton has saved many a farmer from ruin and has made a few fortunes. However, cotton is also a highly risky business with much of the cost of farming up front. If the crop fails or prices plummet, farmers can suffer badly. Partly as a result of this, cotton is also our most sophisticated broad-acre crop, much of it hedged on the futures and options markets internationally.

At this point readers of this book connected with Cotton Australia, the peak industry body, are probably still holding their breaths. For the last ten years, the industry has been in a sort of siege mentality. Why? Because in the past, it has done quite enough to deserve the title of environmental vandal, in both its uncontrolled thirst for water and its cavalier use of chemicals. In recent years Cotton Australia and leaders within the industry believe they have bent over backwards to become better citizens and repair cotton's bad reputation, but they are still highly sensitive. There has also been a major changing of the guard at the organisation's Sydney headquarters. (Ironically, Cotton Australia used to be called the Australian Cotton Foundation, acronym ACF which was the same as the Australian Conservation Foundation. Not surprisingly, the name was changed.)

Just how bad is cotton? On the water front, there's no doubt that cotton is a huge user, most of the crop being flooded or furrow-irrigated—whereby water is sent from the top of a paddock to the bottom in furrows. Cotton areas are also very concentrated. The plant needs to be grown in arid or semi-arid areas, where bugs are fewer than in the tropics but where water is unfortunately limited. As a result, more than 90 per cent of Australia's raw cotton is produced in the Murray–Darling Basin. But it is the way Murray–Darling water is being taken, by harvesting vast amounts of flood water which the wide plains have enjoyed for centuries, that has generated animosity from both Greens and graziers. Growers have invested in massive storage dams, at least one being bigger than Sydney Harbour. In chapter 10, we'll see how in the scramble for water, cotton has created some of the most bitter wars in the country.

Even within the irrigation community, there's criticism. Jeff Parish, head of Central Irrigation in South Australia's wine country, says, 'The only people who frighten me in this business are the cotton growers, because of the chemicals and the blasé attitude. At least when I go to Deniliquin or Goulburn they have just as much concern or care for water and water issues as anyone else, but I can't find that up north.'

Those who defend the industry argue that cotton uses far less water per hectare than rice, maize, soybeans and even some citrus, and they point to how much money is made per meg of water—an average of $670—far more than rice at $230 a meg. This is also true.

There is no worse example of what can happen with irrigated cotton and rice than the Aral Sea, once a massive lake of 65,000 square kilometres divided between Kazakhstan and Uzbekistan and fed by the Amu-Darya and the Syr-Darya Rivers. Just 30 years ago, water was diverted to irrigate millions of acres of land for cotton and rice production in central Asia. The result has been a disaster; the lake has halved in size and lost over 60 per cent of its water. One hundred million tonnes of salt have been mobilised to the surface. Salt in the lake has devastated a once-thriving fishery and exposed useless soil as the water retreated. Salinity has already reduced crop yields. Even the weather has changed, with hot, dry summers and colder, longer winters. Dust storms have blown away up to 40 million tonnes of this newly exposed soil annually, dispersing its salt particles and chemicals, including loads of DDT. Many of the locals have lung diseases and life expectancy in some areas is under forty.

David Farley, the former head of the major Australian cotton producer Colly Cotton, has done some numbers. 'If you understand that Australian farmers need 6.5 megs a hectare to produce 1,800 pounds [about 800 kg] of cotton and Uzbekistan uses 15 megs a hectare to grow 500 pounds [about 220 kg] of cotton, then if we went over to the Aral and put in our practices, in 82 years we would have saved them enough water to refill it, that's even allowing for evaporation. Now wouldn't that be something? If Australian farmers could show them a better way, within one lifetime.'

The politics of the countries involved in the Aral area make this nigh impossible, but the point is that Australia's cotton yields are nearly 2.7 times the world average.

There's no doubt that cotton could be grown with less water, even in Australia, but the cotton industry has been resistant to the idea because of the expense. Pilot schemes working on drip filter irrigation for cotton in areas like the Macquarie River in northern New South Wales are up and running, but most irrigators shake their heads. Today, such investment seems lunacy.

The bug war

By far the biggest criticism of cotton relates to the chemicals used, from organochlorides and organophosphates to pyrethroids and carbonates. In one management plan, the National Registration

Authority recommended: endosulphans, pyrethroids, ovasyn, kelthane, larvin, methomyl, profenofos and comite.[19] As practices and research stand today, without chemicals the industry would not exist. A nasty bug called heliothis would see to that, but there are other pests as well: mites and green myrids.

Chemicals are a mighty touchy subject within the industry. 'There is no such thing as cotton chemicals,' industry literature states hotly, pointing to melons, broccoli and other fruit. One difference, however, is that unlike most fruit and vegetables, cotton involves vast areas, which are sprayed with chemicals from the air, and flood irrigation rather than sprinkler, or drip filter watering, which targets chemical delivery much more effectively.

Some of the chemicals aren't so bad, after all, in summer pyrethrins float off the mozzie zappers plugged into bedrooms all over Australia. It is what the combination of chemicals does to the surroundings that matters. As we've learned, endosulphan kills fish and there have been a number of 'accidents' where spills into rivers have left the industry squirming. Much more serious for the cotton industry has been that its chemicals have drifted off the properties onto grazing land and started to appear in beef.

For this reason, various organochlorides were banned in the late 1980s, but more recent problems have occurred, notably in the summer of 1998–99, a particularly bad year for bugs. By December, 1998, unacceptably high levels of endosulphan were found in 16 cattle on properties bordering cotton farms both in northern New South Wales and on the Darling Downs in Queensland. In February came the news the industry was dreading. Twenty tonnes of beef from Queensland had been turned away by South Korea, our third largest export market, because of high endosulphan levels found in the meat. Relations between the beef industry and the cotton industry have not been the same since. 'The damage the endosulphan thing has done this summer is going to live with us for a long, long time,' Cotton Australia's chief, Gary Punch, said at the time.[20]

As a consequence, Australia's peak chemical regulator (the National Registration Authority for Agricultural and Veterinary Chemicals) warned that 'endo' would be banned if there was any further contamination and demanded mandatory changes to the industry, including a maximum three sprays per season. Rod Eichner from the regulator was reported as saying some farmers were spraying up to 14 times per year.[21]

The incident was even more unfortunate in that the cotton industry was in the middle of launching its new code of 'best management practice' (BMP), a laudable effort that sets both spraying restrictions and new standards on tailings dams which collect any irrigation run-off on the low side of the field and keep chemicals on farm. Growers were also encouraged to plant 'trap' crops alongside cotton and to fight bugs with bugs, so-called 'integrated pest management', as well as rotate crops to give the land a rest.

Unfortunately, BMP is voluntary. 'BMP has always been, and should always remain, voluntary,' said Cotton Australia.[22] As at March 2000, even those promoting the code admit that the take up has been disappointing. Eight per cent of the industry had been audited by the peak body, reflecting both the cost and, importantly, the time it takes farmers to adopt the BMP regime. The peak industry body has called for incentives for farmers.

While farmers' use of endo has fallen by 57 per cent in six years, the area under cotton has expanded. Keeping chemicals on farm isn't always easy when the wind changes direction during spraying or floods take out the tailings dam that has been carefully trapping nasty water. As some consolation, at least there is a dilution factor.

The country's most influential freshwater ecologist, Professor Peter Cullen, has said there's no reason why best management practice should be compulsory: '"Best" is probably a crook word because they can always be improved, but the cotton industry is one where they do make a fair effort, because water is so precious to them. They do try to recycle their tail waters more than anyone else . . . We've got the problem of the odd rogue farmer and I'm not sure whether they're catchable.'

Like all industries, cotton is also only as good as the worst farmer. When the cotton price shoots up and novice dryland cockies move into the crop for a season, many do not have the cash or experience to always do the right thing. Add to that a bad bug year like 1998 and the only way farmers know to fight back is to throw on the pesticide.

Nevertheless, rules on endo use are getting tighter every year. Dr Wayne Meyer, leader of the sustainable agriculture program at CSIRO, believes aerial spraying will eventually be banned worldwide. 'No matter how accurately it's done, there is a public perception now that it's not good.'

As might be expected, more than half the research in Australian cotton is spent on ways to reduce chemicals. A new strain of cotton known as BT (*Bacillus thuringiensis*) Cotton, which has been genetically modified to resist pests, has been welcomed by most groups, despite a rocky start. The Australian heliothis bug turned out to be much tougher than its American counterepart, with early BT Cotton crops giving patchy results. BT's supplier, the US multinational Monsanto, also came in for some flack over the price it charged Australian farmers. However, these issues have been smoothed and BT Cotton's success is growing. Not all Greens are happy with BT, though. 'It's a trade-off for less chemicals, but we worry about GMOs [genetically modified organisms],' says Kathy Ridge of the Nature Conservation Council. 'The German experience is that these gene fragments can move up the food chain.' It's also worth pointing out that Monsanto may be busy working on genetic modification so plants need fewer chemicals; however, it's also working to create plants like 'Roundup Ready' soy, that can withstand higher doses. But a Seattle Man eulogy is a little beyond the brief of this book.

To the relief of those in the industry, the acrid comments made about cotton in the media are lessening. 'I look at *The Land* and country papers all the time,' said Cotton Australia's Ralph Leutton in early 2001. 'Even the Green groups are pretty much leaving us alone.' In fact the Australian Conservation Foundation (ACF), the World Wide Fund for Nature (WWF) and the Nature Conservation Council of New South Wales are all working with the cotton industry. 'But we need to make progress on BMP,' added Ralph Leutton. 'When we have 30 per cent to 40 per cent of growers audited, that's when Tim Fisher and Kathy Ridge will be saying, OK, the cotton industry is really making some progress.'

The Australian Conservation Foundation's Tim Fisher is quite conciliatory about cotton.

> There was bad PR in the cotton industry and it earned it. But the cowboys are fewer now and it's no more or less evil than some of the other crops. I don't want in any way to defend it, but the soils and climate of the Murray–Darling Basin are suited to cotton. After all, they're not as leaky as Coleambally and parts of the Mallee. And compare it with the dairy industry. Seventy-five per cent of the phosphorus into the Gippsland Lakes a couple of years ago was from dairy fertiliser run-off.

A much more serious issue for Wayne Meyer is the impact of years of intensive cotton growing in the long term. There is growing evidence that earthworms and other invertebrates suffer from prolonged cotton cropping, a key indicator of degradation in the form of acid soils and salinity. 'The trouble with cotton is that it has very little biomass that is returned to the soil. Compare it to rice or sugar that's green cane harvested, the organic resources are not there,' explained Wayne Meyer. (Green cane harvesting leaves the 'trash' or sugar cane leaves on the ground.) One of the more startling sights to encounter in the cotton growing region is the newly napalmed look of a field of defoliated cotton; chemicals burn back the crops to sticks supporting the white cotton balls, done to make harvesting easier. 'The real danger is that we get to a stage where [what is done to] the land is not just economically irreversible, but it's bio-geo-chemically irreversible as well,' adds Wayne Meyer.

Yet if this is all true, why has the cotton industry in the United States been rolling along for 200 and in some places 300 years? One reason could be that in Georgia, South Carolina and Mississippi, cotton is generally grown without irrigation. But more importantly, their soils are much deeper and richer, and able to buffer the abuse. This is a point we shall return to—time and time again in Australia, experience is showing that farming practices from the northern hemisphere have but a finite life down under.

So, should we be growing cotton in Australia at all? 'No!' we all cry in our pure cotton T-shirts. (Don't forget, by the way, that cotton garments 'made in Indonesia' could well be Australian. Asia is one of our best export destinations.) Hypocrites we may be, but that does not get the cotton industry off the hook. It's a massive user of water and the chemicals that threaten the landscape are essential for cotton's survival. The problem, according to Wayne Meyer, is the vested interests in the industry. 'It's not the cotton that's the real problem, it's the greed of those who want to grow cotton and the political processes that are not addressing the [fact that water must be provided not only] for the user, but also for the river.'

'Greed' is a word used frequently about the big cotton barons. The most poignant criticism of the industry is that some of the big boys are not setting much of an example. Integrated pest management on the biggest farm in Queensland, Cubbie station, for instance, is practically non-existent. And there's hardly a tree standing, unless it's rotting in a water storage. Kathy Ridge points to a recent corporate expansion near the Macquarie Marshes in New South Wales where even the company's own environmental impact statement admits the land will be seriously degraded within 50 years. She says 'no' to cotton growing, 'not unless it's sustainable, and I mean [in] the very broadest definition, that means water, soils, genetics and pesticides.'

Let's be clear on this. The way most cotton is grown in Australia right now is not sustainable, and even best management practice has room for improvement. This is what the experts, whether at CSIRO or in other parts of the world, are telling us. If we go on growing cotton as we do, our soils will become barren within a lifetime. And it's not just cotton. Sandra Postel, one of the top experts on the impact of irrigation practices around the world, says it is a sad fact that no irrigation era in history has survived beyond a century or so. Changes in irrigation practices must come, as will tough mandatory rules on farm practice, to make our irrigation era as long as possible. It will be an extremely tough transition for farmers and one which they cannot do alone.

So again, is there a future for cotton? 'Oh, yes,' says Wayne Meyer. 'Cotton will probably grow in as big an area, but what will determine the future is the use of less water. And as we get better at gene technology and pest management, I think the crop will move further north and away from the stressed rivers and land it's on at present. But not to the Cooper,' he adds firmly. The story of the Cooper is for chapter 6.

5

Assalted

Water, water everywhere
And all the boards did shrink
Water, water everywhere
Nor any drop to drink
      S.T. Coleridge—The Rime of the Ancient Mariner

The Cap

The ignorance of those in charge of water sharing, and the greed of some opportunists crippled the Murray in just a few decades. By the time the decision makers worked out that the river itself needed a share, it was the early 1990s. Those who were watching the demise of Australia's greatest river were calling for something radical. It happened in June 1995: the Murray–Darling Cap.

But before we go any further, a quick lesson on the bureaucracy, or we'll soon be in deep water.

Management of the Murray–Darling system had changed remarkably little since the job was given to the Murray River Commission in 1917. In 1992 a new agreement, the Murray–Darling Basin Agreement, was signed between the States and Commonwealth government, creating the largest river system management program in the world, known as the Murray–Darling 'Initiative'. To do this, three bodies were created.

The *Council* is made up of ministers from each of the Murray–Darling State governments, the ACT and the Commonwealth, and is the decision making body. The *Commission* carries out the Council's decisions and manages the Murray–Darling system on a day-to-day basis. And the *Community Advisory Committee* advises the Council and is a two-way channel between the Council and the community.

The Cap was the most dramatic step ever taken to fix up the system, the result of a meeting of the Council. It was decided that 'a balance needed to be struck between the significant economic and social benefits that have been obtained from the development of the Basin's water resources on the one hand, and the environmental uses of the water in the rivers on the other.'[1] In other words, more water was needed for the ducks and fish, urgently.

The Cap was radical for three reasons. First, it put a ceiling on the amount of water that could be taken out of river catchments. Brought in originally as 'interim arrangements', water use has never been the same since. The ceiling set for the Cap was also lower than current levels of use. Critically, it involved a big claw back of water from water users, with no compensation whatsoever. The third reason, and a most important achievement, is perhaps sometimes overlooked. Before the Cap, the existing Murray–Darling agreement between the States covered only the main trunk of the river system, including a couple of large lakes and dams. According to this agreement water from this central channel (as measured at Albury) was shared fifty-fifty between New South Wales and Victoria, with South Australia getting a fixed amount of Murray water, 1,850 gigalitres a year (1,850,000 megs). This meant that Victoria and New South Wales were free to extract as much water as they liked from the big tributaries that fed the Murray–Darling. When the Cap came in, for the first time authorities had the power to regulate rivers beyond the main trunk of the Murray. This was momentous. Suddenly everyone in the Basin was affected.

The Cap was set across New South Wales and Victoria at the level of water used in each catchment in the 1993–94 year. In South Australia, the Cap was still 1850 gigs a year (its 1993–94 level). The Cap hit irrigators where it hurt, particularly in New South Wales, but all three States agreed to it because it was clear that the amount of water being extracted from the system was increasing at an alarming rate. The outstanding winner was South Australia. While the State still got its entitlement of 1,850 gigs a year, the average amount that flows to South Australia in any year is much more, around 6,200 gigs. Had the Cap not come in, theoretically all the States upstream could have (and indeed, would have, in time) sucked away the water, right down to the 1,850-gig figure. By agreeing to Cap diversions at 1993–94 levels of development, the States have now locked this water in for South Australia. 'That is not enough water to keep the system healthy, but it is a hell of an advance on 1,850,'[2] was how John Scanlon, a former head of South Australia's Department of Environment and Natural Resources, explained it to a State Parliamentary Select Committee.

The impression one gains from Hansard is that those listening to John Scanlon were rather indignant about this, never imagining

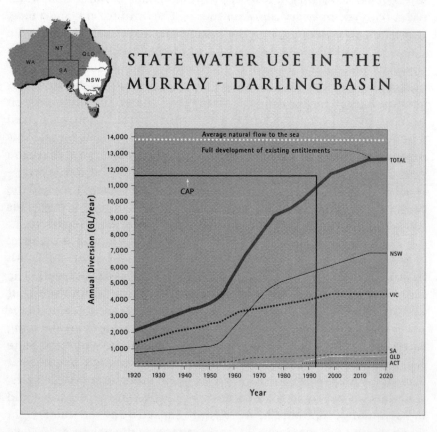

Source: Murray–Darling Basin Commission

South Australia would get *less* water. But the fact that the State has not had its allocation reduced is not lost on Murray Irrigation's George Warne. 'In South Australia the Cap has been set at a level above their record use, which makes a nonsense of 1993–94 levels of development, but given that the Cap is about protecting South Australia, maybe you can forgive them for ignoring it.' John Scanlon insists that though South Australia does receive 1,850 gigs, it has a self-imposed cap of 700 gigs, put in place in 1969. The loudest squawks are not over South Australia's allocation, however. They are about Queensland.

Queensland's prodigal status in the Murray–Darling Basin is notorious. It is no secret that things are done differently in Queensland—the Joh State, where almost one in four people voted for Pauline Hanson in 1998 and close to 10 per cent did in 2001. Yet spend a couple of years in 'Brisvegas', and it is true that the southerners start to look unreal.

Situated at the top of the Basin, Queensland wasn't having a bar of the Cap. Everything is produced in Queensland: tea, coffee, milk, sugar, beef, wheat, cotton, wool, bananas and avocados, but irrigated development has come late. As a result, the Sunshine State takes only about 6 per cent of the total amount of water in the Murray–Darling system, compared to New South Wales' 56 per cent. In its favour too is that the health of rivers up north is better. Environmental research is behind on how current development may impact. So why should the State put a cap on development just when things are taking off? The problem, of course, is that even with the Cap, there is not enough water left in the Murray–Darling system for Queensland to take more than its 6 per cent.

The fight to bring Queensland 'into line' with the Cap and reform is one of the Basin's most passionate stories, which we look at in chapter 9.

Apart from Queensland's behaviour, the Murray–Darling Basin Review of the Cap showed good levels of compliance. (Queensland finally committed to putting Cap levels in place by June 2001 but missed the deadline.) What is less clear is how much the Cap has caused water users to look to ways other than the river to get their water. As the controlling arms of governments extend from the trunk to the tributaries, the irrigators are busy intercepting water closer to home.

Policy makers are not stopping with the Cap, however. The Cap is just the beginning. For now we move to the most serious natural resource issue facing Australia—salt.

The power of Morgan
In water politics the most powerful town in all of the Basin is also one of the smallest. Morgan sits about 160 kilometres northeast of Adelaide, in country covered in twisted mallee trees, just where the

Murray River bends south. It's a picturesque little town, straddling the Murray, which at this point cuts between high limestone cliffs. There's a smattering of little houseboats and a ferry to take local traffic and the odd tourist to and fro.

Right next to the Morgan Bowling Club sits a compound behind high-wire fences, holding some of SA Water Corporation's most precious assets, the Morgan Water Filtration plant and the Morgan Whyalla No. 1 Pumping Station. It is from here that the Murray River supplies 40 per cent of Adelaide's drinking water and all of the water for Whyalla, one of the three towns forming the industrial 'iron triangle' (the others are Port Pirie and Port Lincoln), via concrete pipelines that snake far across the landscape.

While Albury is the town where water quantities are measured for the Cap, Morgan is the benchmark of water quality for the Murray. Whatever happens at Morgan also dictates how much water irrigators and rural towns and ducks in South Australia will receive, because if the quality is below par, healthier water must be added back to the river system.

Older folk on the lower reaches of the Murray, in Morgan or Murray Bridge, can remember seeing the bottom of the river. Today they would be lucky to see the end of their finger if they put a wrist in. But the authorities care little about looks. What they test for at Morgan is not so much pollution but salt.

Below the salt

Farmers have known about salinity for many years: 'Some of the earliest signs occurred within a decade of trees being ringbarked, when railway engineers discovered that the water in reservoirs they built to supply water to locomotives was too salty to use. By 1897, astute observers began to make the connection between clearing the trees and the salting up of creeks on which the pioneer settler so vitally depended.[3]

It is extraordinary, then, that the first real offensive on salt came only in the 1980s. By then it was clear that salt levels in Morgan's water, destined for the people of Adelaide, were on track to top what the World Health Organization deemed the desirable upper limit for drinking water. In parts the river is about as far away from fresh water as a good margarita.

Don Blackmore is the man who runs the Murray–Darling Basin Commission in charge of river management. By the late 1980s he and his team had also become aware of a much more sinister and overwhelming threat, dryland salinity. But that remained firmly in the too hard basket, according to Don Blackmore. 'We said to government, we'll buy you 30 years because no one knows how to deal with what was then, back in 1988, the emerging dryland salinity problem. Now, we knew we had it but we had no idea how big it was and how ugly it was going to be, so we said we'll get on and run a ring around things we can protect now, which is irrigation, which was the big hazard.'

The solution to salinity had to be quick, but the price was that it was an engineering solution rather than a natural one. In 1988, a federal Salinity and Drainage Strategy was adopted to make sure Morgan water stayed below the World Health Organization drinking water level. In industry parlance, that means no more than 800 ECs for 95 per cent of the time. (ECs are electric conductivity units which measure saltiness—much beyond 1,500 ECs and river health and irrigated crop yields fall dramatically and plants use even more water).

The culprits were identified as the irrigators who were busy 'watering the soil and not the plants', raising the watertable and bringing the treacherous salt to the surface to run off into the river. Take George Warne's lot. By 1981, 19,200 hectares of the area serviced by the Murray Irrigation Area near Wakool in New South Wales had a watertable within 1.5 metres of the surface, making more than 2,000 hectares of farmland completely barren.[4] There were attempts in the 1960s to combat the rising water, with pumps erected to extract salty water from underground mechanically. Unfortunately, the salty water was pumped back into the river; there was nowhere else to go. As can be imagined, it did little for river health.

The solution has been to pump this salty groundwater to sacrificial 'out of valley' evaporation basins, dead, white areas often visible from the road. Work began around Wakool in 1987 and today the watertable is back below 2.5 metres from the surface. Murray Irrigation owns and operates the largest salt interception scheme in the Basin, pumping ground water from 54 pump sites. About a third of their work is funded by government. An average of 14,600 megs of ground water is pumped out annually with a salt level of 26,600 EC (compared to irrigable water, remember, of 1,500 EC tops). In 1999, 240,000 tonnes of salt were removed. Basin-wide, over 600,000 tonnes of salt are diverted away from the Murray River each year.[5] That's 600,000 one-tonne trucks of salt.

Salinity levels in water at Morgan have fallen, as planned. But when in October 1999 the Salinity Audit was completed for the Commission it was clear that the problem was much, much bigger. Put bluntly, a 'do nothing' approach will mean that over the next 50 years, salinity in the already affected lower Murray will increase by 50 per cent. More detailed figures provided were quickly interpreted by the media and politicians to mean that by 2020, for two days out of five, Murray water for people in Adelaide will fail World Health Organization standards. Sitting 'below the salt' now conjures up a whole new meaning for the people of Adelaide.

In many of our other big rivers, the Namoi, Macquarie and Lachlan in New South Wales and the Avoca and Loddon in Victoria, salinity is set to rise to levels that threaten their viability as drinking water and even irrigation water over the next 20 to 50 years. Data is poor in Queensland, but scientists expect that inevitably rivers such as the Balonne, Condamine and Warrego will be affected—to say nothing of threatened wetlands and river habitats.[6]

This gloomy news was only part of the Salinity Audit's findings, however. The real horror is that salt problems on a much bigger and broader scale now threaten vast areas of the Basin. What the Murray–Darling Commission found too hard to contemplate in 1988, we all now have to face up to.

The Grim Reaper—dryland salinity

The last thing dryland salinity sounds as though it involves is water. But water is the key to the lurking menace which has taken less than 100 years to appear and will see all of us out. It is the biggest natural disaster in the nation today.

Australia is both blessed and cursed by the fact that it sits on a number of huge underground aquifers. We are blessed because were it not for the Great Artesian Basin and the Murray Basin, there would be no such thing as the Australian settler. Today, there are about 3,700 bores across the land watering the acres that produce our food. But we are cursed because lurking under the ground is salt, laid down millions of years ago, sometimes by dumps of rain laden with salt which was picked up from the sea, and sometimes through invasion of the sea over the land. In many areas, like the Riverland country in South Australia, rock formations contain clusters of tiny shells. It is this salt that has resurfaced to haunt us.

When first explained, the process seems totally illogical. In our dry land when the trees are cut down, water appears and frequently, very salty water. Go to Africa and the reverse is true; the watertable drops away. The reason is the delicate balance in Australia between trees and the underground watertable. Trees are thirsty drinkers and keep the watertable down. But when 15 billion of them are cut down, the delicate balance is broken and the unconsumed water begins to rise. With it comes the salt which has been safely trapped for thousands of years.

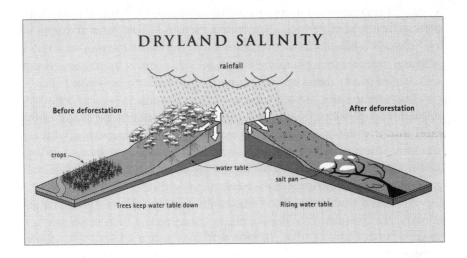

DRYLAND SALINITY

rainfall

Before deforestation

After deforestation

crops

water table

salt pan

Trees keep water table down

Rising water table

The sorts of figures involved are too big to take in properly. Between three and five million hectares of land in the Basin will be salt-affected in the next 100 years, by which time the cost to the taxpayer will be a staggering $1 billion a year, a nice little bill to pass on to the grandchildren.[7] According to the Commonwealth's Australian Dryland Salinity Assessment 2000, some 20,000 kilometres of streams, 52,000 kilometres of major road and 3,600 kilometres of railway could be salt affected by 2050. Farmers don't enjoy talking about tree clearing. A standard response in some areas is, 'No, no, this used to be a floodplain, never a tree here anyway.' In some cases that is true, but for many others, the comment is up there with 'The cheque is in the mail.'

To appreciate the seriousness, it is worth looking beyond the Basin to the other great food bowl in the west, where the nightmare is even greater. On a clear day from the window of an average commercial plane flying between Perth and Adelaide the evidence is indisputable. Gaping pale yellow and green ulcers scatter the landscape, ringed with white, crusty edges, like bad skin on an old leg. Salt is eating up the wheat belt of Western Australia, a massive farming region running from inland of Geraldton, 400 kilometres north of Perth, down to Esperance in the Great Australian Bight.

After a decade working as a hydrologist in the Murray–Darling, Dr Tom Hatton moved to Perth with CSIRO. Nothing in the Basin had prepared him for the scale of the wheat belt's problems. 'In summer it is white with salt, but these big areas that look like bare earth, like a river has flooded and washed everything away, almost all of it was cropped and often that was the most highly productive land, because in these alluvial valleys it was fairly good soil, before it was salinised.'

Hatton continued in an even Canadian accent. 'Actually the area affected goes beyond what you think. When you look at it with the right kinds of bands in the satellite data, you'll see crop productivity affected much more widely than this. This isn't a landscape that's nearly yet in equilibrium; it's going to get much worse than this.'

Beneath the uncleared country in the west salt is typically buried, spread throughout 50 and 70 metres of porous rock, laid down millions of years ago when rain, heavy with sea salt (cyclic salt) from the west, blew across the landmass. Just like in the Murray–Darling Basin, agricultural clearing on a massive scale has unleashed the salt menace as the watertable has risen.

Already 11 per cent of the wheatbelt is affected by salinity, which means that productivity has been at least halved. Scientists reckon that as much as one-third of the belt, six million hectares, will be affected over the next century before any stabilisation is reached.

Those living in the southern corner of the State, full of development promise, around Albany and Denmark, are not spared either. According to Hatton, a salt threat to the Kent, Denmark and Tone Warren river systems is already recognised as limiting potential development. And the concerns in the south go beyond salt. Rising watertables in the west have filled inland valleys with groundwater, which in turn shed water more quickly when it rains. Flood risks along the river systems in the southwest will increase greatly in the next 50 years. 'Storms such as we had in the early 1980s,' continued Hatton, 'two little cyclones came through in summer and caused some very serious flooding along the Blackwood River and Nannup and towns like that, cut the Albany Highway for a few days. If that storm were to happen 20 or 30 years from now, it would have two or three times the peak volume of flow, for the same-sized storm, so the implications for towns along these rivers is pretty serious.'

Back in the Murray–Darling Basin, the worst of the damage is in mallee country, in western Victoria and South Australia. Sixty per cent of the increase in salinity is expected in dryland areas in the lower Murray, and of that more than half will be in the Mallee. Like the wheat belt in the West, broad-acre crops blanket an almost treeless landscape. Despite all the 'world's best practice' farming that is going on, scientists now believe the rainwater that seeps through to ground water systems in broad-acre crop land and grazing areas is often ten and even 20 times what it was under native vegetation.[8] And looking at some of the century-old photos of mallee trees being uprooted it is easy to understand why. Nature made these tough deep-rooted trees some of the best watertable managers in the business.

The irony is bitter. We are now paying for the 15 billion trees that our ancestors determinedly uprooted over the last 150 years trying to bring a wild Australia to bear fruit for their offspring. The news is devastating for farmers. The head of the New State Wales Farmers Federation, Mick Keogh, has admitted, 'We've already got the situation where farmers don't want to know the salinity status of their land because if they want to sell for their retirement in the next five years, it doesn't do them much good to know that they've got high salinity problems.'[9]

The salt circle

Now it's time to bring the crisis full circle back to the Murray. This is because the watertable not only rises with too much water flowing in, it also moves slowly underground in one direction. Ground water follows a gradient, usually towards a river. River salting from ground water is not a new phenomenon. When Charles Sturt came upon a reach of the Darling in 1829, he wrote, 'The men eagerly descended to quench their thirst. Nor shall I ever forget the cry of disappointment that followed their doing so, or the looks of terror and disappointment with which they called out to inform me that the water was so salt as to be unfit to drink . . . On the 6th of February, we discovered some springs in the very bed of the river, with a considerable stream gushing out. They were, however, brine springs.'[10]

Over the southern part of the Murray–Darling system lies the huge Murray Basin, and within it, according to the Commission's Don Blackmore, who has recently spent about $9 million charting the salt menace, lies 100 billion tonnes of salt. 'Just the Murray Basin . . . 100 billion tonnes. Now we don't want to mobilise too much of that, it's a bit naughty,' he muttered.

While ground water is quite sweet around the Murrumbidgee, moving west it confronts aquifers, in some places three times as salty as the sea, and bumps them on towards the Murray River. At the wrong end of this journey, at Loxton in the wine country of the South Australian Riverland is Jeff Parish. Jeff runs nine critical irrigation districts in South Australia, from Renmark to Mypolonga near Murray Bridge. 'The landscape here over a couple of million years was twice a desert and twice inland sea, one of which was sea water, one of which was fresh. There were these highly saline deposits, with the mallee on top and nothing else, it was in equilibrium, but delicate equilibrium.' He interrupted and carefully pointed out a couple of Australia's top vineyards, then continued. 'We came and we cleared. So what used to be only 0.1 millimetres of rain a year got through to the beds. When we cleared it for cereal, we increased the amount of water that gets through to those big saline aquifers up to about 25 millimetres, which doesn't sound much. It's only an inch, but it's a hundred-fold what it used to be, and when you irrigate it, you can probably increase it ten-fold again, if you don't manage it properly. Now when we stuff up irrigation, it's not the water that escapes that causes the problem, it's the fact that the water escaping through hits the saline beds and jets them sideways to the river.'

'Would you like to see?' As we spoke together in his utility, Jeff Parish turned onto a dirt road and there in front of us was a wide, sterile depression. Very wide.

Around half of the salt in the Murray River is added in South Australia. Ground water that enters the Murray from Loxton originates in the Grampians in Victoria, hundreds of kilometres away. It moves perhaps a metre a year, taking millions of years to arrive, in the process supplying magnificent ground water for the southeast and even as far as Pinnaroo on the Victoria–South Australian border. But, as luck would have it, about 50 miles short of Loxton it starts to intercept the ancient seabed aquifers.

Don Rowe from Sunraysia Rural Water at Mildura near the Victorian border agrees. 'Bear in mind we're only 30 metres above sea level, and the river's still got 1,000 kilometres to go, theoretically. You get big surf down at Glenelg and it runs up here, you know!' He grinned at me, and I realised it was too late to check my eyebrows. 'Seriously though,' he added, 'you can have water running down the Murray River at 350, 300 EC, and the water in Lindsay River [flowing into the Murray just east of the South Australian border with Victoria] is 650, 700 EC just from the pressure, and that watertable coming through the top.'

Dryland salinity has taken the fight against salt to a whole new level of sophistication. Ask Don Blackmore why we shouldn't take out all the weirs below Mildura to help the fish and he'd probably see the merits, but he'd also tell you why not. 'The River Murray from Mildura down is a ground water discharge feature. Somebody in 1920, without any understanding of that, built weirs, and they've now raised that mound water level by about another 4 metres. If we pull the weirs out, what will happen is [that] for the next 70 years, we'll have a huge ground water inflow, so unless I want to have a saline river, I then have to put interception in to cut off the ground water created by the weirs. Now, that's the quid pro quo you've got. Personally I think that's a lousy legacy to have been left with.'

Salt goes to town

The angular, glass-sided city council chambers say a great deal about the inland New South Wales town of Wagga Wagga, a regional city determined to keep its head up in rural Australia. Councillor Kevin Wales, the former local police chief, sits comfortably in his design-award-winning chambers. More than a thousand tonnes of salt passes through Wagga each day, not in trucks, but dissolved in the water of the Murrumbidgee.[11] Unfortunately, as in many places, there's also salt underneath Wagga.

The town's salt drama began in 1994, when somebody noticed that the grass in the inner ring of the showground wasn't growing back. 'The showground used to have a beautiful green cover of the centre fields,' explained the mayor, 'and then all of a sudden some ten years ago, the grass wouldn't grow and they were getting salt on the top of the ground.'

The town's watertable was rising and with it salt that would damage some of the town's oldest areas. The salt is not hidden. A quick trip to the back entrance of the base hospital shows the bitumen buckling. Some roads here last only half the time they should do. Around the corner from the hospital, the facades of brick houses are falling apart as the salt oozes through the foundations then crystallises and smashes the brickwork. The postman obviously keeps bogging in one water-sodden driveway. Around 600 homes in Wagga Wagga are affected, with an average repair cost of $20,000 each. By 2020, if nothing is done, that number will reach 7,500. These days, anyone buying real estate is told upfront about the risks.

Yet again, a history of ignorance and a litany of errors has been the issue. First graziers within the 44 kilometre catchment area of Wagga Wagga cleared trees for cattle, thereby raising the watertable. Over the years a ground water mound (a high point in the watertable) right under the town has been created by proud gardeners soaking their gardens. Exacerbating this, the best practice in suburban water management until quite recently was to put a big rubble pit in the back garden to soak up run-off. One estimate of the urban impact from August 2000 is an increase of seepage into the ground water from around 1 millimetre to 50 millimetres.[12]

With the watertable two metres from the ground in some places, Wagga Wagga has had little time to muck around. A State Government analysis found that doing nothing would cost Wagga Wagga over $180 million within 30 years.[13] In 1997 the council threw $3 million at the problem; luckily, it had almost that much from the sale of the local gasworks. By late 1999 nine bores, sunk deep enough to take relatively fresh water from below the saltiest level, were put in to lower the table three to five metres, keeping the salt well down. 'Our underground water, we're pumping that from 60 metres underground,' the mayor said with some satisfaction, 'and then when that comes up, we've reduced the saline level of that water to go into the Murrumbidgee, so it's potable.'

Today, rear block drainage is replacing the old gravel pits, trees are being planted, schools run education programs and garden nurseries carry salt-tolerant plants and literature about how to care for them. At the north end of town on the hills, however, pressure for urban development has created a familiar scene of treeless, concrete nudity and young children tricycling in ever-decreasing circles. You can't win them all. Yet it must be frightening to be a ratepayer in Wagga Wagga. Each time the Department of State and Regional Development does an economic study, the annual cost of fighting salinity to the Wagga community goes up. It has risen from $440,000 to $3.2 million to just over $6 million in the July 2000 report.

Wagga is only one of many towns to suffer salt damage. There are now an estimated 80 towns damaged by salt, including Dubbo, Wellington, Forbes, Yass and Cowra in New South Wales; Bendigo, Kerang and Cohuna in Victoria and others in South Australia. Six council areas in western Sydney are also showing all the signs.[14] At Dubbo, Wellington and Forbes, a mobile desalination plant is now operating.

In Western Australia, predictably, around 30 towns are under attack from rising watertables that are known to flood pub cellars. 'One of the things that is always a surprise to people', explained Tom Hatton, 'is that these inland towns in the wheat belt of WA are actually—most of them—reticulated off coastal water. So these coastal dams that provide resource to Perth and to Kalgoorlie also provide water up to the wheat belt. There's pipes running all over the wheat belt. So you go out to an inland town like Corrigin, Brookton, they've actually been reticulated and are importing water, which is adding to the excess problem.'

All these problems; all over Australia. The message is clear. Water, soil, trees and salt are all inextricably linked, and if any are tampered with, change is more than likely. What should be done? Well, that depends on who you talk to.

The unlikely alliance

Ian Donges and Peter Garrett are chalk and cheese, or I should say, cheese and chalk. As President of the National Farmers Federation Ian Donges runs cattle, rears prime lambs and grows winter cereals in Cowra. He is a quietly spoken, conservative man, a negotiator, a listener.

Peter Garrett, on the other hand, not that far apart in age from his counterpart (which dates us all) is passionate and forthright. He still records for Midnight Oil, the band that took Australian rock to the world in the 1980s and 1990s with protest songs of Aboriginal disempowerment, burning beds, blue sky mines and rivers running red:

So you cut all the trees down
You poisoned the sky and the sea
You've taken what's good from the ground
But you've left precious little for me[15]

This is the man whose bald head and bulging neck veins spelt anger and emotion long before Michael Klim got started. At the closing ceremony of the Olympics, with four billion people watching, his clothes were emblazoned with the word 'Sorry' many times over. What a shame that he didn't have room on one sleeve for the word 'Water'!

It's not often that Greens and farmers are mentioned in the same sentence. But in May 2000, Ian Donges and Australian Conservation Foundation President Peter Garrett sat side by side and hit us with some bad news. It will cost $65 billion over ten years to fix what we have done to Australia. Sixty-five billion dollars, of which about half must come from government and half from the private sector and community partnerships. 'Forty billion trees, ladies and gentlemen who are watching this program, must be planted,' Peter Garrett later told SBS's *Insight* program.

The picture these two groups paint is even worse than that of the Murray–Darling Salinity Audit of the previous October. Salt-affected land will increase from 2.5 to over 15 million hectares, around 30 per cent of Australia's cultivated land. Half of our woodland birds will disappear within decades. And they reckon the cost today of our neglect for the land to be over $2 billion annually, more than half the net annual value of farm production.[16]

Farmers and Greens have joined forces before. 'A decade of Landcare' was a world first, whipping up an enthusiastic volunteer force backed by a few grants to plant trees, fight weeds and raise awareness. 'Thanks to Landcare, the rest of Australia now knows that rural people can be both green and productive when they want to be,' said Leith Boully, chair of the Murray–Darling's Community Advisory Committee.[17]

Unfortunately, the challenge turned out to be far too big for Landcare, with even its founding fathers pondering its failure. The

Salinity Audit agreed. 'At this stage there is no indication that Land-care and other natural resource management programs have altered the "business as usual" trend lines for rising salinity across the Basin predicted by the Audit.'[18]

But that was Landcare then. This is now. The new farmer–Green partnership has raised the stakes to $65 billion, or around $6.5 billion a year for ten years. It also raised some eyebrows in the ivory tower in Canberra.

The salinity strategy

In September 2000, a draft Basin Salinity Management Strategy (the draft Salinity Strategy—not to be confused with 1988 Salinity and Drainage Strategy and Salinity Audit) was produced by the Murray–Darling Basin Council, a 15-year plan to fight the salt problem. It was approved in March 2001. The hope of the Murray–Darling Commission is that once again 'the Strategy will buy time for land-scape change options and incentives to be developed, and for them to start to take effect'.[19]

Morgan in South Australia is to be kept below 800 EC for 95 per cent of the time. Each State has a salt credit balance, a certain level of salinity that it is permitted in its part of the Murray–Darling system. So if new salt problems crop up that affect Morgan, that State has to fix the problem fast. In fact a State credit system has been in force for some time. If, for instance, a South Australian farmer wanted to buy in water from Victoria, which he can do these days, he will be charged a salt levy. Queensland is also committed this time, which covers the Darling River above Menindee Lakes.

The next step was to put the pressure on at the grass roots. For each of the 21 rivers across the Basin, 'end-of-valley' salinity targets have been imposed, a sort of cap on salt pollution. This time it wasn't just for irrigators, but for everyone. The exception, naturally, is Queensland, which will set its targets by 2003. Targets are to be met by planting trees, cutting back water rights, and through on-farm improvements. However, because end-of-valley targets and actions to offset the salt in the Mallee will fall short, State governments will have to put in more salt interception plants to reach the Morgan target.

The good news is that Victoria, New South Wales and South Australia are all still committed to a number at Morgan, so even if things get worse, as they undoubtedly will in some areas, the States will have to fix them up, by adding new artificial salt interceptors. The Achilles heel to the Strategy is its supposed strength: the whole-of-catchment approach. Policy makers talk 'whole of catchment' and 'integrated catchment management' as if they are some kind of Nirvana. The eyes of anyone normal glaze over pretty fast.

The trouble is that to reach the end-of-valley targets, very different interest groups—government, rural communities, broad-acre farmers, the Greens, Aborigines and entrepreneurs wanting new industry development—have to work together. In New South Wales the Nature Conservation Council alone is fielding 440 water and natural vegetation hawks. The aim is to ensure the 'health' of the river, which includes animal and plant habitats. The Strategy itself admits it's a challenge.[20] The comment from Tim Fisher, coordinator of the ACF's Land and Water Program, was simply that 'undue faith has been placed on integrated catchment management as a key delivery mechanism for water quality improvements.'[21]

Making matters worse is new State legislation that has changed all the committees through which catchment management works, mainly to make them greener. Take New South Wales: the Customer Service Committee replaces the old River Advisory Committee and sits under the Water Management Committee, which took over from the 'redneck' River Management Committee, and above this sits the Catchment Management Board which has replaced the Total Catchment Management Committee. That board then reports to the Department for Land and Water Conservation. Frankly, it's not easy to fill the seats. Critically, catchment management organisations have no control over money to compensate landholders. That makes it very hard to be on the front foot in negotiations. Try telling a farmer that he needs to plant trees on his property because it's affecting some bloke that he doesn't even know 100 kilometres downstream. 'Prove it!' he says. Or explain to Mayor Kevin Wales why he may lose the abattoir that employs so many in Wagga Wagga to Victoria because the business wants to expand and New South Wales water rules are tougher than over the border.

Word from the ivory tower

Prime Minister John Howard did respond to the call to arms from Messrs Garrett and Donges, in person, five months later in October 2000. Not with $65 billion over ten years, though; instead, $700 million over seven years, on the strict condition that the money is matched dollar for dollar by the States. It pans out to be 4 per cent of what Peter Garrett and Ian Donges want from government.

Nevertheless, it was a National Action Plan[22] and cash was attached, which pleased those who were pushing the Salinity Strategy. There was also the promise that the money would be targeted to effective outcomes, unlike the $1.5 billion Natural Heritage Trust money which came from the sale of Telstra.

Professor Peter Cullen from the Centre for Freshwater Ecology in Canberra sees the National Action Plan as a rare chance for government to show leadership. 'The onus is now on the States and Federal Government to say what they see as particular values, because all the schemes that we've done so far have reflected local interests. If they don't do that, we're dead. We've got sprinkled around the country a number of Ramsar wetlands and it's an absolute joke to pretend that you can draw a little texta colour round a wetland on a map and pretend you can protect it. We've got to start talking about Ramsar catchments and that means restricting the water extraction from some of those rivers.'

What is extraordinary, however, is that instead of picking up on the Salinity Strategy and its 21 river catchments, the Federal Government selected 20 highly affected catchments *nationwide* and focused on dryland salinity and water quality with programs of revegetation and new environmental river flows. Needless to say, the river catchments were not all the same as those selected by the Salinity Strategy. This is despite a draft of the Salinity Strategy being released by the Murray–Darling Basin Commission one month before the Government made its announcement. Having kicked off the water drama with the Salinity Audit, the Commission naturally felt it had some ownership of the process.

This bureaucratic bungle has caused total confusion over targets and, in the words of former head of SA's Department of Environment and Natural Resources, John Scanlon, 'the National Action Plan even fails to mention the 2000 Draft Basin Salinity Management Strategy. This will inevitably weaken the Murray–Darling Basin Initiative—just at the time when we should be strengthening and reinvigorating it'.[23]

Trees

Trees are in vogue. Farmers, Greens and governments are all talking trees, perhaps billions of them. Trees are being asked to do an awful lot of repair work.

Our top scientists don't have the same faith in trees, particularly those in the west. Tom Hatton, with the CSIRO in Perth, says he's rather tired of being called Hanrahan. 'There's a constant tension between scientific realism and the pressure to be optimistic and to not steal hope from the community,' he said, and then proceeded with the bad news. 'You can't solve [the salinity problem] with trees. Not even if you covered 60 per cent or 70 per cent of the catchment with them. I'm continually frustrated by people's lack of understanding of the scale of intervention required to make almost any change in the degree to which salinity is going to progress. Trees are a waste of time for most of Australia. And they know that. The Commission has been briefed. The government ministers have been briefed. To approach the problem with tree planting would require revegetation on such a massive scale that it's not politically or sociologically or economically feasible.'

Like $65 billion. David Pannell at the University of Western Australia believes 50 per cent of the landscape needs to be replanted to make a difference, but as he says, with a 2,000 hectare farm of which 1,000 hectares need trees at a cost of $1,000 a hectare, this brings the cost to $1 million per farm. $1.4 billion doesn't go very far.

So how has government handled this news? 'It's received with the same processes of grieving that the news of a sick or dead relative is greeted,' according to Tom Hatton. 'First there's denial, then there's anger, and they go through all the steps, and eventually they get to acceptance. If you're lucky. But for the most part we have government ministers and industry leaders [who] are through the denial stage, mostly through the anger stage and now we've seen movement toward some realism—that a few trees up in the back paddock and a few trees down by the creek are not going to fix this problem.'

There is also a nasty side effect of tree planting that has the folks at the Murray–Darling Basin Commission shifting from buttock to buttock. Trees suck up an awful amount of water—irrigator water and fish water. According to the Commission, 'Preliminary regional scale modelling suggests that the longer-term benefits—reducing the area of salinised land and improving water quality—may be more than outweighed by the shorter-term impacts. These impacts include reducing water quality, reducing surface water yield, and reducing revenue, especially where irrigation industries are strongly dependent on surface water supplies.'[24]

Earlier practical work seems to back this up. In Victoria, a 320 hectare catchment in the Kilmore area has been studied since it was sown to pines in 1984. While the pines reduced the salt loads from 0.45 tonnes per hectare to 0.13 tonnes per hectare a year between 1989 and 1993, annual flows in the river dropped from 690 megs to 203 megs.[25] This and other research suggesting that tree planting to address the salt problem might backfire in some instances by reducing river flow led to some rather wild political claims that reafforestation was a solution of despair in the Basin and could stop water reaching Adelaide altogether. It is a sad fact that the number of trees required to fix the damage that cutting them down in the first place has wrought is astronomical, too large to be 'practical'. While tree planting is clearly good for the environment, in the short term it's bad for irrigators. That does not mean we shouldn't bother, however. Even scientists agree that targeted plantings can be very effective.

And did someone mention that dreadful word 'biodiversity'? Trees are critical to bringing back animal and plant life, if nothing else. And most important of all, now we have the knowledge, Australia, particularly Queensland, should not be cutting down any more.

The price of science

In the last few years about $1.2 million a year has been spent on researching different agricultural uses for Australia. 'Peanuts,' as Dr John Williams put it. '"R&D" is a dirty word,' he said, 'you cannot even use the word "research" in briefing Ministry. They don't see that as getting on with the job.' In a good year $150 million is spent in research on broad-acre dryland farming, which makes up 90 per cent of the cleared area in the Basin. The trouble is that most of the money is spent on profitability and production, not on research.[26]

Here we come to one of the most fundamental problems for people who have any interest at all in water. The science is lacking, and the scientists will be the first to tell you so. As Graham Harris, head of the CSIRO's Land and Water Division, has warned, 'Almost none of the environmental engineering community can meet this need. Agricultural agencies have largely deskilled and moved out of extension work and, anyway, they do not understand all the landscape interactions involved. Many other agencies are deskilled and dumbed down.'[27]

The upshot is that the *entire* water debate is being held back because the science cannot tell us what Australians desperately need to know: where and how we can fight back the salt, and just how much environmental flow is needed to keep the rivers healthy. Given that all the rows between water users and the Greens revolve around these issues, our position could not be more unsatisfactory.

According to Peter Cullen, what is holding Australia back is there is not large-scale, long-term experimental science because of government NIMTOO thinking (not in my term of office). Unbelievably, Peter Cullen reports that our water flow monitoring networks have been reduced because the Federal Government has stopped funding the work. 'We don't have a major national instrumented wetland and floodplain site in this country. Other countries have major instrumented catchments. I've been looking at some American ones just recently. There are very distinctive features of the Australian landscape. Floodplains, dryland lakes, terminal wet-lands, all that sort of stuff. Why on earth haven't we got one of those substantially instrumented for long-term data collection so we understand how variability works?'

The Commonwealth Government says it want its National Action Plan targets based on good science and economics. John Williams of the CSIRO and Peter Cullen are yet to be shown the money. 'The difficulty for the scientific community is that [governments] expect that knowledge to be available instantly,' explains Peter Cullen. 'There's no engineers who would build a dam on three months' rainfall. They'd like to have ten years' rainfall as an absolute minimum, normally 100 years' rainfall, and yet ecologically we're expected to comment on how their rivers work with one field visit.'

Peter Cullen's work in the Campaspe River in Victoria is one of the only full river research projects on environmental flow. The Campaspe is a typical dammed river—the natural flow of wet in winter and dry in summer is reversed. The work began a few years ago, but unfortunately it coincided with Victoria's worst drought in almost a century and the dam water was too 'precious' to experiment with. Twenty-five per cent of the dam water was released in May this year and the team is now busy looking for signs of life. 'If it works, it gives us much solid foundation for the environmental flow stuff. If it doesn't work, [if] 25 per cent is not enough, that's a bit gloomy, but we're optimistic.'

Peter Cullen marvels at the ignorance around him. 'Sometimes science hasn't been asked. Sometimes science has got it wrong, and sometimes science has got it right, but has failed to transfer the knowledge to where it can be used. In this context governments have cleverly dismantled the technical services that used to deliver such knowledge and replaced [them] with a new concept known as "wishing".'[28]

For a short time, a few politicians, along with the media, hailed as the answer an amazing machine developed by CSIRO and sitting at the Schofield air base in Adelaide. Using aerial magnetic data, the Tempest, as it is called, can pinpoint where the salt lies. It has been very useful, but what it cannot do is tell you where the water is carrying the salt; the machine is only really useful if all the hydrology is done as well. This hydrology is the time-consuming water research that is lacking.

The Murray–Darling Commission certainly has a wish list: a few inadequately researched areas they've pinpointed include measuring river flow and groundwater salinity monitoring, common protocols for analysis, better and longer historical information, measuring the linkage between improved irrigation and drainage with change in river salinity, the same thing for a change in vegetation cover, and assessing water recharge and run-off for salt mobilisation. And if the research is behind, the capacity of State agencies, catchment management organisations and landholders to access the research and apply it to salinity problems is lagging even farther behind.[29]

White coat revolution

CSIRO did not think much of the government's National Action Plan. Just the other side of the lake from Federal Parliament, John Williams' team had spent months putting together some fairly radical thinking for the government on land use.[30] It seemed that not much had sunk in.

According to John Williams, through its National Action Plan the government will only create welfare-dependent farmers. 'John Howard's Plan just says, "plant trees and we'll compensate you for your income loss,"' he said at the time. He now argues quite differently: rather than compensate farmers, we need to find profitable systems for replacing existing land use. 'The whole process is predicated on a belief that it's a matter of adoption failure rather than having been a failure of something to adopt,' he said, waiting politely while the statement was absorbed. 'That is, that people are not taking up these land uses because they haven't been communicated well enough . . . But the truth is that in every bit of work we've done, farmers can take up very few things that will make them money and address salinity at its cause.'

Tom Hatton agrees. '[A] commercially attractive [area] for tree planting certainly doesn't go below 700 millimetres and probably in most places doesn't go below 800 millimetres of rain. The wheat belt is below 600, mostly below 500. So it's easy to say you need farming systems that don't leak water, but there's nothing that's feasible, economically.' So what sort of revolution are the boffins from Black Mountain talking about?

Unfortunately, the farmers' best produce, annual crops, are at the front of the firing line. Typically, 5 per cent to 15 per cent of long-term average rainfall gets past the roots of annuals and eventually into ground water systems, compared to 1 per cent for native perennials. Cereal growers may already have put *Watershed* down at this point, but there's more.

'We need to pioneer the development of a new landscape,' explains John Williams, 'a mosaic of tree crops driven by large-scale industrial markets such as biomass fuels and high-value annual crops, as well as mixed perennial–annual cropping systems.'[31] John Williams talks of new, more suitable crop strains that may even have 'resurrection genes' (something we're all after!), but these crops could re-sprout after harvest if there were summer rain. Forestry, yes, but only in the 6 per cent of the Basin where rainfall is higher than 800 millimetres a year. Other native shrubs producing nuts and oils should be introduced, and mallee trees can produce charcoal and industrial solvents.

John Williams believes these changes could eventually allow farming to stand on its own without handouts. The bad news is that most of these options need a great deal of research to learn to mimic the bush and produce crops that thrive in a more natural landscape. That means time (between five and 30 years) and, of course, money.

Pilot programs to attract investment in forestry are under way, some even with a bit of government funding, and there is much talk of credits: carbon credits, biodiversity credits, salinity credits. The concept of credits is that they can give the owner (say in the case of salinity credits, an irrigator) the right to create a certain amount of salinity. He can acquire the salinity credits by spending money on a process which is known to reduce salinity. On the Macquarie River, for example, salinity problems are worst up-river from irrigators where the land has been cleared for crops. A local group of irrigators has bought salinity credits based on the amount of water transpired by trees planted upstream, with money passing from irrigators to the farmer doing the planting. The hope is that one day all these credits will have a real value that can be traded, although any decent market in these credits looks to be some way off.

The way forward, then, is unpleasant. There are some tough decisions to make. The research isn't ready, there's no magic bullet. What should Australia do?

According to the Murray–Darling Basin Commission, 'The challenge is to set limits to the degree of degradation we are prepared to accept, and to decide on how much landscape degradation will be worthwhile.'[32] Cut our losses is what this is saying. Focus on what's worth saving. Prioritise. What are these areas likely to be? 'Areas of critical biodiversity, rural towns,' said Tom Hatton, looking moderately optimistic. Recognise fabulous irrigation areas where you put in the salt interceptors? 'Yep, maybe that's where you invest your money, as opposed to the Vegemite approach, which tends to happen for obvious reasons—political ones—put a little bit everywhere so you don't have any impact anywhere.'

Peter Cullen agrees. 'Trees will be part of the answer in some places, engineering works are going to be part of the solution in some places, and retiring people off the land will be a solution, so there is no single panacea. We all know that if you put a lot of trees into the catchment, the economics of that land use are going to change. But equally, leaving things as they are isn't working, and they will get worse. Those people are going to go broke slowly, with a lot of hardship for them and their communities.

'In the PM's National Action Plan for the first time it signalled that the money could be used to purchase land or water if we have to have dramatic land use changes in certain catchments. Now, that's going to need a lot of argument at the end of the day and a bit of muscle to make it happen. But there are places where I suspect we should just give up on farming—in salt-generating areas—and try to stabilise the landscape to minimise the downstream impacts.'

About other parts of Australia, Tom 'Hanrahan' Hatton can become quite excited. We can't beat salt, so we live with it. 'Half the game is coming up with the technologies and strategies for helping people adapt to and profit from salinised resources. Living with it, profiting from it, coping with it. Facilitating society's relationship with living with salinised water, that's where the game is.'

Living with salt

It seems pretty bleak, this idea of living with salt: *Beyond Thunderdome* landscapes dotted with stunted saltbush. It doesn't have to be like that, though. Australian farmers are alive and well and living with salt, and to find some of their best, it's worth heading back to the Coorong, not far from the mouth of the great river, to Duck Island.

James Darling and farmers like him between Keith and the coast turn off some of the better cattle in Australia. Darling's Duck Island property, as its name suggests, was once a natural wetland. Back in the 1920s the land was reclaimed. 'The Department of Lands took 4,000 [acres] from me in 1980 because I wasn't clearing enough land according to them,' said James Darling with an 'I told you so' look.

At some times of the year, one-third of Duck Island can be under water. 'See that?' Darling asked when I visited him on his property, pointing to a patch of white in the bottom of a drain on the property. 'That's about as saline as you'll see. But if you're a metre and a half under the surface here, you're about one and a half times as saline as the sea; two and a half metres, you're up to two and a half times as saline as the sea.'

For James Darling, the key to understanding water is simple. The whole of the Australian continent is a dynamic balance between interconnected watertables. He also believes a huge amount of bad management is blamed on salt. The first lesson on farming with salt, he said, is learning to under-graze. 'If you can do that, you change what might have been considered your least productive land into something that at least rivals the most productive land on your place. By calving twice, you don't bare your ground before opening rains, which was another sort of Western District Victorian habit. [Calving twice and selling off early means the grass does not get pressure from larger yearlings.] People think that if you haven't eaten every blade of grass then you haven't used it properly. Instead of keeping cover, having income coming in all year round, using your bulls twice.'

James Darling's secret is a grass called *Puxinellia*. Originally from Menemen in Western Turkey, *Puxinellia* is magic. It's high in protein, will grow under water and provides great autumn feed and, most importantly, is salt-tolerant. Green paddocks are fenced off on Duck Island. 'It's the one crop that will always be razed to the ground. The cattle love it.' The grass is shallow-rooted and does little to change the high watertable, but then this is living with salt, not fighting it.

Large parts of Duck Island are also left *au naturel*: melaleuca tea trees and samphires alongside mallee. James Darling is almost cocky with his salt. 'People get frightened about salt—if something has salt in it— and yet I find it very clean. It's very good to put it on hills with bad weeds, salt gets rid of weeds. I don't have to use any sprays.'

Other farmers are busy making outback oceans for aquaculture and looking at salt harvesting for commercial sale. The question is: how much of Australia will have to do this sort of farming in future? How many will have the money? And, frankly, how many will want to be bothered?

There are many sites on the Internet about the Basin—one even has a tombstone for the Murray River. Its epitaph reads:

> RIP
> Murray River
> 100,000 BC–2030 AD
> *'SEE—I told you I was sick'*[33]

You don't have to be a raving greenie to care about the Murray-Darling or any other river in Australia, or any other river for that matter. You don't have to love every mosquito that lives along the surface to realise that something is not quite right.

6

Go With the Flow

'And you really live by the river? What a jolly life!'
   'By it and with it and on it and in it,' said the Rat.
'It's brother and sister to me, and aunts, and company,
and food and drink, and (naturally) washing. It's my
world, and I don't want any other. What it hasn't got is
not worth having, and what it doesn't know is not worth
knowing.'

   Kenneth Grahame—Wind in the Willows

The Gondwana inheritance

There would be not too many of the X and Y generations who would have heard of *Silent Spring*.[1] This courageous book, written in 1962 by Rachel Louise Carson, predicted a world where pesticides and other abuse would destroy it to the point where birds stopped singing. The ideas in *Silent Spring* were indeed ahead of their time. That time is now.

   If you speak to professional Greens, some, like Tim Fisher at the Australian Conservation Foundation, are optimistic. But for the most part, there is a deep mistrust, built up over years of being let down by developers with their promises of 'Relax, this won't hurt a bit'. In truth, the hurt of 200 years of colonisation in Australia has been massive, and much of it is irreversible.

   Australia has its own Rachel Carson. Her name is Mary E. White. Mary likes working in the mornings when her encyclopaedic mind is at its clearest. In recent years, this palaeontologist-cum-environmental scientist has devoted many thousands of mornings to writing about Australia's changing land and, more recently, its water. But it is her study of the ancient history of Gondwana,[2] of how this continent came to be the driest vegetated continent, which led her to a stark conclusion. We have completely failed to take into account the prehistoric events which determined how Australia would respond to

European-style agriculture and land use practices. For 200 years, we have been misreading the land.[3] 'I suddenly thought, well for God's sake, with a history like that, how could it ever have responded happily to people coming and treating it like a bit of the northern hemisphere.'

Armed with a good cup of tea, Mary White continued, 'The northern hemisphere's got a completely different history, particularly the ice age. It had had all its soils renewed, deep, fertile new soils created, and it had, after the ice age disappeared, the establishment of reliable climatic cycles where agriculture and land use practice evolved. They were suited to a place in which you could depend on spring when you prepared the soil, you planted, summer the stuff grew, it ripened in the autumn, you harvested it, and it was safe to leave your field fallow in the winter. And these cycles were working with you, the soil was working with you.'

When the European settlers arrived in Australia they had no idea of the country's harsh past. 'A land that had been a wind-blown desert,' expounded Mary White. 'We had the absolutely horrific histories particularly of the last glacial stage of the ice age, at 18,000 years ago, when this continent was twice as dry and twice as windy and therefore twice as fiery. And ancient worn-out soils—the last time they were renewed by ice marching over the landscape grinding up new rock, 300 million years ago.'

It should be no surprise that water was different on this continent too. As pioneers like Sturt and Hume became painfully aware, Australian rivers were not at all like the one on which Ratty and Mole enjoyed nothing better than messing about in boats. In Europe, water in rivers starts in the hills and flows to the sea, for almost all of it a journey of just a few days. In contrast, most of the rivers in Australia's arid, flat land run, not to the sea, but inland, many petering out altogether. When it is dry, the land is tough, and when it is wet, it is unmanageable.

The secret to how our land of extremes works is that life has built up around the floodplain. Australia is covered in wide plains of often rich earth (although not that deep) deposited many tens of thousands of years ago. Rivers meander quietly across them for months, perhaps years, until heavy rains set off floods which trigger a burst of life over the plains as the floods recede. That's how it was: a land of extremes to which nature, including Aborigines, adapted perfectly. To settlers, however, it presented a brutal life of insecurity.

White man's dams, weirs and intensive agriculture are designed to achieve precisely the opposite environment: security and reliability. Engineering know-how, pesticides and fertilisers took the nation even further to boom times. 'I remember those boom years, when there was that lovely feeling after the war,' said Mary White. 'It was relief, it was all over, the war had been so absolutely bloody generally, and you were feeling so nice about things and you really thought mankind was so very clever, it doesn't matter if the population doubles. And then you're pulled up sharply because you haven't taken into account some of the most fundamental parts of the whole thing.'

In the short term, it has been a miracle. In the long term, according to Mary White, to the CSIRO and almost anyone who has read the history books on irrigation through the centuries, it is not sustainable. As we have seen already in Australia, nature's amber light is flashing: the Murray–Darling, the wheat belt of the west, the denuded rivers tipping tonnes of silt onto our great reef.

Mary White has but one solution and it accords pretty much with what we've been hearing from the CSIRO. We have to work with the variability of Australia's environment, not against it. That means picking winners. 'We're not far off the stage where we've got to make a very deliberate assessment and say these areas of Australia, if they are managed with the best technology and the greatest wisdom, are reasonably sustainable—nothing is completely sustainable. But a large number of the areas we are using are not suitable for the use to which they've been put.' That may be a big ask of farmers and graziers but if we don't do it we will lose those areas anyway. But just how far are people prepared to go to save the environment? Many of us need to heed the wake-up calls of a Mary White.

Maurice Strong, the Canadian who led the Rio Earth Summit in 1992, has spoken of Australia as a potential environmental powerhouse in a region that runs from the South Pacific right down to Antarctica. 'What Australia does or fails to do will make a profound difference to this region and this region will make a profound difference to shaping our environmental future as a planet.'[4] In this chapter, we look through Green eyes at what Australia is doing and failing to do, and at some of the battles to protect our living treasures. In chapter 7, we'll turn to the plight of the farmers and why they feel, somewhat justifiably, that they are unfairly having to shoulder responsibility for Australia's water problems.

Where the Greening began

There is nowhere in Australia 'Greener' than Tasmania. While the recent drought is doing its best to refute this, the Apple Isle is the country's environmental conscience.

Tasmania puts the wild into wilderness, for the 'empty' quarter of the State is wilderness in the true sense of the word. And let's not kid ourselves, it is only wilderness today because the land in the south-west corner of the island is utterly impenetrable. It has been argued that had the Tasmanian Aborigines been pushed into this region instead of rounded up and banished to Flinders Island, they might have found some sanctuary and still be a population today. To settlers, the wilderness was useless, save for the copper and gold in its hills. Areas like Queenstown are now scars on the landscape where years of acidic sulphurous fumes have poisoned the surroundings.

It was this unique wilderness that kicked off Australia's green revolution, marked by two major events: the drowning of Lake Pedder and the fight to stop the Franklin Dam.

While Australia is notorious for its coal mining and greenhouse gas emissions, Tasmania is self-sufficient in what most would see as 'green energy', hydro-electric power. It is one of the great paradoxes that to many in Tasmania, 'Hydro' is an environmental vandal. The State Government's vision was that hydro would be the economic saviour. Unlimited power would attract industry from the mainland and invigorate the economy.

'In the 1930s Labor came to power here with the belief and the propaganda that we would become the Ruhr Valley of Australia and that the hydro-engineers would be the modern Moses, who would lead us out of the wilderness,' explained lobbyist and author Richard Flanagan. A self-described watermelon green (green on the outside and pink in the middle), he grew up with the campaigns for Pedder and Franklin. 'Central to that was the idea of the natural world being entirely subservient to man and being transformed into these heavy industrial forms.'[5] As it turned out, the tyranny of distance, this time the Bass Strait, proved too much of a disincentive for industry.

At about the same time as the Snowy Scheme was being proposed, the first hydro-electric station in Tasmania was commissioned in 1938 at Tarraleah on the Derwent River, using the river's water and that of Lake St Clair in the national park. Every year between the late 1940s and the 1980s, the State-owned Hydro Electric Corporation (Hydro) built or expanded a major storage. Today, Hydro dams together now have a capacity of 25 million megs, easily meeting the State's power demands.[6]

From an engineering standpoint, the State is perfect for hydro. Over 80 per cent of the rain falls in the wild west where fierce rivers run through steep mountain ranges. It is not unusual for some parts of the southwest to get 4 metres of rain a year! The other 'beauty' was that hardly a soul lived in these areas, so unlike the rest of the

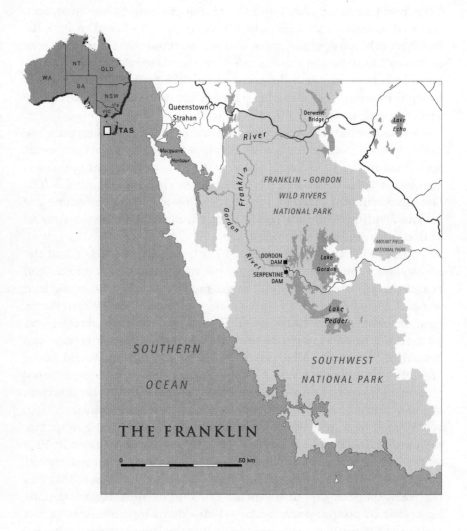

Queenstown
Strahan

Derwent
Bridge

Lake
Echo

River

Macquarie
Harbour

FRANKLIN - GORDON
WILD RIVERS
NATIONAL PARK

Franklin

Gordon

MOUNT FIELD
NATIONAL PARK

Gordon River

GORDON
DAM

Lake
Gordon

SERPENTINE
DAM

Lake
Pedder

SOUTHERN

OCEAN

SOUTHWEST

NATIONAL PARK

THE FRANKLIN

0   50 km

world, issues of human displacement were non-existent. In fact, so untouched are many of the national parks where the lakes have been created that the only way to reach them is to walk.

While damming continued rapidly in the 1950s and 1960s, the conservation movement had also been gathering pace. The last straw for Greens was the drowning of Lake Pedder in 1972, as part of the Gordon power scheme. Before it was drowned, Pedder had been 9 square kilometres, a smallish lake of unique beauty with a broad shining pink beach running 3 kilometres around it. In 1955 it had been proclaimed national park. Even that could not save it, however. Nor did the many photos that protestor Olegas Truchanas pain-stakingly took of the area. Truchanas drowned in a tragic accident while filming the Gordon River in 1972 and became a martyr within the conservation movement. However the determination of then Premier Eric Reece ('Electric Eric') was unmatched. When the Greens established that in drowning the lake Hydro was breaching the *National Parks and Wildlife Act,* Premier Reece passed the Gordon River Doubt Removal Bill, which allowed flooding retrospectively.[7]

In every way, Pedder was a flop. The plan for Pedder was not to be a power scheme itself, but a storage for Lake Gordon. Despite the flooding of 240 square kilometres in 1972, that storage fell well below expectations. Pedder today is eerily black in some places, the tannin from drowned button grass leaching like tea into the water. But the whole debacle unleashed its own environmental tiger. The nascent green movement and the world's first Green movement, the United Tasmania Group, was formed, and for the next battle, which would go down in world environmental history, they were prepared.

It is not difficult to see parts of the Franklin; the quiet upper reaches of the river are crossed by the Lyell Highway. But where Hydro proposed to flood a 35-kilometre stretch of the Franklin is well beyond the reach of most Australians. This is downstream, deep in the National Park. Here the river powers through the Franklin Gorge, which Dr Bob Brown brought to prominence when he kayaked down it in 1976 and formed the Wilderness Society.

Today, Senator Bob Brown is Australia's most famous green politi-cian. He has more opinions than the press gallery sometimes has patience for, but on green issues and anything that impacts on green issues, this is the man. Looking back, he remembers the campaign for the Franklin was very different from Pedder. 'First, we had colour TV for the first time so people could see those magnificent pictures of wild scenery and take some ownership for themselves. We also knew we had to argue about the economy and jobs because the Hydro had scared everyone over Pedder. And there was a lot of debate about peaceful protest. We decided on direct action and intervention against the machinery. We even had secret weekend camps and looked very carefully at strategies to manage large crowds.'

By the late 1970s, Hydro's plans for the Franklin were clear. When work began, so did the now famous rubber duckie blockade of the lower Gordon, which the Franklin runs into, near Strahan (see the map on page 116). Everything was organised, from food drops in base camps to weekend training for protestors. 'The Franklin blockade became pivotal in galvanising public support. The first bulldozer came into Strahan the day of the blockade and police cut the phone lines in town, but missed one underground line and I was able to talk to [ABC's] *PM* on radio.'

At the same time, the first pieces of equipment for the dam were being barged up and protestors in rubber duckies and on foot got as far as 40 kilometres from Strahan. According to Brown, special laws were invoked which allowed the police to arrest protestors for trespass. In total, 1,272 people were arrested and 450 of those went to Risdon Jail. Bob Brown was among them and spent 19 days behind bars.

The white knight for the Greens was the incoming Federal Labor Government, which opposed the Franklin Dam. By this time, Labor was beginning to gain its own green tinge, as environmental concerns began to appear in the polls. In 1983, Federal Parliament moved to place southwest Tasmania on the World Heritage list, preventing development proceeding. Even the air force was sent over to make sure work on the dam had stopped. Outraged, the Tasmanian Government appealed to the High Court, but in July 1983 lost the battle in one of the most important States' rights battles in our history. 'In the end, the High Court ruled, four to three,' said Bob Brown. 'It was a States' rights issue in terms of managing the land and environment versus a Federal obligation to uphold an international treaty, the World Heritage Convention.'

On the day of his release, Bob Brown was elected as the first Green to enter Tasmania's Parliament and on 30 August 1992, he officially launched the Australian Greens Party in Sydney.

In some ways, it is quite bizarre that an area which precious few Tasmanians let alone mainlanders will ever see became such a focus for the Greens. In summer, the 'empty' quarter of Tasmania is far harsher than the bush of New South Wales, although there are the same biting horseflies as can be found high in the Snowies. It's rugged terrain, dense forests and often bleak climate have claimed the lives of a number of Australians. When asked why it would matter to dam the Franklin when there was so much wilderness, Bob Brown responded, 'It would be like putting a scratch on the *Mona Lisa*.'

Dams and dam busters

The Franklin win marked the rise of the anti-dam era. Good and bad, dams have made rural Australia what it is today. Each has a story, from the Sugarloaf Dam built in the early 1900s at Eildon in Victoria, which ended up drowning the farms of some of the settlers who helped build it, to the massive Argyle Dam in Western Australia, which still promises so much.

Dams are a perfect focus for lobbyists because they can do so much damage. 'Australia's rivers have a heartbeat,' wrote Tim Fisher. 'They beat to the natural rhythm of wets and dries, of flood and droughts. Dams are designed to regulate and exploit water resources. But they also deaden the heartbeat of our rivers.' Barriers which stop fish breeding, water at icy temperatures and starved of oxygen because it is released from deep in the dam, and, perhaps most importantly, changes to the size of water flow and an end to seasons—is it any wonder that fish such as the Murray cod and perch are in trouble? As Mary White reminds us, the sacred relationship between the river and the floodplain has been desecrated.

Around 45,000 major dams cover over 400,000 square kilometres of the world. It sounds unbelievable at first, but where the river lovers have been more successful than anywhere is the United States. There the pendulum has swung back so hard that dams are now coming down in serious numbers. Big dams. And even more amazing, the issue driving the dam busting is predominantly fish.

Only one major river, the Yellowstone, in the whole of the United States flows freely. Lucky Yogi. Americans have had their fair share of environmental disasters with dams, one of the most well known being the Glen Canyon Dam on the Colorado, built in 1963, which ruined river flow through the Grand Canyon, wiped out native fish species and prevented the majestic river from ever reaching the sea in Mexico in most years.

One of the earliest public protests on dams in the United States came in 1981 with the cracking of Glen Canyon Dam by radical group Earth First!. The group draped a black plastic 'crack' over the side of the dam in its first, and highly effective, exploit. The move was both backed and inspired by Edward Abbey, the author of the cult novel *The Monkey Wrench Gang*,[8] in which the Glen Canyon Dam was sabotaged to free the Colorado.

Through the 1990s, public interest in the damage dams cause

grew, fuelled by Paul Reisner's book *Cadillac Desert,* published in 1993. But the breakthrough came in 1999 with the busting of the Edwards Dam on the Kennebec River in Maine. Built in the 1830s to give power to local factories, the dam was old and inefficient and a major target for environmentalists. Despite the protest from the power company, the Federal Energy Regulatory Commission refused to renew the licence and on 1 July 1999, to the delight of onlookers, a huge hole was cut through the dam. The beneficiaries were nine species of native fish. Incredibly, Atlantic salmon were reported as returning to their old waters within weeks of the demolition, the first time they had been seen in over 150 years.

Dam busting has now become a national sport in the United States. According to the International Rivers Network, by May 2001 nearly 500 known dams had been removed (admittedly mostly small ones) with a further 100 dams either committed or under considera-tion for removal.

There has also been growing criticism of the role of organisations such as the World Bank for creating situations, particularly in Third World countries, in which greed has been allowed to influence devel-opment with the outcome of 'bad dams'. In 2000 the chair of the World Commission on Dams, Professor Kader Asmal, complained, 'Over the last century we collectively bought, on average, one large dam per day. There have been precious few, if any, comprehensive, inde-pendent analyses as to why dams came about, how dams perform over time, whether we are getting a fair return from our US$2 trillion investment.'

That analysis has now been done. Following a two-year review worldwide, the Commission, comprising representatives from the World Bank and the World Conservation Union, reported its findings to the United Nations in November 2000. While dams had made a sig-nificant contribution to human development, the Commission felt that in too many cases the social and environmental costs have been unacceptable and often unnecessary. It called for a new decision-making process that looks beyond a simple cost-benefit analysis to rights and risks, including the importance of environmental flow.[9] Just how much impact the report will have on dam building remains to be seen, but as Nelson Mandela said at the unveiling, 'It is one thing to find fault with an existing system. It is another thing altogether, a more difficult task, to replace it with an approach that is better.'[10]

Unlike many Third World nations which suffer from general cor-ruption and the strong arm of big multinational companies seeking hydro-electric power development, a level of scrutiny exists now in

Australia which means that building new dams will be much harder to pull off. In our attempt to drought-proof Australia, we have built around 375 major dams. Dam building reached a frenzy in the 1970s when seven storages were commissioned—the Gordon, Ord, Dartmouth, Fairbairn, Copeton and Wyangala, with a combined capacity of about 40 per cent of Australia's total storage capacity.[11]

Hunter Landale's fight for the Dartmouth revealed just what political dynamite dams have been—and still are, particularly in Queensland. 'When the National Party came in after Goss went,' explained water activist Jenifer Simpson, 'they came up with this thing called the Water Infrastructure Task Force. So everybody was asked, "Would you like a dam dear?" You know, only a couple of hundred million, and everybody put their hand up. You and I could be quite popular if we were standing on the street corner handing out million dollar bills, wouldn't we?'

It is this macho dam-building mindset entrenched in bureaucracy for years that Jenifer Simpson believes has been so damaging—the 'damn dam builders' as she calls them, or dinosaurs with their dinosaur technology. According to Jenifer Simpson, it is possible to identify those suffering from so called 'Dinosaur Technology Syndrome' at 20 paces.

> Dinosaurs are most likely to be male. They are short-sighted and their vision is often further impaired with dollar signs. They are cold blooded, have uncertain tempers and are given to irrational rages and tantrums if they don't get their own way. They are very large, powerful animals and have been in the habit of using this bulk to enforce their wishes. Political dinosaurs show cyclical behaviour; the cycles are of three or four years, depending on the species.[12]

Northern Australia has its own reasons for looking at dams, not least because the States up north are Johnny-come-latelies. In 2000, a new dam was announced for the Burnett region in Queensland and prospects for the Nathan Dam on the Dawson River, also in Queensland, are still unclear. But the good news for the Greens is that in the southern States, new dams are off the agenda. The plight of the Murray–Darling has now made it politically incorrect to even talk about dams.

Can we see dams coming down in Australia? 'It's happening already,' said ACF's Tim Fisher enthusiastically. 'The one at Wellington near Dubbo is coming down, it's silted up and has no value,' adding cagily, 'and there are a couple of others being considered.'

In 1994, a campaign to restore Lake Pedder was launched which led to a Federal Government inquiry the following year. The inquiry found it quite feasible to reverse Lake Pedder, but according to Helen Gee, the convenor of the Pedder 2000 campaign, the cost of the project as estimated by the Tasmanian Government ranged up to a massive $850 million. The inquiry wimped out. 'The Tasmanian Government quotes—they threw in everything they could, removal of every part of the dam, which we hadn't asked for,' said Helen Gee. 'There was a state of petrified hysteria in the lower house [of Parliament] at the time, as though they thought the Feds were going to override them again. It was a huge case of self-pride. But the Friends of Lake Pedder will never give up.'

One fixture Greens are targeting is the barrages on Lake Alexandrina at the mouth of the Murray. Pulling down the barrages and replacing them with a weir at the top of the lake around Wellington would allow the shallow freshwater lake to once again become tidal, helping the Coorong, and savings on evaporation by in effect shortening the length of the river, would be huge. Murray–Darling Basin Commission chief Don Blackmore is busy spending hundreds of thousands of dollars on the idea, but believes that the Greens' demands are far too simplistic. 'The environment may be better, but it will be a managed environment because under natural conditions we used to have on average 15 million megs going out the Murray mouth and now we have 4.7 million. So it's not going to be natural. That means the huge area of Lake Alexandrina will be under tidal influence and because there's very little fresh water it will be much more saline than it ever was naturally. So go into it with your eyes open.'

Don Blackmore also has problems with where the evaporation savings would be stored. 'Fifty thousand megs [the suggested savings on evaportion] could become a *safe yield* [a useable quantity] in terms of the way you operate a storage. But Menindee Lakes is full as we speak, Lake Victoria's full. I've got nowhere above that in the Basin to store it, unless I go into aquifer storage, which is possible, technically.' Even if he stores it (or its equivalent) somewhere upstream, such as Menindee Lakes, Don Blackmore sees another problem. 'That water has been flowing down the river all summer to get [to Lake Alexandrina], which is a very important environmental outcome for the River Murray, and then I stop that happening and then I've got a salinity problem.'

The Green response? Bugger the storage, that water saving is for the ducks and the fish.

The last free rivers

As in Tasmania, it is only in tough country that mainland rivers still run free, many of them up in the Gulf. There are a few, however, which are just within reach of irrigators and once again tensions have emerged about the prospects of tampering with these last pristine waterways. Only one free river is left in the Murray–Darling Basin, the Paroo.

Paroo country is as it sounds: 800 kilometres west of Brisbane, this is the end of the honey trail, the last stop west for apiarists in search of flowering gums. It's a flat land of red earth, Cunnamulla fellas and opals.

The huge Paroo catchment extends some 64,800 square kilometres,[13] the Paroo River rising in central Queensland and flowing almost due south over 600 kilometres into New South Wales. Rainfall varies enormously, from less than 1 metre to over 7, but the big floods come just once in 15 years. 'I reckon it's only flooded through to the Darling four times in my lifetime,' said NSW grazier Robert Bartlett, whose family has been 'keeping the Paroo free' for four generations.

In terms of wildlife, the Paroo is the most exciting area in the Basin. The jewel is the Ramsar-listed Currawinya Lakes, where 100,000 pairs of birds have been counted at one sitting, including the freckled duck, one of the rarest waterfowl in the world. The freckled duck is more closely related to the swan than other duck species, with flute-like calls and courtship displays, but it also takes after the goose, with a windpipe outside its breast bone. Unfortunately, these differences aren't enough and the duck is a frequent casualty of the shooting season, despite being protected.[14]

The Paroo is also one of the few rivers where native fish seem to be holding their own against carp, which don't like the natural boom–bust cycle. One of the most experienced and outspoken ecologists in Australia is Dr Richard Kingsford from the NSW Parks and Wildlife Service. His position has not stopped him from doing a great deal of work in Queensland too. 'The first time I visited the Paroo,' said Richard Kingsford, 'I could smell the dead carp from the plane from three or four kilometres away, rotting away as the lakes dried out. The theory is that carp can't find their way back to the river like the natives.'

The threat to the Paroo is only small—a couple of families near Eulo, just north of the Queensland–New South Wales border, wanting to irrigate for hemp and other crops, not cotton. But graziers south of the border were up in arms when the proposal was made. 'One licence is one too many for us, because we lose the whole status of the river you see. It's the thin end of the wedge,' explained Rob Bartlett. 'And one of the families up there is getting on, they may want to sell out, and if they sell a licence on to new people, who knows what they'll grow.'

For the moment, the Paroo looks safe, thanks to loud squawks from the World Wide Fund, which lobbied for the Paroo along with Richard Kingsford. As for the folks involved, there's a Mexican stand-off between the would-be irrigators in Queensland and the graziers from New South Wales. 'They haven't shifted an inch and we haven't shifted an inch,' said Rob Bartlett. Interestingly, despite its 'free' status, the Paroo is one of the more damaged river systems in an important new river survey completed by Peter Cullen's team at the Centre for Freshwater Ecology. The survey focuses on aquatic macro-invertebrates in rivers Australia-wide and will form part of the State of the Environment Report for 2002 and should give the best yet indicator of river health. 'The States have collected invertebrate samples right across Australia. They've identified rivers that they believe are relatively undamaged for that type of river, then we've looked at the invertebrates and created models which show how far below every site is from what we expect from an undamaged site. And most of Australia is fairly damaged.' Peter Cullen can't explain why the Paroo appears to be suffering.

The battle for the Cooper
In outback terms, the Paroo is suburbia really. Australia's best kept secret is another 650 kilometres northwest, just over three hours from Birdsville and well beyond the average Australian's desire for outback travel.

The Channel Country is magnificent and still works like Mary White's old Australia. It was floods on the Cooper Creek and its sister the Diamantina that brought Lake Eyre in South Australia to life in 1999. These far western rivers are not really rivers in the conventional sense. When they flood, they are giant, slow-moving wetlands and the only place to be is in the air. The flat land forces the Cooper into myriad braided channels. Just south of Windorah, Queensland, the river can be 80 kilometres wide and locals reckon there might be two or three thousand channels crossing the floodplain of over 100,000 square kilometres. As a result, the Channel Country is the best cattle fattening territory in Australia. Yet, because it was first discovered in the dry, possibly our most important dryland river is a 'creek' and very likely the only creek in the world supplied by rivers.[15]

The fight for the Cooper began when a group of highly successful

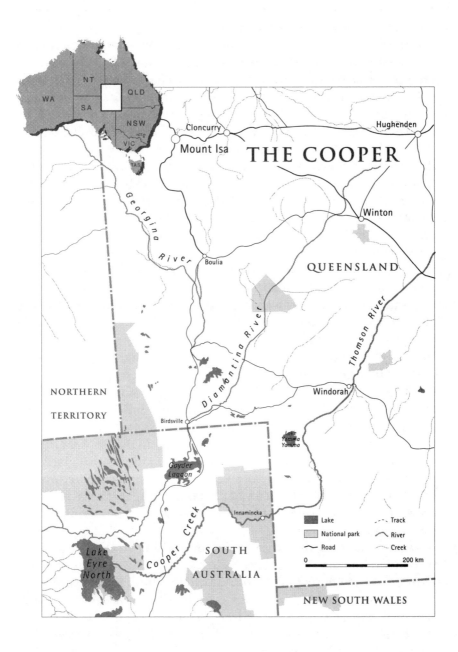

cotton growers from the Macquarie Valley in New South Wales took a fancy to the Cooper and the idea of damming some of its massive floodwater to create the next site for cotton. Their attempts to convince residents of the outback town of Windorah (population 63) and surrounding properties of the merits of cotton is almost Keystone Cops material. Like many areas of outback Australia, the Channel Country is home to some very colourful characters who were not going to take the cotton cockies lying down.

Windorah is a long way from a cappuccino. Mobiles drop out after Charleville, 450 kilometres away. This is presumably the area former Prime Minister Paul Keating referred to as the arse-end of the world, the land best flown over. Mulgas give way to hardier acacias, the fences disappear and the approach is single-lane bitumen splattered with dead roos mown down by road trains. As the day closes, the pig shooters take to the road with the day's bounty, huge, black, hairy bodies strung up by the trotters on the backs of utes, bound for German delis.

Up here, country people are masters of the understatement. Bob Morrish's property, Flodden Downs, described as 'just the other side of Windorah, mate' was another good hour and a half of dirt. The Morrish family is not flush. They live simply, dividing their time between Flodden and Bob's other station, Springfield ('just the other side of Windorah'). Bob borrowed to buy Flodden, a property of 142,000 hectares, although there's little commercial use made of 40,000 to 50,000 hectares of it because there's no water. Power comes from a rattly generator and cattle mustering consists of a chap in a chopper, Bob's wife in the truck and Bob on a horse.

Cattle is Bob Morrish's day job. He moonlights as head of the Cooper Creek Protection Group. He sports a mane of white hair like Bob Hawke, only thicker, and a wicked twinkle in his eye. After a comfy swag and good steak-and-gravy breakfast, and over measurement-taking around the cattle yards, Bob Morrish's tale of Cooper Creek was unfurled.

On 5 September 1995, the cotton growers came to a public meeting in Windorah to put their proposals to the community. Once the community was onside, council approval would follow. It can't have been an easy audience, but one of the mistakes they made, according to Bob, was to wheel in their marketing man, a fellow called Hans Woldring. 'He arrived at this public meeting to tell everyone what a wonderful thing they were going to do, you see,' said Bob. 'I think they expected all the locals to come along and clap or something . . . anyway Woldring arrived there looking very out of place and he apologised for a coffee stain on his tie. You know, none of the rest of the audience *had* ties. This is Windorah, and everyone was sort of saying, well, Christ, get on with it. No one gives a fuck about your tie.'

By a number of accounts, Hans Woldring underestimated Windorah. According to Bob Morrish, he informed the audience that he had been advising the Russians on irrigating from the Aral Sea. Presumably he was trying to solve some of the horrific environmental problems with better farm practices. But this was not how his pitch was taken and in fact if backfired. 'People weren't as silly as he thought,' explained Bob Morrish. 'People said, "We saw that on the TV, that Aral Sea, that's a bloody mess. What do you think you're talking about you bloody idiot? If you're part of that we want nothing to do with it, you hear." Country people have a built-in detector for bullshit.' The situation was not helped by Richard Kingsford following, quite by accident, with a presentation on the death in Aral.

Another speaker from Cotton Australia attempted to explain the facts about cotton and chemicals. Unfortunately, he too, apparently hit a raw nerve when he answered a question from the floor about endosulphan. Bob Morrish continued, 'And he must really not have done his homework, or not realised where he was standing, in Windorah, virtually on the banks of the Cooper. He said, "Oh, well endosulphan's totally harmless, except that it's pretty toxic to fish." And these guys planning to grow cotton up to 10 metres from the bank of this big water hole where a lot of the local Windorah population supplement their diet with yellow-belly. And he just lost it. I don't think he realised it. But most of the town people turned against him on that basis alone.'

It didn't help that the cotton growers pressed their good management of riverbanks and tailings dams, which would keep the chemicals contained. 'What they'd forgotten about,' said Bob with some satisfaction, 'was the famous western Queensland dust storms, and the Channel Country has the highest natural rate of wind erosion of anywhere in the world. The average sediment load in the western Queensland dust storms is something like 16 million tonnes. So if you've got a dust storm running past a cotton farm, you're going to pick up the chemical out of the soil. And their big argument was that endosulphan and the big organo-chlorines and organo-phosphates chemically bond and stick to dust particles! They just shot themselves in the foot because that's going to be blown all over the country. In fact, the big dust storm that emanated out of Birdsville in the early sixties ended up covering the New Zealand Alps a red colour, that's how far it went.'

As soon as the cotton group had packed up and left the stage, the Cooper Creek Protection Group was formed, 35 passionate outbackers spread from Windorah down to Lake Eyre.

At this point it should be mentioned that Cooper Creek was not quite the David and Goliath story it might at first seem, because along the Cooper lie Australia's most powerful pastoral companies: Stanbroke, Kidmans, AA (the Australian Agricultural Company) and, at that stage, Janet Holmes à Court's Heytesbury. 'We were a little bit disappointed in Janet Holmes à Court, but she was tied up with her theatres in London,' said Bob Morrish, still measuring the old yard gates.

According to Bob Morrish, Channel Country just on the Queensland side of the border would turn off conservatively 100,000 fat bullocks a year.

> Lake Yamma Yamma is a terminal branch of the Cooper, a black-soil lake, 800 square kilometres of wetland, on Tanbar [a Stanbroke property]. When Yamma Yamma gets full, it takes a year or so for the waters to recede back and as they recede back it leaves huge pasture growth along the edges. Tanbar can turn off almost unlimited fat bullocks. That's really big fat bullocks. One of the economists that works for Stanbroke's estimated that the annual average export value of Cooper alone, bullocks off the Cooper, is something like 130 million bucks. We were able to use that as part of the argument when these cotton growers were telling the Queensland Government how much extra money they were going to inject.

The Cooper has another important friend in the grazing business. OBE Beef is Australia's largest organic beef organisation and it thrives in the dry climate of the Channel Country, which is tick free. OBE has seven million hectares quarantined as pesticide free. Cattle can be raised without any pesticides, and naturally the last thing all the pastoralists supplying cattle to the run want is cotton anywhere near the herds.

It was not the first, nor will it be the last, time that the Greens and the graziers spoke with one voice. Bob Morrish recalled what he saw as a 'divide and conquer' visit that some of the Macquarie cotton group paid on Bill Scott, livestock manager for Stanbroke's and one of Australia's top stockmen. 'Barbara and he made them a cup of tea, and they said to Bill, "Now look, we appreciate you're businessmen and you think the way we do, but we're a bit worried you've got yourself a bad chairman there." And then Woldring said his secret agenda is to get rid of every bullock on the Cooper, he's a secret raging greenie you see. And Bill said, hang on a minute, why would he have just got himself in debt up to his bloody ears to buy another property if he wants to get rid of all the cattle? And Bill rang me, quite agitated, and said, "Look old man, I owe you an apology. We tipped the tea over their heads and have thrown them out straight away".'

A couple of weeks later, Woldring apparently felt the cooler side of Bill Scott. 'Very straightforward sort of a bloke, very impressive character, Bill—tall, skinny with an eye patch. Woldring bounced up to Bill at the pub to shake hands. Bill fixed on him with that one eye and he stood there like a military colonel with both hands behind his back and stared at this hand shaking the air. Bill eventually says, "You can put that hand away Woldring".'

It seems that while Woldring was left to squirm, the cotton grower who led the charge, John O'Brien at least earned a modicum of respect. 'O'Brien is a tough bastard,' Bob Morrish explained over smoko, 'but O'Brien doesn't bullshit. Bill said to O'Brien in a meeting, "John I'd just like to know: what are you really here for?" And really quickly O'Brien said, "To make money Bill, what are you here for?" Sort of floored Bill a little bit.'

The cotton growers originally planned for 3,000 hectares and wanted to build 25,000 megs of storage, already a challenge as the annual evaporation rate for Windorah is 3.66 metres a year and earth dams over 5 metres need special permission in Queensland. They weren't giving up easily, however. The Cooper Creek Protection Group may have had the big pastoralists on side but they were only as strong as its weakest link. The cotton growers were able to buy one of only three small properties on the Cooper, 500 hectares at Currareva, close to Windorah, and with it, a couple of small irrigation licences.

In the mid-1990s, Queensland's bureaucracy was still gung-ho for development, as was the then Minister for Natural Resources, Howard Hobbs, despite promises not to touch the Cooper. 'Welford has a greater intellectual grasp than Hobbs,' Bob Morrish explained, referring to the current Minister, 'And that's the kindest thing I can say about Hobbs. Hobbs said to us a number of times—he thought he was talking to the converted—he said, "Hey watch out for the bloody Greens, they'll take all your country away." And Bill Scott and I would say, "Hang on a minute Hobbs, these greenies are our mates, you know, you can't make unsubstantiated comments like that, justify them."'

One of Windorah's best known locals is pilot Sandy Kidd, who doubles as the district's emergency lifeline. Everyone who meets Sandy Kidd comes home with a story. 'We took a flight with Sandy to look at some agistment,' said one farmer, 'and I asked him if he ever had to carry any dead people out. "Seventy-nine of 'em," he said, "all sitting where you're sitting".'

Sandy Kidd made things very clear for Howard Hobbs when he dropped in on Windorah (part of his electorate) to test the water on irrigation around the Cooper. 'Sandy is a very rough-and-ready looking bloke, shirt tails hanging out. He looks even scruffier than me most of the time. Hobbs brought a whole bunch of people out from Brisbane and made a speech over breakfast at the Windorah Hall. Sandy interrupted three times trying to change the subject, scratched his head, lit up a cigarette . . .'

By the time he'd finished the speech, Hobb's patience had clearly run out. 'He said, "Well Sandy, what do you think of that?" Sandy fired back, "All I can say to you is that if you take so much as one drop of water out of the Cooper, you c--t, I'll stand against you at the next election."'

A year of tense meetings between the locals and the Resources Department and catchment advisory board followed. Then came a brilliant piece of PR work that for the first time brought the battle to the attention of the general public: a mass demonstration on horse-back at Windorah—in 1996. According to Bob Morrish, 'Windorah is normally 65 people, and there were 128 people just riding horses that day. All the way from the South Australian border, right through to Windorah, and every Stanbroke station in the Channel Country brought their horses in. Tanbar alone brought two truckloads. All the stations were carrying banners. It must have been quite a sight, 128 mounted protesters marching through the main street of Windorah. Not surprisingly, it attracted national media.' The event raised $10,000, which is now in the bank for when the need arises.

At the same time, Bob Morrish got the scientists going. Not your run-of-the-mill grazier, Bob picked up a PhD in psycho-physiology in his early days and while the emotional antics of the fight must have been fascinating, so was data collection and mixing with the men in white coats. He spent half of 1996 organising a week-long scientific workshop with a dozen arid zone and river ecologists including, of course, Richard Kingsford, to document the value of the Cooper habitat. One of these is the Ramsar-listed Coongie Lakes of 20,000 square kilometres. 'For the first time,' reported Richard Kingsford, 'the scientists had the opportunity to put their hand up for a river system, based on their understanding of the future, predicting what would happen if we went with irrigation.'

The Cooper was a feast for any ecologist. Productivity in the muddy Cooper water holes is about a hundred times the productivity of the clear coastal streams, a feature of the classic boom–bust process in which the natural species have evolved over millions of years to capitalise on the high water events and explode into life. And what life? 'Yellow-belly, catfish, black bream, heaps and heaps of bony bream, spangled perch, rainbow fish.' Bob's enthusiasm is unrelenting. 'The major birds—pelicans, cormorants, ducks—that's in the aquatic areas . . . and then herons, egrets and the ibis, a few sacred ibis and thousands of straw-necked ibis. Yamma Yamma last year had 50,000 pelicans on it.'

The Cooper also floods Lake Eyre. In 1990, waterbirds on the lake numbered 500,000, and at the end of the flood in 1975, about 40 million dead fish lay around the lake for another five months providing a feast for bird life.[16]

To the Cooper Creek protectionists, water regulation that comes with irrigation spells disaster for the wetlands and risks turning land into desert. 'You've got to realise that once you get to the south of Windorah, the sand hills are poised this side of the Cooper,' Bob Morrish continued. 'The sand hills already encroach down on the South Australian side and you need a big flood every ten or fifteen years to actually reroute the channels through to Lake Eyre. You would destroy one of the most magnificent arid land systems in the world.'

Howard Hobbs thought little of the workshop. According to Bob Morrish, he branded it a rent-a-crowd affair, full of one-eyed greenies. It was to Morrish's delight that just a few days before the horsemen's parade, the Fifth International Ecological Congress in Perth voted 430 to nil on a resolution to tell the Queensland Government not to allocate any water from Cooper Creek to irrigation. Bob Morrish was suitably vitriolic to Howard Hobbs at the horsemen's parade. 'Armed with that, I tore strips off him. And a few days later, Hobbs capitulated and decided there shouldn't be cotton in the Cooper after all,' he chuckled.

As a green grazier, Bob Morrish has little time for the cotton argument that beef's returns are not economic. 'Let's face it, economics is hopeless when it comes to ecological values anyway. We boldly advance the argument that the pastoral system in the Cooper has been sustainable for a hundred years—the greenies reckon it's still worth applying for World Heritage listing for the aquatic systems of the Cooper and in fact the Channel Country generally and Lake Eyre.'

The battle for the Cooper is not over, however. As Hans Woldring said when the cotton decision came down, 'Our understanding is that the existing licences we hold are able to be utilised. That's why we are a little surprised that our opponents to development in the Cooper system appear to be quite happy.'[17] The Macquarie group's latest proposals have been to grow tea trees, grapes, mangoes and citrus—intensive horticulture, which uses a suite of agricultural chemicals that could be even stronger than for cotton, including endosulphan.

The Macquarie group still has two hurdles. First Minister Rod Welford has allowed the group to irrigate under its licence but not store water for irrigation. As a result the growers sued the Queensland Government. But second, it needs to convince the local shire council that a change in land use is desirable. According to Bob Morrish, of about a hundred ratepayers in the shire, a survey on whether the proposal to irrigate the crops should go ahead revealed that 96 per cent said absolutely no, 1 per cent said yes and 3 per cent were undecided.

The Cooper protectionists believe they will hold out against the developers for as long as they can keep the public focused. 'The Cooper seems to have a magic and mythological quality to Australians generally,' concluded Bob Morrish. 'We've had phone calls from people in Sydney and Melbourne, from people who've never seen the Cooper, but they offered us support to protect it because they like to know it's there. Basically it will come down to a choice of whether the Australian public still want a relatively ecologically integrated arid interior or they want it stuffed up like the rest of Australia.'

The battle for the Cooper is won, but not the war. As ecologist Peter Cullen said, 'There are already pressures outside the Murray–Darling Basin because of the Cap, that's why we're getting pressure on the Cooper and elsewhere. They're going to continue. I fear the most likely outcome is that we get more cowboy developments in those northern rivers and we'll make a whole raft of new mistakes, as we're seeing proposals all the time.' They'll have to get past the real cowboys, however, Bob Morrish, Sandy Kidd and Bill Scott.

The birds and the bees

If there is one area that has generally taken a back seat in water issues, it is biodiversity. Although the Council of Australian Governments (COAG—see chapter 2) talks about environmental flow and water quality, it doesn't talk about habitat or landscapes.

Green groups must get very tired of telling the public about how special Australia is. Our Gondwanaland inheritance has meant we have a unique range of species which, for over 50 million years, has been allowed to develop without interference. This, of course, has all changed in 200 years (0.0004 per cent of the period).

Harsh facts on extinctions are housed appropriately at the Australian Museum.[18] Since 1788, Australia has lost 20 species of mammals, 20 species of birds and 68 species of plants. Missing, presumed extinct. Many more are on the critically endangered list. Populations and gene pools are disappearing before our eyes.

Nowhere is this more obvious than in the wheat belt of Western Australia, now under threat from salinity. Doing much of the work in this area is botanist Greg Keighery at the West Australian Conservation and Land Management (CALM) offices in Perth. Greg Keighery has said that of the 4,000 flowering plant species in the belt, about 60 per cent grow nowhere else in the world. 'Our guesstimate is that 450 plant species could go extinct in the next 100 years.' Compare that with 68 species so far.

But biodiversity is not just about looking after endangered species. It's about preserving the entire ecosystem that props everything up. So far, we have managed to lose 75 per cent of our rainforests, nearly half of all forests and nearly 90 per cent of temperate woodland and mallee.[19] Most of us learned enough about food chains at school to know that the loss of a rather insignificant plant or invertebrate may be what Greg Keighery calls the mine canary, spelling doom for many more plants and animals. 'Would we be happy to randomly delete human genes from the gene pool as we do species from the biosphere?' asked Graham Harris, CSIRO's head of Land and Water. 'I doubt it. Yet we eliminate species and reduce biodiversity with impunity, and even fail to train any taxonomists who might even be able to identify what we are losing.'

Western Australia is the world centre for micro-crustacea. 'I say to farmers they're full stops with hairy legs,' explained Greg Keighery. 'Twenty per cent of the invertebrates are threatened in wheat belt wetlands and a third of the birds will go because there's nowhere for them to nest when the trees die.'

Getting people to give a damn is the issue. 'Only 6 per cent of the Australian population ever goes to the wheat belt. In the past, people had relations there, my grandfather was a farmer, but today very few have links with the farming community,' mused Keighery. 'Up until now,' added Richard Kingsford, 'even in the Murray–Darling, we haven't come to grips with the loss of biodiversity. There are such lag effects and most development in the Murray–Darling is post-1970s in terms of water diversions. In young development like the Condamine–Balonne and the Darling, we might be 20 years off, because animals can live for a long time without water.'

Occasionally there are spectacular successes such as the Green Olympics being forced to spend millions accommodating a frog in old brick pits in Sydney, or the case of the giant dragonfly that has managed to survive 190 million years at Wingecaribee Swamp and which caused the peat mining business to shut down months before the swamp imploded. Local residents have never seen the 15 centimetre dragonfly (the second largest in the world) but the local National Parks and Wildlife officer was adamant. 'If mining continues,' Noel Plump was reported to have said, 'it will drastically affect the life cycle of the dragonfly, which can include 30 years buried beneath the peat. Obviously we do not enjoy the fact that jobs have been lost but there are some jobs which cost the earth.'[20]

But it is a major concern of the environmental movement that they can't often prove the damage until it's far too late. This is why in June 1992, one of the key principles agreed to at the Rio Summit was the 'precautionary approach'. 'Where there are threats of serious or irreversible damage, lack of full scientific certainty shall not be used as a reason for postponing cost-effective measures to prevent environmental degradation.'[21]

Of all the items on the Green checklist, perhaps the one paid least attention to is the need to mimic seasonal water flows in our rivers. Governments and the Murray–Darling Basin Council are all out just squeezing a minimum environmental flow. Yet this, as Mary White, Tim Fisher and Richard Kingsford all agree, is literally vital for Australia's water life.

Wetland welfare

The most vital areas for Australia's freshwater life are our precious but withering wetlands—these giant natural filters which slow and cleanse the rivers. Very few Australians have seen a wetland in flood, a fact that most environmentalists would not be keen to change. 'They are spectacular,' said Richard Kingsford simply. 'The life of a river. That's where you see the full range of biodiversity that depends on a river system, from small crustaceans to pelicans. What can you say? We would be much the poorer if we lost them.'

This book touches on several wetlands, but one of the most controversial is where cotton grower John O'Brien developed his business, on the Macquarie River in northwest New South Wales.

The Macquarie Marshes cover about 150,000 hectares near Coonamble in inland northern New South Wales. Just over 18,000 of these hectares are Ramsar-listed. The marshes are fed by the Macquarie River, which was dammed in the 1960s (Burrendong) and the early 1980s (Windamere). Success in cotton farming during the 1980s along the Macquarie saw the industry increase fourfold,[22] and it soon became clear that the marshes were suffering as a result and grazing pasture might also be under threat.

An extra 50,000 megs of effectively guaranteed annual duckwater was organised for the wetland in 1986, changing the median flow to around 315,000 megs a year. That compared with a peak irrigator use of over 500,000 megs. By 1996, the State Government was recognising that '. . . in the past the water requirements of the environment have not always been given adequate consideration, and while the 1986 plan was an important first step, it has proven to have some significant deficiencies.'[23] Another 75,000 megs per year of water has been clawed back from irrigators and attempts made to represent more natural flows.

Today, it is estimated that roughly half the marshes have been lost to irrigation. On the other hand, through lobbying, half the marshes have been saved with community agreement between the Greens and irrigator's co-op, Macquarie Food and Fibre. Michelle Ward, from MFF, believes farmers feel both relief and a bit of pride at the outcome. The marshes are half full, not half empty. 'After this, we're more ahead in the water reform process than other rivers. No one likes change and change was forced on Macquarie irrigators ten years ago.'

Tim Fisher says it's too early to say whether the rot has stopped, and while he accepts the marshes now have greater security, the much bigger threat in future years is salinity. 'In 20 or 30 years, the marshes could be a wasteland.'

Can wetland destruction ever be reversed? If there is a will, perhaps. There is no better example of this than what is currently happening with the Florida Everglades. Champion of the Everglades cause was Marjory Stoneman Douglas, whose book[24] in 1947 raised public awareness of the free-flowing 'river of grass' that was home to storks, alligators, panthers and flocks of wildfowl. Dredging had begun in 1905 and roads put in during the 1920s. In 1948, a State Government scheme canalised and drained the wetlands, intending to protect the environment, but the impact was devastating. The scheme allowed the human population to grow from 500,000 to six million and opened the area to major agricultural development. Marjory Stoneman Douglas died in 1998, well conserved at 108, but she just missed the announcement in December 2000 that US$16 billion is to be spent on restoration of the Everglades in a historic partnership between Federal, State, local and tribal leaders. Ironically, the US Army Corps of Engineers, which was responsible for draining the Everglades, now has a second bite at the cherry engineering the new scheme.

Grey green
Back in Australia, one man is attempting his own wetland rejuvenation, down on the beleaguered Coorong at the Murray's mouth. It's a

tale of determination, but also one which reveals a problem endemic in most big movements and the Greens are no exception: not everyone agrees on the solution. And in this case, indeed, they do not even agree about the history of the area.

To the northeast of the watery stretch of the Coorong is an area which was originally a rich wetland.

In 1864 the first channels were cut to drain the wetland south towards the sea, and in 1912 a stock bank was constructed that stopped water reaching the Coorong altogether. Much like the Everglades, it made farms like James Darling's Duck Island (see chapter 5) possible, but also dried out the wetland. During the latter half of the twentieth century, concerns grew that the Coorong might have been damaged from the early constructions.

However arguments about the Coorong have raged for over 100 years. In 1983, the seminal Cardwell Buckingham Report argued that it was unlikely any major flows from the wetland had drained into the Coorong since white settlement. This conclusion is one that local Robert England has been working for many years to disprove, with the help of old archives. 'In 1847,' he explained, 'the police commissioner at the time was trying to find a better route from Adelaide to Mount Gambier. There were times when he found himself grabbing the tail of his horse and swimming behind him.' Flows in ephemeral lagoons were apparently running into Tilly Swamp and then north into the Coorong. In 1863, the year before the swamps were drained, 'the surveyor, G.W. Goyder, received notice that it was dangerous to cross Salt Creek even with the height of a stage coach—that would be well over a metre,' said Rob England.

The Coorong is full of passionate locals. Tom Brinkworth, BRW rich-list greenie and one of the most successful farmers in the area, set about reversing the drainage process on his land. He has spent a small fortune digging over 50 kilometres of channels to create new wetlands, sometimes flying in the face of drainage board regulation in the interests of getting the job done. The result is magnificent— some say a South Australian Kakadu—which has delighted many ornithologists. It has also divided the scientific community and farmers in the area. For while the scheme has successfully drained some of Tom Brinkworth's farming land to create the wetlands, because of local politics at the time, a critical outlet channel to drain the wetland (beyond Tom Brinkworth's property) into the end of the Coorong wasn't built. In an area where farmers are busting a gut to manage massive salinity problems by keeping the watertable down, many believe Tom Brinkworth's heroism created an environmental nightmare, backing up water in surrounding areas and adding to the watertable.

James Darling on Duck Island is livid. He says a bridge on the road along the coast between Duck Island and Salt Creek has been raised four times since Tom Brinkworth moved in. The watertable in some areas has moved up 3 metres bringing more loads of salt, and bare hills replace what was once lucerne. To James Darling, those supporting the new wetland miss the point—long-term sustainability means management of the whole watertable. 'If people understood that, they wouldn't advocate ponding on the one hand and deep drainage on the other in the same region—and have the same person doing both. From our point of view in the northern catchment it is a scandal, an out-and-out scandal. The calculation is 70,000 tonnes of salt a year into his wetlands and environmentally it is not a sustainable thing.'

Late in 2000, permission was finally granted for a channel to be dug through to the Coorong. It's not fresh water from the swamps, but it is a lot less saline than the southern lagoon of the Coorong (three to five times saltier than sea water) and should appease the Wetlands and Wildlife organisation, which believes that this will at least stop the area turning into shallow pink lakes and salt pans.

The dispute has not ended there, however. There are others, like scientist Dr David Paton from the University of Adelaide, who worry about the change that a dump of what he believes to be 40,000 megs of water a year will now have on wildlife adapted to the saltier state of the Coorong. He argues that the southern lagoon of the Coorong has always been salty. 'The modelling shows that within seven years, the area will become an estuarine system, destroying a rare piece of biodiversity,' he says. According to David Paton, the State Government decision to allow a drain into the Coorong was a quick fix to solve dryland salinity problems. The real long-term solution for that land, targeted revegetation, was never addressed. He believes birdlife in the southern lagoon is under threat: banded stilts will suffer as their aquatic weed dies out, as will fish-eating birds like hoary-headed grebes and fairy terns that feed on whitebait, enjoying little competition in these salty waters. 'Very importantly,' he says, 'there's no monitoring program in place with a contingency program if there is a crisis.' This is not the view of Rob England, who maintains that other fish species have died out over the years as the lagoon became saltier. 'There was another wet year in 1957 and the old fishermen say that there was not just duckweed but red berry weed in the Coorong, and when the mullet used to grind the seeds you could see the tails thrashing around in the water. There were 50,000 breeding pairs of swans, I don't think you'd find two there today.'

Rob England believes that the birds are slowly starving because duckweed and red berry weed have gone. It's a time of grey green compromise.

The greatest enemy of environmentalists, however, is apathy. Compared to most parts of the world, Australia is unbelievably pristine. Why worry? It is apathy from the public and, despite being the 'guardians' of the land, apathy from many farmers, even if they do have other priorities. As time ticks on, history is proving the Greens right again and again—like that great protest song says, 'When will they ever learn?'

Thank goodness for Ian Kiernan, Richard Flanagan, Peter Garrett and other high-profile people who carry on the cause. Australia needs high-profile people in the Green movement to get the message through. It needs more of them.

Weather we like it or not

A report by the World Commission on Water in 2000 put our water situation into perspective.

> Only 2.5 per cent of the world's water is not salty, and of that, two-thirds are locked up in ice caps and glaciers. Of the remaining amount subject to the continuous hydrological cycle, some 20 per cent is in areas too remote for human access, and of the remaining 80 per cent, about three-quarters comes at the wrong time and place—monsoons and floods—and is not captured for use by people. The remainder is less than 0.08 of 1 per cent of the total water on the planet.[25]

In February 2001, the International Panel on Climate Change announced grim news. Over the last century, the world's temperature has moved up 0.6 degrees Celsius. 'Globally,' says the IPCC report, 'it is very likely that the 1990s was the warmest decade and 1998 the warmest year in the instrumental record, since 1861.'

New reports on Greenhouse with doomsday scenarios for various parts of the world seem to be appearing almost monthly, but the big fear for water in Australia is that if the claims are correct, we face drier droughts and bigger floods than this land of extremes has experienced for many thousands of years. Some say 25 years from now, we won't be able to insure for floods or storm damage.

El Nino is the driver of droughts in Australia, certainly for the eastern side of the country. Every three to six years, the waters off the Peruvian coast become unusually warm. In South America this leads to heavy rainfall which causes widespread flooding and because this happens around Christmas time, it is known as El Niño, 'the boy child'. In Australia the impact can be far from joyous, reducing chances of receiving 'normal' rainfall.

If El Nino's droughts do worsen as the greenhouse effect increases, the consequences are indeed dire for Australia.

Sensitive work done by CSIRO recently (sensitive because of the Federal Government's roguish position on greenhouse) has been picked up by Greenpeace and the Nature Conservation Council. The work predicts in New South Wales an average temperature increase of 0.5 to 2.7 degrees by 2050. By 2030, dairy cows without shade cover will suffer from an average milk loss of 280 litres a cow and wheat yields will also fall with increase in temperature. As if there weren't enough problems, water is predicted to become scarcer. Less rainfall in the system acts as a double whammy: not only does it mean a shortage of water in our already stressed southern rivers, but it also could have a serious impact on the recharging of aquifiers, underground basins like the Great Artesian Basin.

The last word goes to Graham Harris, CSIRO's chief of Land and Water, who believes that whether you believe in global warming or not, greenhouse is the least of our worries. 'I don't give a stuff about climate change, if we've stuffed up the land before that. The sort of problems that a 4-degree change will have are 50 to 100 years away.'

7

Irate Irrigators

'Faites qu'il pleut, mon Dieu! Faites qu'il pleut!'
Jean de Florette in Claude Berri's Jean de Florette

It's a rotten feeling to hit a galah. But this particular road in central New South Wales is the home of the nation's slowest galahs. Suddenly it is clear where the great Aussie epithet originated.

Just the other side of Forbes, famous for gold in the 1860s, Ben Hall and now increasingly for urban salinity, lies the Lachlan Valley. The Lachlan River runs over 900 kilometres from east to west. There are about 800 water licences to extract Lachlan water, mostly belonging to family farms growing lucerne, wheat, hay and in recent years maize. Here on the Lachlan, lives Australia's most frustrated irrigator, Robert Caldwell.

Put your farming hat on for this chapter. As the President of the National Council of Irrigation and Drainage (ANCID), Stephen Mills has observed, 'Irrigation is seen as a destroyer of a resource. However, no one is presenting the view of the productive use of water. There is a good news story that needs to be told.'[1] And as already noted, irrigation is one of the few, very few, bright lights for farmers in Australia. Covering just 2 per cent of the Murray–Darling Basin, it produces products worth over $4 billion a year. Milk, fruit and vegies, wine, beef, rice and cotton: they're in the supermarket, day in day out, and they're really very cheap. Thank you irrigators.

The Caldwells have been on the land for four generations. Robert Caldwell has spent 30 years building the farm around water. His mother lives on the farm and grows camellias that would take out first prize at any show in the country. Robert says that, because of what has happened to irrigation in the last five years, he's unlikely to survive unless he grows cut flowers himself.

Robert Caldwell is intense and difficult to understand sometimes, which is why even some farmers in the district consider him a bit manic. But that is only because he has spent many years now getting down to the hard facts and figures of what water reform in Australia is doing to him and farmers like him. He has to, to understand the jargon and prove his point to the boffins and politicians who want to brush him off.

'Robert Caldwell has done his homework,' said CSIRO's John Williams. 'I admire what he has done. Now, he is a fair person and I think what he wants to know is "I don't mind losing my water as long as it's bloody right." At the end of the day, we're going to have to realise that there's been some fairly average science done in some of these things.'

There's an urgency to Robert Caldwell, because he's seen the writing on the wall. He's written to everyone, the Prime Minister, top scientists, the bureaucracy and the media, in carefully printed writing. And most of his letters go something like this:

Dear Sir,

I own a modest sized irrigation farm which has been made unworkable and unviable by national competition policy, full cost recovery and water reforms.

My water allocation in the good years has been halved without compensation. This effectively halves my ability to generate income and reduces the value of my farm by 30 per cent. To add insult to injury, the Department plans to release up to 350,000 megs from Wyangala Dam, 29 per cent of dam capacity, under the pretext of environmental flows. I believe the purpose of this release is to deny irrigators half their rightful entitlement and so implement the Murray–Darling Basin Cap. The fact that irrigators only divert 13 per cent to 19 per cent of the resource has been ignored.

It means in the non-wet years, the reliability of the remaining half of my water allocation has fallen from one year in 50 with zero allocation of water to now 20 years of zero allocation in fifty.

Water prices have risen threefold in the last four years and more multiples are expected. Full cost recovery and water to its highest value use will see me priced out of production. All of my water at Forbes will go to Hillston for cotton growing within two years. The fact that delivery efficiency to Hillston is only 25 per cent compared to 90 per cent at Forbes is not considered.

I feel betrayed by the Federal National Party.

Yours sincerely

Even if you've missed some of the jargon, the gist of Robert's predicament is pretty clear.

We'll come back to Robert and his predicament. To understand it, however, we must take a deep breath and turn to water reform. (On more than one occasion in researching *Watershed* I received a rather curious smile from water experts. 'Ah, yes, of course, you're going to explain water reform!')

The physical side of irrigation is simple. Irrigators are responsible for the costs of all on-farm irrigation equipment such as pipes and sprinklers. For most irrigators, delivery of water to the farm is the responsibility of the irrigation company. Water from rivers and in some cases, dams, is delivered in open channels or pipes right to the farm gate, the irrigator being charged a delivery and maintenance fee. A step closer to the source, dams and weirs are typically the property of the State or Commonwealth government.

Tally ho—water reform

Water reform is complicated—around the traps, people say there's a conspiracy to make it deliberately so. Yet even without the conspiracy, water comes to us from many sources: it falls as rain, runs off as surface water into rivers, lakes and storages and it lives underground, in ancient aquifers, fed by annual seepage of rain. The challenge is to regulate it and make it efficient.

The horrendous problems in the water business these days boil down to one thing: the water resource is over-allocated. As we saw in chapter 3, pressure for water reform built up in the early 1990s. In rural areas, it had become clear that there was more land to be farmed than there was water to irrigate it, and claims for water had been increasing. New claims came from farmers (encouraged by governments) being pressed out of tough markets like beef and wool. At the same time those who had always had water licences wanted to continue using as much as they ever did. Add to that a new share for the environment we now believe to be so important for the health of our rivers and the over-allocation of water has just blown the end off the scale.

Reform was prompted by other issues as well. For reasons already noted, irrigation infrastructure such as dams and channels has been built and maintained largely by governments to catalyse growth in agriculture and rural communities. Unfortunately in recent years, the States have not been putting money aside to keep this infrastructure in good nick.

Decision makers also began to look closely at watering practices. Historically, water has been very cheap for farmers. As it dawned on everyone that water was actually a scarce resource, those grape growers using high tech and expensive gear to water their crops frugally looked much better corporate citizens than graziers who flooded paddocks to fatten cattle, sheep or keep dairy herds.

These issues festered in the 1990s in a climate of tally-ho reform, led by ecorat, Treasurer Peter Costello and flanked by Fred Hilmer with his reformist zeal. For better and worse, as it turns out, water reform was inevitable. In 1993 South Australia, Victoria and NSW began pushing for reform through COAG. The Federal Government had its own reasons for joining the move. Apart from environmental concerns, much of the water infrastructure was Commonwealth owned, and viewed as a sunk cost by the Feds, who were wooed by the prospect of making the industry pay for itself. The sparkly solution was that the *market* would solve everything.

The logic ran like this. Even though there seemed to be an increasing demand for water, if that demand were managed, and a decent chunk was put aside to look after river health, water's true value would be realised. Water users would be able to trade water and that way it would naturally end up with those prepared to pay the most for it, the most efficient businesses. In effect, water would have to pay its own way, with a series of charges to water users to maintain their delivery systems. Reform was to be finished by 2001, and so it was that in February 1994, COAG (being all the States, Territories and the Federal Government) signed up to a strategic framework for water reform, for both the cities and country Australia.

On the rural side it meant setting aside water for the environment, pricing 'appropriately' and separating water rights from land holdings so that water could be traded. Enter ghastly phrases like 'full cost recovery' and 'highest value use' which we saw Robert Caldwell use with such familiarity and are now part of the rural lexicon.

Importantly, there was a sting in the tail for the States, which are in charge of water in Australia. In 1995, worried about a slowdown in reform (a phenomenon known as 'COAGulation'), the Commonwealth slipped the water reform agenda into National Competition Policy and put the financial screws on by making payments from the Commonwealth to the States and Territories over eight years of reform from the 1997/98 financial year dependent on progress. A total of $5.5 billion has been recommended by the National Competition Council in return for reform in water, gas and electricity. The States, particularly Queensland, which at one point had $15 million held back, have not enjoyed the experience.

These Commonwealth payments are in three tranches. The first tranche was for urban reform. The second tranche is for rural reform —a $1.106 billion incentive to get the process started. To receive money from the third tranche worth $600 million, the States have to show the reforms are up and running.

The challenge for reformers was that there were both economic and environmental goals. You might wonder what the Greens think about all this. The Australian Conservation Foundation's (ACF) not-so-secret weapon on this matter is Tim Fisher, based at the Green headquarters in Melbourne. Pony-tailed and much taller than his television appearances suggest, he sits not in a chair, but on one of those inflatable balls that are otherwise found in gyms. 'Have a read of this, see what you think,' he says nonchalantly and hands over a draft of his *Water: Lessons from Australia's first Practical Experiment in Integrated Microeconomic and Environmental Reform*. (Like Robert Caldwell, this guy does his research and the ACF is very lucky to have him.)

To Tim Fisher, the Murray–Darling is saltier, greener and bluer than ever. He has issues with the reforms because the focus on legislation is on access and use, not the environment. Nevertheless, the Greens accepted the overall package as it coincided with their view that irrigators had benefited from subsidised water prices that did little to encourage efficient use. As the document Tim Fisher presented says:

> Surely a pricing policy of fully recovering your costs will lead to the more efficient use of water. Water trade, too, could assist by tending to transfer water from inefficient leaky and high run-off irrigators to more efficient and less polluting irrigators. And we asked ourselves, how many dam and weir proposals could pass a twin test of economic viability and ecological sustainability?[2]

The criticism against irrigators over water subsidies is one of the more controversial areas in *Watershed*. Irrigators are infuriated at the 'freeloader' connotation it brings and say the reality is far from easy street. This chapter sets out to explain why irrigators are irate.

The first water rights

Since the first Europeans arrived, and indeed long before that, water has always been attached to the land. With white settlement, however, came the doctrine of *terra nullius*, the assumption that Australia was empty when whites arrived. Britain immediately acquired sovereignty over Australia and its water. The first rights of Australian settlers to water flowed directly from the Common Law of England. Put simply, a man who owned land with river frontage had reasonable use of the water. He had the right to have that river run through his land unimpaired and its quality undiminished *except* where it resulted from reasonable use of the river upstream. There was quite a lot of argument during the nineteenth century over what was 'reasonable'. Tensions were exacerbated by the discovery that the rivers of Australia were rather different from the rivers of the motherland. As all Australians know, it never stops raining in England.

Fights of the sort described in chapter 4 around Billabong Creek inevitably followed. In the 1880s, H.G. McKinney, engineer for a Royal Commission to examine the water resources of New South Wales under Sir William Lyne declared with some exasperation, 'As a means of stifling enterprise by preventing the utilization of the natural water supply of the country, the British Law . . . could scarcely be excelled.'[3]

Alfred Deakin got his teeth into the problem. As a result of the Deakin Royal Commission's report in 1884, legislation to clarify a man's right to water was enacted first through the *1886 Irrigation Act* in Victoria and then the *1896 Water Rights Act* in New South Wales (the latter prompted by a major drought).

The new laws vested all flowing water in the Crown (which at that time was the State). From then on, stock and domestic rights were private ones, but dams and irrigation needed a licence, which was attached to the land. There were several brawls, culminating in a landmark case in 1900, the judge pronouncing, 'If this Act does not aim to take the old common law rights from the riparian owners and vest them in the Crown, then I do not know what it was passed for nor what it means . . . I do not think the language of the Act could be clearer, and plainly the rights of the riparian owners were divested and vested in the Crown.'[4]

Water leaves the land

For almost a century, water rights remained essentially unchanged. The big move came in the 1990s when reformers decided to give water its own property right, so that it would have a separate value from land and, most importantly, could be traded without the land. This, they argued, would mean that water would flow to where it would be valued most . . . as the price of water went up, the market would make sure that big wasteful users of water would no longer be able to hang onto it, and we would get this wonderfully efficient 'highest value use' for our liquid gold.

It might appear that with irrigators having the water and the value of that water going up, separating out water rights could only be good news for these fellows. There is a hitch, however. Irrigators' water rights are wobbly. And when you're looking to buy and sell them, firm rights are absolutely fundamental. It cannot be stressed enough how serious this is for water users. COAG, at least on paper, understands the situation clearly. 'In order to facilitate trading, governments will need to ensure that property rights to water are clearly defined and specified in terms of ownership, volume, reliability, environmental flows and tenure.'[5] The bottom line is that the ownership, volume, reliability, environmental flows and tenure are, frankly, all a bit unclear. So, what does a water licence grant its holder?

The licence

In chapter 1, we established that rural water users in Australia don't actually pay for water as a commodity. What they do pay for is the cost of delivering that water and the cost of a licence which gives a right to take water. You can't do much in Australia without a water licence. In rare cases, such as in the Northern Territory, a licence may be almost free, but in most of the country, they cost dearly, up to $1.5 million each.

In the old days, licences were area-based. Each licence in the Lachlan area was attached to 400 acres of land on which you could apply as much water as you needed, provided that you arranged the pump in the river and pipes to the property. Most licences in the Basin have now been converted from area licences to volumetric ones, to encourage better water usage. But how much water the licence *allows* an irrigator and how much he actually *gets* are two completely different things.

Anyone growing up on an irrigated farm quickly learns how licences work, because a licence is a livelihood. But a lesson in minutes is a brain teaser. A *licence* today will give you an entitlement to a set amount of water, say 600 megs a year. You probably won't get it though, because at the start of the year, the State water authorities decide how much is going to be available for everybody (depending on what is in the dam) and gives you your annual *allocation*. In some set-ups, like the cotton growing areas of the Gwydir, dams were built and licences were given out in the full knowledge that irrigators would never get 100 per cent of their entitlement. Before the reforms, Robert Caldwell's allocation was 70 per cent of his licence.

But that's not the end of it. Once the allocation is sorted out, the next question is: how *reliable* is that allocation? To put this in perspective, cities have 100 per cent reliability, there will always be water. Gardens have about 97 per cent reliability, 97 years in 100, there'll be water for the garden, but in three there'll be drought. Before

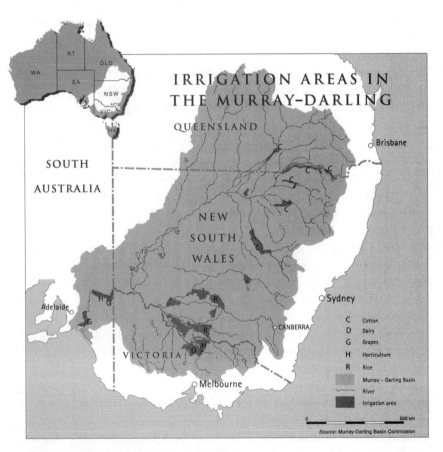

Source: Murray–Darling Basin Commission

the reforms, Robert Caldwell could rely on receiving his full allocation of water 71 years in 100. The other years, he would get a lower allocation and for four very dry years in 100, he had to live with no water at all—zero allocation.

So when Robert tells the PM that as a result of these new reforms, his allocation is now half what it was in the good years and he calculates that for not four, but 40 years in 100, he'll have no water at all (and bear in mind, historically dry years come in clumps, and he could well have several years together of zero allocation) you can see why he's a little upset.

At each stage therefore, from a farmer's entitlement under the licence to his allocation and then the final amount received depending on his reliability rating, water is getting chipped away from him. As one irrigator put it, he feels like a boiling frog, not realising when he's cooked until it's too late.

Still with it? Good, because what follows is the story of how Robert Caldwell, and many others have been done over, both in the allocation and in the reliability of their vital water. For this, a focus on the Murray–Darling Basin is inevitable, so apologies to the Ord, the Burnett, the Derwent and all other river systems outside the Basin.

Zap the Cap

Remember what a watershed the Cap was for the Murray–Darling Basin—how for the first time, it allowed the Murray–Darling Basin Commission to control not just the main channel, but all the major tributaries into the Murray and the Darling? The Cap even controls the Lachlan, which Robert Caldwell is quick to point out does not even reach the Murray, except perhaps in one in 20 years when water floods into the Murrumbidgee.

As can well be imagined, in June 1995 when the Cap was announced, the reaction in rural Australia was furious. Even city dwellers heard about it. By the time Tim Fischer (not the Green one, but the then National Party leader) got to Wakool in the Riverina for a talk, the locals wanted blood. Fischer promised it: 'Zap the Cap' became a rural catch cry.

At one level, it was a brilliant and classic two-minute Tim solution. Just like 'bucket loads of extinguishment', there was 'Zap the Cap'. National Party President Helen Dickie smiled when I asked her about Tim Fischer's idiosyncrasies. 'Journalists liked the catch phrases. Tim always walked away and left you thinking about them.'

The National Party dropped its opposition to the Cap with far less fanfare than it adopted it. Regardless of environmental concerns, it had become clear to the Nats that without a cap, irrigator security was under threat. Robert Caldwell wrote in one of his letters, 'Deputy Prime Minister Fischer said one month prior to the last election in Wakool that he would "Zap the Cap". It is now apparent from policy that he had no intention of fulfilling this statement.'

To be fair, six years later, most irrigators see the logic of the Cap. For a one-off measure, it has been very effective at cutting back ad hoc developments which would only have hit their reliability.[6] If there is a criticism of the Cap, it is that it is a blunt instrument. Rather than make a judgment based on historic water use in the area, the Cap was set at whatever the total amount of water used was in the year 1993/94. As a result, some catchments were hit much worse than others. The Lachlan didn't do too well from the Cap. It's an undeveloped river anyway, with only about half its licensed water used on average. Worse, it was a wet year in 1993/94, and the Cap is set at 40 per cent of the licensed entitlement.

'Surplus' water

One of the areas to be hit worst was rice country, back around Deniliquin, where George Warne's Murray Irrigation diverts water to its farmers through the 160-kilometre-long Mulwala Canal. Here development of irrigated farming had been done on the basis of extra, unallocated water in the Murray in addition to the allocated stuff. This water is much less reliable, but is also very significant. In New South Wales, it's known as 'off-allocation' water. (Victoria has a little of the same sort of water, but it calls it 'off quota'.) Irrigators also call it surplus water, but you can imagine what the Greens have to say about that.

Depending on how you look at it, NSW irrigators live either dangerously or opportunistically. It means the State doesn't leave its dams very full, but it does maximise the use of water by taking it quickly out of the dam for irrigation and at the same time creating room in the dam for the next downpour—the 'use it or lose it' approach, George Warne calls it. In this way, irrigators can take advantage of more water than they have been formally allocated—hence off-allocation water—and it works well, provided there is no drought. In Victoria and South Australia, it's the opposite: squirrel management. These States store enough water in dams to ensure almost all water is so-called 'high security water'. For most water in the dairy State of Victoria, full supply is expected 96 years out of a hundred.

According to George Warne, in 18 of the last 20 years, he has used more water than Victorian irrigators from the same storage potential.

Then again, in August 1998 the Murray Valley's allocation in New South Wales for normal security water (most water under the irrigator's licence) was zero and Victorian irrigators sharing the same dams got 60 per cent to 70 per cent of their normal supplies.[7] It's a different philosophy, some say because of what is grown: New South Wales grows annual crops more opportunistically but Victoria has dairy and a perennial crop, grapes, which need more security. That said, George believes the difference is just history.

The way Warne's irrigators have done their watering may seem cheeky at first. Off-allocation is after all 'high risk' water. But irrigators say they've been encouraged to develop using this water and now it is under threat. Victoria is the same. When the big Dartmouth Dam was built in the early 1980s, Victoria's allocations were not increased. Instead, farmers were told 'non-secure' water would be there on a regular basis and large-scale dairy farms moved in and spent capital. Now water availability is looking more dodgy.[8]

In Deniliquin, most of George Warne's farmers had a very high water use. With off-allocation water, they were using an average 110 per cent of their allocation and had committed capital and expanded their farms accordingly. Unfortunately for these chaps, about one-third of water users in the same Murray catchment area were low users (taking perhaps 40 per cent of their allocation), pumping direct from the river rather than getting fed through the Mulwala Canal. The Cap level decided upon for the area was the average of the high users and low users—one level for all at 87 per cent of each farmer's allocation. Ouch if you've built a farm based on 110 per cent, 120 per cent or even 200 per cent.

But wait, there's more . . . for the ducks

The Cap was only the beginning of the cut-backs to create more water for river health: fish, ducks and water skiers. Ominous for irrigators, the board reviewing the Cap's operation for the Murray–Darling Basin Commission noted, 'The Cap is not an end in itself, but rather a first step towards achieving the long-term objective of the Initiative (their rather grand name for the steps taken in the interests of river health).'[9]

The uncertainty over the moving fish-water line, for which once again there is no compensation, has created even more resentment. The speech of irrigator and former Elders IXL chief John Elliott at the 1997 ANCID conference is now legendary. 'He had the South Australians trying to get up on the stage and hit him,' said George Warne. 'He wouldn't let up, he's a fantastic after-dinner speaker. I think his quote was "Birds don't pay taxes, fish don't pay taxes", and an old

cynic at a nearby table said, "And from what I gather, neither do you".'
John Elliott's point hit home to most present, however.

Historically, environmental rights had been residual to development. Under the water reforms initiated by COAG and implemented
after the Cap, they are prior rights. These days, many irrigators see
COAG as the Committee Of Anti-farming Greenies, or just another four
letter word. As part of water reforms, COAG requires each State to
decide on environmental flows. According to the ACF's Tim Fisher, the
only State that has paid more than lip service to COAG's demands is
New South Wales, so you can guess where most of the tensions are. In
New South Wales, a claw-back of *up to* a maximum 10 per cent of water
from irrigators on *top* of the Cap was put in place by the State Government in 1995, in the name of healthy rivers. The actual amount of
duckwater is determined on a catchment-by-catchment basis. After
what must have been some beaut disputes, a handful of river communities have agreed to a set level for duckwater and all communities are
required to have a ten-year proposal by September 2001.

Robert Caldwell is less happy about the Lachlan and he has a conspiracy theory. If the Lachlan Valley irrigators' allocation under the
Cap of 253,000 megs is taken away from its original entitlement of
630,000 megs annually, it comes to 377,000 megs. For Robert, that's
awfully close to the 350,000 megs he believes was plucked out of the
air for duckwater by the State Department of Land and Water Conservation (DLWC). Far from the amount of duckwater needed being
based upon science, he reckons it is more than coincidence that the
figure is almost all the remaining water from the irrigators' original
entitlement. The duckwater allocated has priority over irrigation
water. There are conflicting reports regarding Robert Caldwell's suspicions,[10] but the fact that almost every document you read on water
stresses the mess that authorities are in over what correct duckwater
levels should be doesn't harm Robert's theory.

Irrigators everywhere would be appeased just a little bit if there
was some certainty in life. Farmers get enough excitement from the
weather. It's one of the reasons why a much loved idea of 'capacity
sharing' took off, pushed by cotton growers, such as Bernie George
from Togo Station in northern New South Wales. This is a way for all
water users, irrigators, urban users and the environment to share the
water in storages. With this idea, every water user gets an agreed share
at the start of the year and it's up to each one to manage their share
most effectively. If they are caught short, they have to buy water off
the other users.

The capacity-sharing idea was an attempt to give all water users a secure right. That works very well for cotton growers in the Gwydir and Namoi Valleys, who are sophisticated enough to tell down to the last bucket how much water they use. They are the professionals at water management. But for the Greens it was less than satisfactory. It assumes the environment has its own fellow with funds to buy and sell water, and raises big questions about accountability. More importantly, the Greens don't want a secure right, they want the right to claim more if they need it.

It is all academic, however. Despite the then NSW Minister, Kim Yeadon, agreeing to capacity sharing trials, the DLWC has since refused to implement the system.[11]

Sleepers and dozers

Now we arrive at perhaps the most sinister characters in reform: sleepers and dozers. They sound innocent enough, but they are a long way from Snow White. Sleepers and dozers are the water licences in Australia that are either owned by people but not used (sleepers), or only partially used (dozers). It is difficult to know just what volume of water in the Murray–Darling Basin is represented by sleepers and dozers, but in some rivers, according to the Commission, it's over 30 per cent of licensed allocation. Many are also sitting on entitlements to water in undammed rivers and over groundwater.

The creation of water rights and the encouragement of trading has suddenly given these licences enormous and instant value, and they are being bought, transferred and activated at a cracking pace. Sleepers and dozers have caused at least as much chaos up and down the Murray–Darling as the Cap, because their awakening simply winds back the amount of water available to existing licence holders. So much for 'increased efficiency' leading to less use of water by irrigators, and all this in a river system where existing water allocations and fish water are already under stress.

Water reform would have been the perfect opportunity to take these licences out of the system. COAG was warned before it set off on the reform track: 'Dormant rights should be substantially reviewed before the creation of a system of tradable water entitlements.'[12] It seems mind boggling that governments could not bring themselves to grasp the nettle and pay out sleepers and dozers early on—the earlier in the piece, the cheaper they would have been. But no.

The Murray–Darling Basin Commission's Don Blackmore agrees it's a mess. In his view the sleepers and dozers should have been frozen and owners should not have been allowed to sell them. Only the original owners should have been allowed to develop them. 'If you'd increased your use because the kid's coming home from ag college or whatever the excuse was, and put the capital into that development, well you should have been able to take that up.' That way, according to Blackmore, would have been fair, and change would have been slow. Instead, by allowing owners to trade sleepers, they don't have to put in a dollar of capital. 'We've given them access to the market at the marginal value, so they've got a huge free kick, because they've paid a very small holding fee for the sleeper licence.'

But even without the trading, sleepers have caused ructions. A classic example has been enormous expansion of cotton irrigation around Bourke on the Barwon and Darling Rivers, controlled by the UK Swire group. In the early 1990s the DLWC allocated special licences which allowed growers to harvest floodwaters. Unfortunately, the Department completely underestimated the number of licences that would be activated. The reason for this was that these were only licences to capture water when the river was in flood, so naturally licence holders would need to spend quite a bit of money on a dam to hold their water. Most licences therefore would remain as sleepers. It was another error of judgment. 'They didn't expect more than 10 per cent to be used, [otherwise] they would have demanded large amounts of capital by people to build big dams and pick up those flows,' explained Don Blackmore. 'Humans are bastards, they actually thought it was an opportunity. Cotton was profitable, a few good years and they put in a storage, and there was rapid growth past the 1993/94 Cap level of diversion.'

By 2000, the Bourke expansion had caused such outrage that South Australia threatened to support court action against it by the NSW Environment Defenders Office but had to back off because the project was legal. Despite the excesses, development for the whole of the Darling catchment, if averaged out in terms of water use, from the Queensland border to Wentworth, is still under the Cap level. Clearly the Cap rules covered individual catchments but, due to a bit of clever negotiating on the part of New South Wales that's allowed under the Cap agreement.

However, in February 2001 the NSW Environment Protection Authority State of the Environment Report caused such alarm about damage to the river that a ban on all further development was enforced, covering 1,500 irrigators along the 1,400-kilometre stretch of the Darling. Needless to say, other growers in the Darling are pretty annoyed that one group should snatch all the development upside.

The question is: should Bourke's development go back to its 1993/94 levels, or is it fair that these people exercised their rights?

Landsat photos are being accumulated by all sides. 'It will be quite frankly a huge stoush between the upper Darling and lower Darling people, around Menindee Lakes and below,' warned Don Blackmore.

Irrigators generally are being told the solution to their reduction in water rights is to buy up sleepers and dozers to maintain their previous real allocations. At a cost of between $300,000 to $1.5 million per licence, that's really helpful for family farms.

Farmers do not like paying for water beyond their licence. Take the infamous motorbike ballot-box hold-up of 1999. George Warne's irrigators had had their worst allocation ever, largely due to drought. For the first time, the private irrigation company decided to enter the market on behalf of its irrigators to buy water. They did a rather complex deal with the Snowy Mountains Hydro-electric Authority to buy an option on the right to the following year's water stored in Snowy dams (to be delivered that year, 1999), and went back to the irrigators with the offer of putting up $12 per meg for the option and a further $12 if the farmer exercised the option.

The farmers were furious. 'Yeah,' George Warne grinned, 'because in the past in previous droughts, the government had agreed to extra releases from the Snowy, but now as the power company is becoming corporatised, they're just sticking to their digs and saying no, we're contracted to provide this minimum amount for electricity, if you want more, you're going to have to come and talk to us about some sort of commercial arrangement.'

When the day came to decide on the issue, Murray Irrigation sent out a letter requesting nominations to buy options by noon. Farmers were asked to put them in the ballot box in the town hall so that the company could apply in bulk to the Snowy. It was all too much for one farmer, according to George. 'We got a tip-off and the poor bugger, at sort of 12 o'clock, he came in, in his motorbike helmet and he grabbed the ballot box and ran out the door and unfortunately he made the escape in his own truck with a couple of merino rams in the back. And we'd already emptied the box, except that there was one cheque and one application that went in from one of our tougher farmers and he had his two very tough sons with him and they charged out after. So they grabbed him, got the box and brought him back.'

Who owns the rain?

So far we've been talking rivers. In July 1999 a Victorian farmer, Julian Kaye, got the shock of his life. The local water authority demanded that he pay $30,000, because the rain that landed on his property did not belong to him. Julian Kaye had decided to grow a few grapes just west of Ballarat and put in a dam.

Fanned by the media, the news of privatised rain caused uproar in both Victoria and New South Wales. For some years now, struggling pastoralists have been diversifying into irrigated crops—horticulture, grapes and olives—and they've been encouraged by governments to do so. These pastoralists, typically upstream in the water catchments who had traditionally used no more than their basic right to water for stock and domestic, started damming up the watershed.

With a permanent water supply costing $1,000 a meg in some parts, dams went up fast. In Victoria alone, there are said to be about 90,000 dams, and for every meg held in the dam, between one and three megs are lost to the system in evaporation. Not surprisingly, this started affecting Victoria's compliance with the Cap.[13]

Then in New South Wales, government made a ridiculous decision. You would think that with all the discomfort they had been through over tightening allocations, it would be impossible for any government to increase allocations of water to farmers. But no. This is what New South Wales did. It was policy on the run at its worst. Under new laws, farmers are allowed 10 per cent of the water that falls on their land for free. Any more and they need a licence. More squeals followed. The State was privatising 90 per cent of a farmer's rain—rain which was rightfully theirs! A frightful row has ensued. People like ecologist Peter Cullen argued that the farmers' claims of rain ownership are outrageous. 'They don't feel quite the same way when they have floods. If it was privatised and they flooded me, I'd sue them for letting their water onto my property, so they need to get a bit clearer as to which way they want it. They can't have it theirs under some conditions and an act of God under others!'

Farmers may not have been happy with 90 per cent of their rain snatched away, but looked at from the other direction the result was even worse. What the NSW Farm Dams Policy did overnight was to give farmers 10 per cent of the water that fell on their land for free, without a licence, unmetered and unmonitored, water which in the past had been running down the catchment.

Former chief of the NSW Irrigators Council Gary Donovan has made it clear that the State Government knew the risks.

> The NSW Government's proposed Farm Dams Policy will allow new water for harvesting for irrigation without licensing, metering or enforcement. The government has reported, 'There is significant concern in the MDBC [Murray–Darling Basin Commission], as with

licensed users and conservation interests, that the Farm Dams
Policy will result in significant growth in water use and thereby
could impact on either the Cap, the security of supply to other
users and/or downstream flows. While New South Wales has not
reserved a volume for this purpose, if growth occurs then New
South Wales will maintain the Cap by adjusting licensed diversions
as necessary.[14]

That translates to more cut-backs for irrigators with licences.

'If you go forward in ten years' time,' said a firm George Warne,
'and you said we've just done an audit of your farm and the rainfall
figures here, and you're way over the dam policy limit, the farmer
would say, "I've got 70 acres of apple orchard here now, I've got 12
people employed. What are you going to do about it?", and you'll find
they're going to do not much. That's the history of water management
in Australia.'

Over the border, thankfully Victoria didn't go for the 10 per cent
rule, but there are certainly rain squabbles. The State had already had
one go at fixing the problem in 1989 with laws that allowed farmers
to build a dam, so long as it was not on a 'waterway'. The trouble was,
despite seven paragraphs of legislative definition, no one could define
what a 'waterway' actually was. Eddie Ostarvic, an olive grower from
Stuart Mill near Avoca in Victoria, spent $140,000 building a dam in
the belief that he was not on a waterway. The authorities disagreed,
saying he should be paying $20,000 for the water.[15]

These woolly laws have set farmer against farmer, not helped by
a four-year drought in Victoria. Those on the traditionally irrigated
country who also enjoyed a bit of sales water (extra low-security water
for free) felt this would be threatened by developments upstream. Yet
in the high country, farmers believed that they had already paid for
the high rainfall because it was reflected in their property prices.
Highlanders also had to pay to put in their dams and pipes when
traditional irrigators had it sweet, with all the systems built for them.

The dairy industry in Victoria has had to deal with confusion too,
over tailings dams versus storage dams. The industry has been openly
attacked for its poor record on nutrient run-off into the Thomson and
Macalister River, which has caused havoc in the Gippsland Lakes with
fish kills and algal blooms. Dr Peter Fisher reported, 'In the Macalister
irrigation district, properties are approaching feedlots with no store
cattle and no calves. There is no treatment of the effluent that flushes
from the paddocks into water courses in either catchment. In Lake
Victoria there have been 12 recent fish kills due to deoxygenated

water.'[16] Farmers are busy working on drainage and tailings systems, but in some cases, that has led to confusion over whether the tailings dam is purely for nutrient run-off collection or for water storage (and heaven forbid if it was on a waterway). As dairy farmer Max Fehring said, 'When is a tailings dam not a tailings dam?'

A new policy for Victorian dams was due to be announced by the State Government in June 2001,[17] but was delayed. By the time *Watershed* is on the shelves, the 'waterway' definition will no doubt have gone. All new dams will need permission and any new water that is needed in already capped catchments will need to be bought in the market. And out will go any concept of private rights to water.

'They've got no private rights to it,' said Don Blackmore, who headed up the review. 'The only private right existed before 1886; [the 1886] Act vested all water in the Crown. We've also said that the first priority is to protect existing users. There's no benefit in Victoria transferring wealth from one group to another by having a dopey dams policy which is just a surrogate social policy. If you're going to do that. Then do it with your eyes open.'

Everywhere, water users are trapping water higher up the creek and floodplain management remains a hot issue. In the 75,000 square-kilometre Condamine River catchment on the Darling Downs in Queensland, large earth-walled dams known as ring tanks had been sprouting like daisies until a moratorium last year. Worse, throughout the State, a dam has to be 5 metres high before it is 'referable' to authorities. Ecologist Peter Cullen despairs. 'So there are these piles of dirt that go 4.9 metres high, that go 30 kilometres along floodplains, harvesting every drop of water that runs off that floodplain and it goes into a storage.'

State rules—thanks for nothing

We can try to keep up with water reform, as each of the States beds in its own legislation, but it makes for very dry reading, especially as Victoria and South Australia had already made progress before the reforms.

Watershed is certainly not going down every rabbit hole on legislation. But with a water user's hat on there are two hot clauses in State legislation. They spell out the numbers of years the actual licence runs and the number of years that governments give themselves before they can change the rules again on environmental flow. With the Murray–Darling Basin Commission making it very clear that we may not have seen the last of the duckwater claw-backs, State law will determine when, whether and how much money a farmer can borrow against his water. If water is removed from land, in most places there's not much left to lend against in land value terms. It's difficult to predict what will happen with the banks. Banks will generally lend against licences. However, State laws differ in relation to licence terms. The interesting question is how banks will behave when the term of a licence nears, and just prior to the time of a possible new claw-back of water by governments. This is what keeps irrigators awake at night. All this reform has led irrigators to question whether they actually own the water right at all. And if so, given the lack of security, what is it they do own?

Robert Caldwell is fond of reminding people of his State's *1912 Water Act*, Section 17:

> The person holding a licence in respect of any work shall have absolutely, during his lawful occupation of the work, the quiet enjoyment and the sole and exclusive use of the work as against all other persons, whomsoever, including the Crown and the Ministerial Corporation, and shall be entitled to take, use, and dispose of any water contained therein or conserved or obtained thereby to the extent and in respect of the land and in the manner specified in the licence.[18]

Robert's 'quiet enjoyment' is over. Section 17 was repealed in November 2000. It's all a bit irrelevant now. The new legislation has moved the goalposts and halved the distance between them.

Trading places — new world order

It is here that we turn to the most exciting part of the reforms (from a reformer's point of view). Already water trading is said to have increased production by $100 million a year.[19]

The idea behind trading is that through buying and selling in the market, water will naturally find its highest value use. This 'highest value use' mantra has been taken up by reformists, politicians, regulators—and many farmers at the *higher* value end of the chain. Indeed, to sell the reforms to Government, the supporters argued they could repair environmental degradation by clawing back water from users, with no loss to the economy, simply by allowing water to be traded to its highest value use.

At its simplest level, the theory is very clean. Different crops and other farm produce like milk and beef have very different water use patterns. Some are annual, and some are permanent and so need guaranteed supplies each year. More important is the dramatic difference in how thirsty each of the sectors can be. And we should not forget mining as a major competitor for water. If you compare sectors *purely* in terms of average amounts of water used per hectare, the numbers are dramatic as illustrated in the table below for New South Wales.

Irrigation requirements for crops in New South Wales

Product	Average Irrigation Requirement[20] (per hectare/per annum)
Cereal	4
Grapes	4
Pasture—extensive grazing	5
Horticulture	6
Pasture—dairy	7
Sugarcane	8
Cotton	10
Rice	13 (can go to 16)

Source: Department of Land and Water Conservation 2000
Note: Irrigation requirement after rainfall and assuming 30% loss of water on farm.

Predictably, those most enthusiastic about trading are the efficient users. And at the other end, pressure is on farmers to move out of their businesses. Another favourite set of statistics is jobs created from traded water. At one extreme, horticulture offers 30 jobs for every 1,000 megs; at the other, grazing offers just 0.62 jobs. Across the country there are now well over 100,000 hectares of wine grapes under cultivation. And early assessments of trading show that irrigated horticulture is a clear winner from water trading overall, well affording the high water prices.[21]

There are, however, juicy counter arguments to the highest value use theory, the most obvious being that we can't all grow grapes. And to those who might argue that the market will kick out other 'non-performing' products, this too only works within limits. Milk, for example, may not become as expensive as wine, but nor will water reform destroy the industry. Australians need the 'legendary' stuff.

'It needs to be shouted as loud as it can,' said former New South Wales irrigators head Gary Donovan. 'A range of factors determine agricultural production and best economic use. What is considered high value one day, may be in overproduction or may be overtaken by technology shortly thereafter. The target is not water use efficiency, but economic efficiency.'[22]

In fact, the whole idea that trading is environmentally friendly is hotly disputed. George Warne's Murray Irrigation happens to be the biggest water exchange in Australia. 'This is one of my favourites.' He smiled. 'Trading doesn't lead to environmental good. Trading leads to commercial good, but don't confuse it with environmental good. It might bring environmental good but it might not.'

George's rice irrigators are the biggest and by one measure some of the least efficient users of water in the Basin. Logically, they should all be going out of business sometime soon. 'On the ground, the exact opposite is happening,' said George defiantly. 'We're the biggest importer of water in Australia.' The reason, simply, is that water is but one of the factors driving profitability and therefore what a farmer might grow. For instance, one farmer might apply about 2,000 megs per family business in rice but labour costs in rice are tiny. For a horticultural operation a farmer might only need 100 megs of water, but there are many other labour-intensive jobs that will have to be costed.

Rice growers also happen to own the growers' cooperative, which provides a single selling desk for rice. This sort of leverage means rice growers are extracting about double the world market price for their product. The farmer owns the packets on the supermarket shelf and doesn't get screwed by middle men, that means he gets 45 per cent of the shelf price compared with the wheat grower's 3 per cent for the loaf of bread.

There are others not enamoured with trading. Take Robert Caldwell's great fear that as new cotton strains become hardier and the industry creeps south into the Lachlan, he will be out-priced on water. From wheat to cotton, that's hardly a win for the fish.

Even the Greens accept that higher prices won't always help the environment. But it is odd that they don't seem too concerned about the threat of a major shift into cotton. 'Well it's just a complete loss to me,' continued George Warne, and then with an eye to the bigger picture of a move to higher value uses, he added matter-of-factly, 'Maybe it's got something to do with the fact that they are being sponsored by Southcorp, and that Southcorp is involved in big vineyards right on the Murray.' These are the sorts of suggestions that the Greens may have to wear if they are going into partnership with business.

Efficient water use in the new trading world does have a downside, according to agri-businessman Doug Shears, whose Berri Limited suppliers produce 150,000 to 160,000 tonnes of fruit a year. Predictably, Doug says reform should be welcomed, but he also believes the portability of water licences creates a frightening conundrum. It means concentrated areas of thirsty crops like cotton are expanding and water supply to places like Hillston on the Lachlan River can only further disrupt natural flows. 'Restructuring the licensing system down the track is inevitable.' Even if it means upsetting powerful players like cotton companies and Doug Shears' Berri Limited? 'We might be powerful players, but we're not the majority of the electorate,' was Doug's considered reply.

Pressure from the big smoke

If water trading is taken to its logical conclusion, the efficient irrigators might not be that happy either. In terms of returns per meg, city water is by far the most productive. The price of water in the city is up to $1,000 a meg. In rural areas water is $20 to $60 a meg, based on bulk water supply arrangements. If it ever came to competition over water, there is an awful lot of buying power in the city, and as US water expert Sandra Postel notes, 'There's an old adage that water flows uphill towards money.'

Interestingly, this is where the supposedly clean and clear game of trading becomes decidedly unbalanced. If this seems an extreme idea, president of United Dairy Farmers of Victoria, and a dairyman himself, Max Fehring will assure you that it is not. The Thomson River that supplies dairy farmers all the way to the Gippsland Lakes also supplies Melbourne.

'The reality is that if a city community wants water, they will get it', explained Max. 'They are hypocritical in telling everyone else how to run their resources and yet when the time comes they'll say, "Don't you criticise my use, or I'll come and buy yours off you." When Melbourne is tight, they'll just come in and get it. California's given us all the history on this. Melbourne has four years' security in water. It's not a threat, it's a reality.'

According to Sandra Postel, American cities are 'buying water, water rights or land that comes with water rights in parts of Arizona, California, Colorado and elsewhere'. In 1989 the massive Imperial Irrigation District in California provided 130 million cubic metres (130,000 megs) of water a year to Los Angeles in return for an investment in water efficiency improvements. In 1992, the city convinced Palo Verde irrigators on the Colorado River to leave fallow part of their land and pass the water savings to Los Angeles.[23]

Max Fehring gets fairly exasperated with the continued criticism of greedy irrigators from city media. 'Look at Adelaide,' he said. 'Adelaide is a classic case of hypocrisy. It takes three-quarters of the water in a catchment that it doesn't even live in. It's not even in the Murray–Darling catchment.' It is true that the town of Murray Bridge is a good 70 kilometres from central Adelaide, but this is all academic for those in the city who see the Murray as 'their river'.

David Farley believes that one day Adelaide will be a major beneficiary from water trading. Farley is the man who spent 20 years putting the energy into Colly Cotton, now a $1.7 billion business bigger than both Elders and Wesfarmers, and who has spent a great many hours thinking about water trading in current and forward markets. 'Water is an unrepeatable resource after all. Most cities in Australia are the other side of the catchment, so they couldn't benefit from the western flows, but Adelaide is an obvious player.'

It's already happening in California, according to David. In some cases, farmers are now being paid to 'rest' their land by not farming at all. They then sell the water which they don't need for that year to Los Angeles or San Diego. 'Not even heroin would give you that sort of return!' he remarked.

'If Melbourne does wish to come north,' added Max Fehring, 'maybe they will be able to use the capital to improve our systems and we can better use the water.' Precisely this sort of experiment is happening to obtain the extra water for the Snowy River (see chapter 8).

The exchange

Trading has been going on within Victoria since 1988, but a decade later, government pilot programs commenced between three States. The perfect region for trade is where the Murray runs through western New South Wales, Victoria and South Australia. In Swan Hill, unemployment has dropped from over 13 per cent to under 4 per cent on the back of new business.

Trades happen through exchanges, but also via stock and station agents and private agreements. There are two types of water traded, which must not be confused or there is awful trouble. Sales of *permanent water* mean selling your water in perpetuity; sales of *temporary water* transfer water on an annual basis.

Temporary water trading is where the action is, at about 850,000 megs a year. There's too much uncertainty over the future of permanent trade, which sits at about 100,000 megs a year.[24] People don't want to do it, according to George Warne, because they want to hang on to the capital value. 'It's like having a house in Paddington, a building in a boom. You don't want to sell it, because you don't know what it could be worth tomorrow, whereas you're very happy to rent it out and cop the high rents. And that's what people are doing.'

What permanent trade there is, though, is very telling. Water is moving, *permanently* to the wine industry and frequently away from grazing country. If you go to Renmark in South Australia or Mildura in Victoria, the buying capacity for water can be seen from the highway, in the massive wineries and infrastructure.

At Don Rowe's Sunraysia Irrigation around Mildura there's no funny off-allocation water. Irrigators get exactly 9.114 megs per hectare for their grapes. So to expand, they need to buy. In the three years to 2000, about 9,000 megs a year were bought by Sunraysia mostly from Murray-Goulburn Irrigation Company and, according to Don Rowe, 'from the fat cattle people upstream—those that are trying to make money growing grass. They're getting out, or in some cases, guys from this area are going up and buying the property and then just moving the water.'

In the wine-making areas of the Riverland in South Australia, the story is the same. In fact, South Australia has turned out to be the big winner from trading, both in terms of developing its irrigation industries and having the extra water as a dilution flow down river. 'Tandou [one of Australia's bigger agricultural companies] alone has got at least another 5,000 to 5,500 megs annually buying from farmers upstream of Torrumbarry, Pyramid Hill, up as far as the Barmah Choke,' explained Don. It is apparently not wise to go beyond the Barmah Choke, because the water might not get through. Near Echuca, the Choke was formed squillions of years ago when a land fault moved the Murray.

There's a catch to permanent trading. It's to do with the Cap. Take the case of Mildura, perched just the Victorian side of the Murray. If a farmer owns a property there and buys 100 megs of permanent water from Buronga or Gol Gol, just the other side of the river, in New South Wales, that 100 megs comes off New South Wales' Cap and goes onto Victoria's. It's been legitimately traded. 'That's why people scream about certain people who throw up huge ring tanks in Queensland and don't pay anything for the right to it in the first place,' explained Don Rowe.

Ironically, the Cap itself threatens both interstate and inter-valley trade. Why? Because of sleepers. If a sleeper sells his water out of the valley, it means his next-door neighbour can't use it and the Cap is adjusted accordingly. So communities in these valleys have ring fenced their trading, not allowing permanent sales of water outside the valley. It makes buying water quite challenging, as George Warne found out. Murray Irrigators needed water after being hacked back by the Cap. 'Government said, "Don't worry Murray Irrigators, you'll be able to trade your way out",' says George, shaking his head. It's turned out that neighbouring valleys have been rather unwilling to sell permanent supplies.

The problem was acknowledged in a report to COAG: 'Some communities are concerned about the permanent loss of water to a district that might be caused by trading. This emphasises the need to effectively incorporate consideration of the social, physical and ecological constraints into water trading policies.'[25]

There are a couple of other touchy issues about the much-touted trading system. No environmental clearances are required for temporary trading, the area in which the bulk of trading takes place.[26] Another issue that has been pushed into the too hard basket is whether a meg of water upstream is worth the same as a meg downstream. The Productivity Commission certainly took the point, quoting Robert Caldwell about the Lachlan:

not true! refer to the water trade exchange rates!

> If a farmer at the bottom end of the system makes a request for 1 meg of water, up to 6 megs must be released from the dam. However, despite such transmission losses, the downstream farmer is presently paying the same price for a given quantity of water as an upstream counterpart. Such inefficiencies in water could increase as trade in water entitlements becomes more widespread.[27]

And just to throw in a final bombshell, filed under 'Hard Yakka' in one expert report—trading in groundwater is well behind.[28] As river users turn to groundwater to make up for water losses, the whole set-up is likely to become even more tricky to regulate.

The money makers

With the price of a rural water licence going in only one direction these days, there's an obvious question: isn't somebody busy making an awful lot of money from this?

Even before trading started on any scale, water licences had been rocketing in value. Up until about 1980 in the Lachlan you could get a water licence for $150. Today they're worth around $300,000. But around Dubbo, try $1.2 million, and Moree $1.5 million. What's driving the price is cotton, and as this wealthy industry expands, it can surely only benefit from trading. Enter a new player: the water hoarder.

'If in 1992 you'd bought in, you'd be pretty wealthy now,' agreed George Warne, 'because it's grown a lot better than any super fund. You buy the farms with the water and you get the land for free. In fact one fellow had a house block at Mildura which he had a lot of water rights stacked up on, thousands of megs, under the guise of land and water even if the land was only an acre, so it's happened.'

It's happening on a much larger scale than that. Take the multi-millionaire land owner John Kahlbetzer and his company Twynam. He is certainly betting that H_2O has a long way to go, with massive water interests in New South Wales. In 1999, he took over Colly Cotton and a swag of water licences. Aside from the private irrigation companies, the converted co-ops, John Kahlbetzer has the largest water holding in Australia, at almost 160,000 megs a year. It includes a huge allocation off the Murrumbidgee for his one million acres of Riverina properties, and equally impressive interests on the Gwydir, Barwon and Macquarie Rivers. All these rivers eventually find their way into the Murray. Now there's some healthy assets for future trading with Adelaide.

Hot on the heals of Twynam is the Queensland company Cubbie, founded by entrepreneur Des Stevenson, with up to 150,000 megs a year. Cubbie Station, located near the top of the Murray–Darling System is the subject of chapter 10. It's difficult to know the exact pecking order of the top water barons; partly because some, like the Harris family in New South Wales, have water licences in different family companies, others also have holdings within the major irrigation companies and may have groundwater, as well as river water. Suffice to say that most of the big ones, like Auscott, Tandou, Sundown Pastoral and RMI are into cotton.

Not all water accumulators are farmers, however. Take Collin Bell of Bell Commodities, a stock broker who recently bought the Murdoch irrigated pasture properties and licences on the Riverina. These add to his already substantial investments at Hay in rice. But the biggest rice grower is John D. Elliott. 'Well he claims he is,' said George Warne playfully, 'although I heard the other day that a farmer rang him one day and said "Tell the second biggest rice grower, that the biggest rice grower wants to talk to him," but he's got the largest area of rice in.' Victorian property developer Harry Trigguboff is also said to have more than a passing interest in water.

Amazingly, even though the new water rights are probably worth a great deal more than the land they were once attached to, governments do not seem that interested in a proper register of who owns what. This is despite recommendations from their consultants.[29]

Water magnates worry people on the land. There are fears that some Pitt Street farmer will buy up all the water around them and skew the market. The situation became so tense in Victoria, the government has insisted that water has to be bought with some land. It's a nice contradiction that an original aim of reform was for water to be a separately tradable right.

All this is paranoia according to the Murray–Darling Basin Commission's Don Blackmore. 'I designed the original water trading environment in Victoria in 1984, and when I went around the community that question was asked all the time: "Am I going to be raped?" But how are they going to do this? There's no compunction to sell, you're presumably selling because it's in your economic interest.' Tell that to John Hill, a grazier downstream from Cubbie Station in Queensland, who was outplayed for water licences and whose land was devalued as a result. Simultaneously, another family business has piled up licences and is now worth a large fortune.

If water hoarding ever did become a problem, Don Blackmore believes the irrigation community is too smart and would refuse to buy the water from the licence hoarder. Instead they would wait for it to be redistributed as sales water when the seller couldn't find a market. But there are genuine fears. As farmers have their entitlements literally 'watered down', it is much easier for big companies to buy up new licences to keep their allocations high. And this is just rural Australia. In chapter 12, we will see how the money in water for mining purposes has led to very interesting corporate behaviour.

Some folk do make money purely from temporary trading. 'I can take you to a block where there are six farmers in the same region and they don't put a drop of water on their property. They sit there and dryland farm and they just wait for the stock market to suit them,' said Don Blackmore. Last year the early spot price was about $90 a meg for temporary trade, compared with an average of $50, because of the demand from the horticultural industry around Shepparton. 'There are room for opportunities but not for big blocks,' Don insisted.

Full cost recovery

If there are those making a tidy killing in water, reform has meant for many more there's a fair bit to pay. COAG has declared that the industry must be fully recovering the costs of delivery of water to irrigators. If there are any subsidies remaining, these should be made clear. It was always expected that rural reform would lag urban reform but the National Competition Council—the body in charge of overseeing water reform—wanted full cost recovery in place and on time by 2001. It tried pretty hard to slip in this requirement as one of the conditions for Commonwealth tranche payments to the States.[30] However, even COAG saw the political impracticalities of this one.[31] Full cost recovery, adopted whole hog, would put many of Australia's farmers out of business.

So just how far should full cost recovery go? For the Greens, it's simple. User pays. Subsidised water brings both overuse and inefficient use of water, and it means more dams and weirs than absolutely necessary.

While pricing to account for *management* of water is changing, paying to fix up dams, pipes and channels (and many are aging rapidly) is another matter entirely. David Farley (formerly of Colly Cotton) finds this attitude unbelievable. Society has simply forgotten about agricultural infrastructure. Money has been spent on improving agronomic practices, we've re-engineered world trade and yet, now faced with the challenge of upgrading our water infrastructure to be more sustainable and green friendly, the State wants no part of it. 'Shouldn't every citizen pay for the maintenance of an Australian airport, or painting the Sydney Harbour Bridge, or just those who fly or drive over it?' he asked. Investment in the industry is critical.

Tim Fisher disagrees and what's more is rather frustrated at the current level of State support. Government largesse, as he calls it, is responsible for the Ord, Burdekin, Snowy, Eildon, Dartmouth, and channel systems in the Riverina and the Goulburn Valley. He points out that when you start dividing dam costs among the farmers they serve, it seems outrageous.

> ACF's assessment of the Teemburra Creek Dam near Mackay is that capital subsidies equate to $250,000 to *each* of 60 cane growers downstream from the dam. A proposed (but never constructed) off-river storage near St George in Queensland would have provided a capital subsidy to around three dozen irrigators, equivalent to between $300,000 and $500,000 each. That's a lot of money for an individual or modest farming enterprise. When the Hume Dam required major repairs in the late 1990s, 100 per cent of the funding was paid for by the public sector, including $12 million that was misappropriated from the Natural Heritage Trust. In all other industry sectors, economic viability means commercial viability. Is there some reason why irrigated agriculture developments are different?'[32]

Robert Caldwell's response is blunt. 'If they sent us the bill, there wouldn't be an irrigator left irrigating.' Charges in the Lachlan have gone up threefold in the last four years, and Robert certainly believes they could end up doubling the trading price of temporary water. Bumping up fixed charges for licence holders also has the impact of getting sleepers to wake up and to sell. Another consequence of higher charges ($30 a meg used) is that farmers are being forced into horticulture, which Ted Morgan, in charge of Jemalong Irrigation on the Lachlan, reckons is a disaster. He says Lachlan farmers are graziers with little capital investment or expertise in irrigation practices. 'You know, if you grow 40 hectares of corn and miss your watering by two days, it costs you 50 per cent of your yield. What they'll do is create peasant farming.' As one farmer explained bitterly, the whole idea of full cost recovery sticks in his throat when he gets a visit from the extension officer in his shiny new government 4WD.

The dairy farmers of the Murray–Darling are another group under pressure. Already reeling from dairy deregulation (for which they were handed a $1.6 billion adjustment package), it is expected that many smaller farmers will leave the land. This can only be to the detriment of society's rural fabric. Yet if dairy farmers are to stay, as one industry chief put it, can we continue to flood irrigate pastures and have fertilisers running into our waterways? And if not, who is going to pay for the best management practice that demands we convert to sprinkler irrigation? [33]

That 'collateral damage'

When the reformers talk about full cost recovery, however, they no longer actually mean FULL cost recovery. This awful phrase was originally supposed to include what are known as 'externalities', the environmental cost downstream of what a farmer might be doing to the water. Reducing it, polluting it, halving the number of fish. A progress report on reform confirmed what everyone suspected: 'A gap exists between current decisions by jurisdictional price regulators and the requirement to factor in the full costs of resource management and environmental degradation.'[34] In other words, pricing is not taking account of externalities. In terms of healthy rivers, externalities are precisely what governments should be addressing. Unfortunately, the damage is hard to calculate and far too politically sensitive. Who wants to ask what the cotton industry and the melon industry really cost Australia? With growth in both cotton and horticulture, these are questions that should be asked and answered. Not to address external costs is, in Tim Fisher's words, a 'cop out'. While Tim's activism continues, unfortunately far too little agitating is going on, largely because the work is 'all too hard'.

Compensation

It should be no shock that reform is tightening the noose around many an irrigator's neck. Public awareness is growing about the plight of our river and groundwater systems, even if few seem to appreciate their contribution as consumers to that plight. Land and water resources departments all over Australia have been shedding their developers' clothes and putting on shiny new green ones. And if there's one subject Macca on ABC Radio talks about more than almost anything else, it's water.

Cutting back off-allocation water and improving the efficiency of water use are now accepted as good management practices. And that there would be casualties was also clearly envisaged by those designing the reforms; there would be ordinary people whose livelihoods would be battered about and in some cases destroyed. Usually, these people are the small family farmers.

Yet perhaps a step back is needed. Do people on the land need to be compensated? Aren't they rich enough already? The answer is no. In 1997 the Murray–Darling Commission surveyed farmers and graziers in the Loddon–Campaspe catchment in Victoria, seen as fairly typical for the State. Leith Boully, chair of Murray–Darling Basin Community Advisory Committee, reported its findings to the International Landcare Council. It found that 'In many dryland, mixed farming districts more than 40 per cent of farms were generating disposable incomes of less than $20,000 per annum. The Australian Bureau of Agricultural and Resource Economics (ABARE) survey upon which [the Commission's] paper was based defined $50,000 as the income level required to allow reasonable reinvestment back into the farm to maintain its assets.' In these areas, the average age of farmers rose four years over the last decade to around 55 years old. The message from this is that people don't have the faith to hand on the family farm. [35]

Peter Garrett and Ian Donges recognised that assistance is needed. It is factored into their $65 billion. There's any number of other signs of 'good intention'. Water trading, according to economic expert Marsden Jacob, may bring adverse environmental impacts to the selling property, possible loss of income, stranded irrigation supplies and higher costs for remaining irrigators. It advises that 'rather than stop water trading to prevent structural adjustment, government will, on occasion, need to provide structural adjustment assistance to prevent trade and growth being impeded.' [36]

And the Federal Government's National Action Plan also pointed out that 'Governments need to ensure buying back or 'clawing back' ground and surface water allocations with compensation to promote adjustment to affected individuals where appropriate.'[37] The Federal Government is quick to point out that compensation was the prime role of States and Territories as owners of Australia's water, although it might be 'prepared to consider a contribution towards appropriate compensation to promote adjustment'.[38]

There is no doubt that COAG appreciated what sort of impact reform would have on some farmers, even if it did have to use that Orwellian newspeak term 'structural adjustment': 'The speed and extent of water industry reform and the adjustment process will be dependent on the availability of financial resources to facilitate structural adjustment and asset refurbishment.'[39] Deputy Prime Minister John Anderson is left exasperated by it all. 'You know, the 1995 COAG reform obliges States to properly recognise water rights. And that's not happening,' he explained. 'It's time for the Commonwealth to play hardball. We must get serious as a nation and recognise property rights where they exist and legitimate expectation where it exists. And that's why this has become so acrimonious. To your suburban readers, I would say that they need only consider how they would react if in the interest of sounder environmental concerns, their backyard was changed into a nature reserve in a way that it not only impacts on lifestyle but dramatically reduces value of their assets. Well, of course they would expect some recompense.'

'Structural adjustment' is a horrid phrase born of the bowels of the bureaucracy. Most farmers see it as a euphemism for being squeezed off the land, throwing in the towel and moving into Toowoomba or Albury-Wodonga. In many instances they are right.

To a farmer, there's something very irritating about being told what to do by a suit who probably couldn't grow alfalfa on his back windowbox if he tried. But there is also a growing feeling of unease in the farming community that not only are they going through a very painful change, but that the structural adjustment fairy will disappear. Indeed, the potential for any magic wand waving in the future looks very slim. State legislation is shamefully coy when it comes to compo. Loss of income may be hard to assess but to avoid it is iniquitous behaviour, given that cash from the $5.5 billion in tranche payments from the Commonwealth was supposed to be put towards assisting social upheaval.

In South Australia, the head of Central Irrigation, Jeff Parish, managed to get money out of governments for irrigation improvements, a remarkable feat, but none of it was for structural adjustment. In fact the money handed over was conditional upon Central Irrigation carrying out most of the 'adjustment'. 'We actually did a lot of restructuring, we looked at 300 farms and took 130 out of production, not a lot of fun I can tell you, when they're starting to take out double-page adds on you.'

Dairy chief Max Fehring is at the end of his tether on all this. 'We need all the rights and securities defined. Set out clearly. Then if society wished to change those affected, fine. But I am sick to death of the community being forced to change. We need to stop changing for a moment, stop disenfranchising people and compensate those people so that they can then go and be productive in another part of the economy.'

Tim Fisher is rather unsympathetic. On this particular front, economic rationalism is popular with the Greens: ha! Another subsidised industry.

> Given governments actually own water resources, why should they give away a compensatable right unnecessarily when the market has already shown that it is prepared to invest in licences that are annually renewable? One could reasonably expect that some growers in commodities (e.g. grapes) could well afford to pay higher water infrastructure costs than those that could be afforded by other growers in different commodities (e.g. sheep). Requiring a positive rate of return on existing infrastructure could potentially accelerate the speed of structural reform in the irrigation sector by encouraging a more rapid transition to high value uses for water.[40]

We all know why government purse strings are drawn tight. Treasurers will always behave like the congregation member who looks into his wallet and can find nothing smaller than a lobster. We know that unless a politician is bombarded each night by ugly pictures of the car factories in Adelaide or John Howard's brother's factory in the Hunter, spending the taxpayer's money on subsidies or compensation is very unappealing. Yet this time it is the States, not the Federal Government, that should by rights be the farmer's first recourse for compensation. That they are turning their backs is a disgrace.

Easy water

There have been many imaginative novels and films about what happens when water gets scarce and expensive. In reality there may be one very real consequence of water reform we have not touched upon. It is what happens when money takes over. Poaching.

Just how much water poaching has gone on over the years is very hard to know. Back in 1947 H.O. Monteith made the following observation to the Institution of Surveyors: 'The pump often works long hours with great efficiency. Housed in a little tin shed, down on the river bank, its importance is inclined to be overlooked. I think Surveyors should know more about it—where it is to be found, how it comes to be there, and what it is doing to our streams.'[41]

Water pinching clearly goes on, although as you'd expect, it's a touchy subject with farmers. The mayor of Toowoomba, Dianne Thorley, is from a farming family and pulls no punches. 'Come on, I come from the farming community, we always stretch the boundaries a little bit. [Yet] it seems like there is an inability to understand that if you abuse this and we do the multiplier effect all the way down the Murray–Darling, on every tributary, on every available water, and if we all abuse it to even half a meg a year, what sort of effect does that have?'

There are various ways to nick water. Traditionally, people fiddled with the old Dethridge wheel, which acted like paddles on a steamer, clocking up the water used as each blade went round. Sometimes a frozen fish would be jammed in and as it thawed out, so did the evidence.

'They used to, they don't do that now.' George Warne looked amused at the interest. 'If you're really going to steal water, the frozen fish way is not enough, you've got to get the backhoe down there and lift the wheel out. Like the story of the farmer that went down there one night on his own and lifted the wheel out, they're quite heavy, and he went back at 5 a.m., he had it going all night, and during the night the thing had blown into the channel, and he was out there in his underpants in the freezing cold August day trying to get the wheel back up the bank.'

Clearly there is some educating to do, just on a day-to-day basis. In Don Rowe's quarterly magazine for Sunraysia irrigators, farmers were reminded about the importance of monitoring their water use.

> Over the last four weeks, I have made a number of calls to cus-
> tomers who are close to using their water right or in fact have
> gone over it. Most of the responses have been, 'What do you mean
> I can't use more than that?' The question has been asked, 'How is
> the Authority going to stop me anyway?'[42]

While Sunraysia irrigators are today more closely monitored, questions like these from some farmers are a little concerning.

The climate for poaching is ripening. One Queensland irrigator from the Condamine–Balonne made matters clear. 'If you cut someone back until it's a question of losing the farm and his shirt, or taking a little more on the quiet out of the river, what would you expect him to do?' And with new rules like the *Farm Dams Act* in New South Wales encouraging farmers to use dams, but with no effective maintaining of their use, the opportunity for opportunism is there.

Ecologist Peter Cullen is not surprised. 'The value of water is going up and there's going to be more theft. I think you'll also find there'll be quite high levels of community policing. I mean they steal water from their neighbours rather than a wider community. I also think there's a whole lot of satellite technology and other things that let us identify people having significant breaches. Police use them for marijuana crops, already.'

It may be that in years to come, Australia has a water police force. Not the Rats on the harbour, but a privatised organisation with employees who work like the wheel clampers in London—zealots who are incentivised to catch as many illegally parked culprits as possible. And the penalties could be high. 'I wouldn't charge them $1,000. I'd take 10 per cent of their water rights which might smarten the game up,' said Peter Cullen.

'Yes, we will need a whole enforcement army,' agreed Justice Peter McClellan of Sydney water inquiry fame. Privatised? 'Well in Victoria, I'd probably say private.' He smiled. 'In New South Wales, it would be government. And depending on the circumstances in which it bites— in drought or in plenty—I can see significant unrest.' If this sounds bizarre, it can surely only be because water doesn't cost enough . . . yet.

The rub

Is water reform a good thing on balance? Overall—yes. Yes, mainly because the rate of growth in irrigation in the Murray–Darling in particular was at a level that would have hurt existing farmers as well as the river itself. Irrigation can improve. There is still more watering of the soil than the plant, with only 30 per cent to 50 per cent of harvested water used to produce dry matter under irrigation. Given that in total most water used in Australia is in intensive agriculture, even small savings should make real differences.[43]

Governments' unwillingness to provide clear rights has come back to bite them on the backside, because permanent trading is hampered, and moves from flood irrigation to drip, from fat cattle to oranges, aren't happening as quickly as they should.

Of all the downsides of water reform summed up in the progress report to COAG on water reform implementation,[44] by far the greatest challenge is measuring this 'ruddy duckwater', the environmental flow. It's pretty hard to get a handle on water skiing water as well. No one knows how much water the leisure industry should rightfully demand and irrigators simply do not accept that the environment has some God-given right to be vague. As already discussed, science's inability to lock in a long-term flow rate means the Greens have no interest in secure amounts, unless those amounts are massive. The 'precautionary principle' at work.

If you believe CSIRO's Sustainable Agriculture leader Dr Wayne Meyer, irrigators are not just competing with the environment, but with themselves. 'The notion that improved water efficiency will free water for environmental purposes is simplistic and will not work. On the contrary, the demand for irrigation use is likely to increase with water savings unless there is some explicit mechanism to encourage saved water to be reallocated.'[45]

At the end, as with this whole water struggle, the answers should lie in someone attempting to quantify fairness. Because right now, it's unfair. 'Farmers have always considered they were in control,' said Berrigan rice grower Ian Mason, 'and in water, it's gone and they're bewildered. They are even turning against each other.'

8

Snowy's Revenge

I am the one that rode beside the man from Snowy River.
*Ian Mudie—*They'll Tell You About Me

The Mighty Snowy

By far the most famous river in Australia is the Snowy. Waters of myth and legend. In 2000 the Snowy, the trickle, won the right to run free once again, or at least about a quarter of it.

The Snowy story is the story of two of Australia's greatest icons pitched against each other. In one corner, the great Snowy Mountains Scheme which drought-proofed the inland and became the most important symbol for multicultural Australia; in the other, the legendary Snowy River itself, inspiration for one of the country's most famous poems.

The Snowy was the final straw which broke the Liberal Government in Victoria in 1999, but perhaps most importantly, the October 2000 decision signifies a change in thinking about rivers and the environment. Even in the 1970s and 1980s, as the coal and gas power stations around Australia continued to burn, the Snowy Scheme was heralded as a magnificent clean green energy. Some years later, many people are revising their views on just how clean and green the project really is.

The Snowy is only a short river. It rises high in New South Wales' white-capped Snowy Mountains and runs 400 kilometres down through Victoria to Orbost and the Tasman Sea. Before the big dams at Jindabyne and Eucumbene backed up the water, it ran closer to 500 kilometres, and each spring when the snow melted vast amounts of water crashed through steep rugged country in spectacular torrents, a sight that earned it the title of the Mighty Snowy.

At the time Banjo Paterson wrote his famous poem immortalising the man from Snowy River, the river was in its element.

He hails from Snowy River, up by Kosciusko's side,
Where the hills are twice as steep and twice as rough,
Where a horse's hoofs strike firelight from the flint stones every
  stride,
The man that holds his own is good enough.
And the Snowy River riders on the mountains make their home,
Where the river runs those giant hills between;
I have seen full many horsemen since I first commenced to roam,
But nowhere yet such horsemen have I seen.

While Paterson devoted few words to the river itself, the famous yarn of how the stripling on a small and weedy beast brought the mob home was enough to transform the river into a household name.

Much of the farming area along the Snowy was developed as soldier settlement after the First World War. Up near the source, 'So clear and so pure are those waters of melted snow that the wild cattle of the ranges, craving salt, will come to the call of a human voice like a flock of chickens.'[1] Those who live along the Snowy, in places like Dalgety and south along the coast, still remember it as a fishing Mecca, one of the best bass rivers in Australia, and full of grayling and eel. Some remember having to scramble to the roof of their house in a flood.

Today, however, in the country just below Jindabyne the river is but a trickle, 1 per cent of its original flow. This is how it stays for another 80 kilometres before other tributaries kick in. Countless pictures of notable pollies jumping across it have been taken over the years and such was the river's demise that the 1982 film *The Man from Snowy River,* with Tom Burlinson and Sigrid Thornton, had to be filmed on the upper Murray. The lack of water has allowed salty tidal water to run inland 15 kilometres to Orbost, and farmers say their land is now choked with sand.

The Snowy River was the big casualty in the great mountain scheme for electricity and irrigation. Its powers were to be harnessed for what at the time were seen as much more worthy causes.

The vision

In the 1940s, ideas about a big drought-proofing scheme were not new. A member of the Lyne Royal Commission on conserving water in Australia in the 1880s, civil engineer, F.B. Gipps, had recommended use of Snowy waters for hydro-electric power. And by 1913 a Victorian engineer, Gregory, had suggested 30-mile long tunnels to divert the waters to the Murray. The challenge was technology.[2]

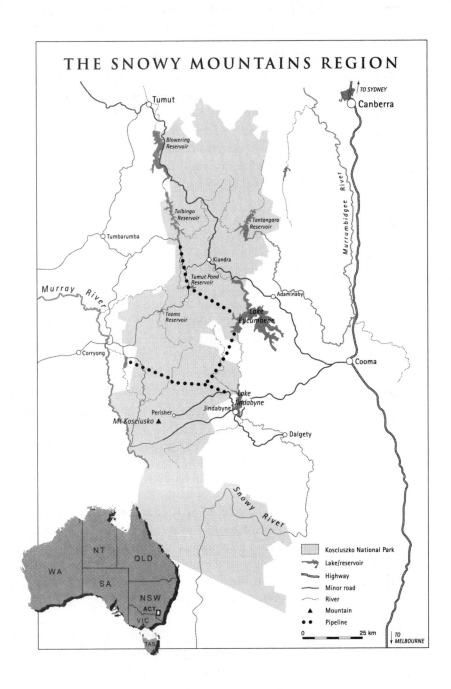

THE SNOWY MOUNTAINS REGION

TO SYDNEY
Canberra
Tumut
Blowering
Reservoir
Talbingo
Reservoir
Tantangara
Reservoir
Tumbarumba
Kiandra
Murrumbidgee River
Tumut Pond
Reservoir
Adaminaby
Murray River
Tooms
Reservoir
Lake
Eucumbene
Corryong
Cooma
Lake
Jindabyne
Perisher
Jindabyne
Mt Kosciusko ▲
Dalgety
Snowy River

Kosciuszko National Park
Lake/reservoir
Highway
Minor road
River
▲ Mountain
• • Pipeline

NT
QLD
WA
SA
NSW
ACT
VIC
TAS

0   25 km

TO
MELBOURNE

Even by today's standards, the scale of the Snowy Scheme is so huge that as a government project, it is never likely to be repeated. At the time it was commissioned, in the late 1940s, it must have been truly awesome. The aim of the Snowy Scheme was twofold: to harness the power of the Snowy as it dropped down through the mountain ranges for hydro-electricity and to divert the waters of the Snowy River inland down the Murrumbidgee and the Murray to provide secure farming water for thousands of growers.

The first rock was blasted at Adaminaby on 17 October 1949, with Prime Minister Ben Chifley, Governor General Sir William McKell and Commissioner of the scheme, Sir William Hudson, looking on. The scheme would take 25 years to complete and come in on time and within budget.

The scheme's vital statistics are staggering. Right at the heart lies the man-made Lake Eucumbene with a capacity of 4,800 gigs of water, or more than nine Sydney Harbours. The dam walls are 116 metres high. The whole system connects fifteen other dams. Seven power stations, some hundreds of metres under the ground, constructed within giant underground caves, which look like Bond action-movie sets, were carved out. In total, 145 kilometres of tunnels and 80 kilometres of aqueducts connect the 16 dams which can store 7,000 gigs, or 13 Sydney Harbours.[3] Ninety-nine per cent of the Snowy's water was diverted. It meant every little tributary tapped for the scheme was sent west for the farmers on the Murray and Murrumbidgee. In 1969 the American Society of Engineers recognised it appropriately as one of the seven engineering wonders of the world.

How this massive infrastructure was assembled is the most incredible feat of physical and social engineering Australia has ever seen. Between 1949 and 1974, 100,000 people from over 30 countries helped build the dams and tunnels. Many were grateful refugees from the Second World War, commandeered for the project, who camped in prefab units through two decades of winters high in the mountains. Suddenly, the Snowy Scheme was nation building in another dimension, as former enemies discovered mateship and overcame language differences, the skills and trades they learned setting many up for later life. Archival pictures taken by the authority have thankfully preserved an extraordinary record.

There were casualties. Two Snowy Mountain towns (Adaminaby and Jindabyne) had to be physically moved. In all, 121 men died in the construction, most on the roads. The most horrific accident came just before the Christmas break in 1963 at Island Bend Dam. Men working on deep shafts within the dam wall were knocked over into quick drying cement, one up to his waist. He died of shock as doctors desperately tried to amputate.[4]

In 1999 a campaign to find as many of the workers as possible was launched in the media and a massive 50-year anniversary party reunited 1,000 of them on 17 October that year, an occasion which brought home to some younger Australians just what a human commitment the scheme had been.

The politics

The challenges were not all technical. Federation had not anticipated the Snowy Scheme and had left both water and electricity controlled by the States, not the Federal Government. In order to get around this, in 1949 Prime Minister Ben Chifley used the constitutional powers of defence to enable the Federal Government to wrest control from the States and create a statutory authority to carry out the works. This was clever politics, around the time of the Berlin airlift and the drawing of the iron curtain.

According to one engineer on the scheme, credit for the idea should go to Ben Chifley's Minister for Works and Housing, Nelson Lemmon. This is Lemmon's account:

> I went to Chifley . . . and I said, 'There's only one way to handle this . . . Put the whole thing under the *Defence Act* . . . and we'll be the boss.'
>
> He said, 'WHAT? Your name's Nelson Lemmon, not Ned Kelly—you can't do that.'
>
> So I said, 'Why can't I?'
>
> 'Well,' he said, 'you tell me how you can!'
>
> So I said, 'Listen! You had subs in the Harbour. The way we're building everything now, all they want is a decent cruiser and they could sneak through the guard and they could blow all your power stations out without an effort! You've got Bunnerong built on the water, you've got the big one at Wollongong built on the water, they could blow all your damned electricity out in one night's shooting! Where'll you produce the arms? Where'll your production be with all the power of New South Wales buggered?'
>
> Chif says, 'You might get away with it. If you can get Evatt [the then Defence Minister) to agree with it—and if there's a case he'll have to fight it in the High Court—if you can get Evatt to agree, I'll go all the way with you!'[5]

That was not the end of it, though. In the reverse of the Franklin Dam situation, in which Tasmania was stopped by Federal intervention, the Federal Snowy Scheme was almost kyboshed by the States. Construction had already commenced when both Victoria and New South Wales cried foul on the basis that electricity generation was a primary State responsibility. Court action was only averted by the incoming Menzies Government in 1950, which negotiated with the NSW and Victorian Governments to pass enabling legislation at State level.

Even today there is argument as to whether power or water had priority. The man instrumental in returning water to the Snowy, Victorian Independent MP Craig Ingram, has no doubts. 'Originally the scheme was a power scheme, a concept developed in the early days of Federation. They were looking for a site for the new capital and Dalgety near the headwaters of the Snowy was a front runner. Legislation was passed which gave the Commonwealth power to use the water to develop hydro-electric power and for domestic water.'

When Canberra was selected as the site, the Act was still in place. 'In fact the first option was to put the hydro-station on the Snowy and not divert water [inland] into New South Wales', explained Craig Ingram, 'which would have created more power. But then New South Wales came in to suggest water be diverted into the Tumut [River] and the scheme developed from there.'

Before the greening

In all the preparation and implementation, one issue was clearly absent—any consideration for the environment. This is easy to say with hindsight. At the time the omission was surely understandable. Even the word 'environment' was barely used, except by town planners as a rather stuffy word for the area around a city. Thousands of people, not fish, had died in the previous decade and no one had ever heard of an environmental impact statement. The factors which were driving decisions were the demand for power, as hungry industry expanded causing blackouts, and the value-added production in agriculture and improvement to rural communities which water could bring[6]—in the words of ecologist Professor Sam Lake, from Monash University, 'No one ever asked what it would do to the river. The rhetoric of Chifley and Hudson was of wealth and prosperity, not a word about the environment.'

One exception to this was the South Australian Premier, Sir Thomas Playford, as demonstrated in this fascinating exchange with Robert Menzies in 1958, in which Sir Thomas showed rather good foresight about the Murray's future.[7] At the time, South Australia was taking action in the High Court against the Commonwealth and its powers to build the Snowy Mountains Scheme. At what must have been quite a tense ministerial conference, the main beneficiaries of the scheme, New South Wales and Victoria, were arguing that the new Snowy water would fill the Hume Dam and not affect South Australia's water allocation from the Murray. Sir Thomas retorted:

> Now that they know there will be additional amounts of water, I have not the slightest doubt—and I do not offer any criticism in this regard—that they will develop their use of water much more extensively. I do not accept for one moment the argument that because more water will be coming in there will be less likelihood of restrictions. Why are we arguing about water?

The Prime Minister spoke next:

> Could I try to put your view in my own words to see if I am clear of it? The level of the Hume will not necessarily be enhanced by the inflow of the Snowy and might in fact be reduced by the heavier consumption lower down, and in those circumstances, the periods of restriction may not be less frequent, but may conceivably be more frequent.

'Yes,' endorsed Sir Thomas.

Shortly afterwards, the PM took the conference with the ministers in camera, politely suggesting:

> Do you think it might be useful, just for half an hour to have a look at this without the presence of other people? I ask that because there are some political angles involved and I never like to embarrass distinguished civil servants by introducing them into political arguments.

In the end, High Court proceedings were discontinued, although South Australia made some gains through the Snowy Agreement,[8] which sets out water allocation and security post the Snowy Scheme construction.

Movement at the station

About half of the Snowy water from the scheme was sent down the Murray, with entitlements to be shared equally between Victoria and New South Wales and the other half to the Murrumbidgee for New South Wales only. Overall the split was 75 per cent/ 25 per cent.

The change that Snowy waters made to areas like the Riverina in southern New South Wales was phenomenal. Rural populations thrived and the Murray–Darling Basin became what it is today, the supplier of 90 per cent of Australia's irrigated agriculture. However, by 1970 farmers were becoming suspicious about the operation of the Snowy. In their view, the scheme that had been built with the number one goal of irrigation was now dancing to a different tune: supplying power. The force of falling water is essential to drive the turbines which generate electricity. Around 75 per cent of the Snowy Scheme's water storage capacity can be used for electricity generation, the seven power stations generating an average 5,100 gigawatt hours of electricity each year, roughly 10 per cent of the State's total energy consumption.

This is important because surges in demand for power do not necessarily coincide with demand for water. Unfortunately, the original deal that had been struck with the States was that they would pay for power, but that water would effectively be 'free'. As the years went by there was increasing pressure for the Snowy Scheme to pay for itself by releasing water for electricity when needed, ahead of irrigation needs.

[handwritten margin note: Another example of competitive pressures overcoming property rights.]

By early 1970 the level of concern was enough to get Hunter Landale (after his success with Dartmouth Dam) into action again. 'I organised another debate in the town hall in Deniliquin, between Sir William Hudson on the one side, and my cousin David Fairbairn, the Minister for National Development, on the stage in the town hall,' Hunter recalled. 'The agenda for that was one thing and one thing only—what was the priority for that water, in the Snowy? David wouldn't do releases unless you could make electricity, and we were short of water at that time.'

So worried was Hunter Landale that there might not be interest in this, he had asked the local headmaster to give the schoolboys a half day to be bums on seats. As it turned out, 150 farmers left little room for the boys. There was apparently some confusion about whether the honorary guests had been informed of the confrontational nature of the debate, which according to Hunter led to a previously admiring David Fairbairn clearly quite unnerved, to say to him accusingly, 'You set me up!'

Corporatisation

Irrigator cynicism about Snowy Scheme priorities continued over the next two decades, and their fears were certainly not appeased by talk of corporatisation of the Snowy Hydro Authority. With competition reform in the 1990s came the creation of a national electricity grid

and electricity trading between States. Governments agreed to corporatisation of the Authority in 1993 and complementary State
legislation was passed in 1997 to ensure that the Authority could
operate on the same basis as other power companies.

Until this time, the Snowy Scheme's assets and liabilities had been
owned by the Commonwealth. Electricity rights had been split 58 per
cent to New South Wales, 29 per cent Victoria and 13 per cent to the
Commonwealth. Importantly, the Scheme had not been set up to
make money, rather to recover costs of operating and servicing the
almost $1 billion loan. Income was received from the Victorian, New
South Wales and Australian Capital Territory power companies. After
corporatisation, the plan was for Snowy Hydro to be responsible for
its own financial management, organise its own financing and focus
on making profits.

The announcement of this plan triggered new concern, not from
the Murray and Murrumbidgee irrigators, but for people living along
the Snowy River. Snowy supporters had been living in hope that one
day the water flows would be reversed and the Snowy River would once
again be mighty. There were real fears that a fixation on profits would
demand maximum electricity generation and therefore set the diversion of the river even more firmly with the Snowy flow stuck at 1 per
cent for the entire 75-year licence period granted to the new corporation. It was this fear that kicked off a new campaign to free the Snowy.

The rise of the Snowy lobby

In 1996, the first scientific report into the Snowy was commissioned
by the Snowy Genoa Catchment Management Group. The study was
driven by Professor Sam Lake. Professor Lake was a good choice. He'd
known the Snowy since childhood and worked on the scheme each
summer during the 1960s. His report concluded that the hydro scheme
had turned a mighty river into a drain. Water quality had declined
with the bottom of river pools devoid of oxygen and uninhabitable for
most creatures. There was a build up of sediment, sand and other
organic materials which would have been cleared out by floods in the
old days.

The second task for Sam Lake and a team of hydrologists was much
harder: to put a figure on what flow was needed to repair the Snowy
and recommend a flood regime. 'Engineers were then asking, "How do
you build a flood?" They've never had to think about it before,' said
Sam in an office which oozed research papers from every shelf. The
figure they came up with was a return of 28 per cent of the original
flow. Furthermore, water releases should mimic natural seasonal flow.

At about the same time the report was released, a vocal lobby group, the Snowy River Alliance, representing aggrieved farmers and communities down the river, began agitating for increased flows. Calls for a 28 per cent return of flow grew louder from the Alliance, so loud that in April 1998, corporatisation of the Authority was post-poned and a full Federal inquiry was launched to produce a range of costed options on water flow. It was to study environmental, social, economic and heritage impacts.[9]

A rally was held in Dalgety to support the inquiry, with actor Tom Burlinson pushing all the right buttons. 'In my travels internationally,' he said, 'the Snowy River is now world famous and so there is more than just the environmental issue at stake here. It's our national reputation.'

The inquiry favoured a 15 per cent increase in flow, which its Commissioner, Robert Webster, recommended as the best compro-mise with a minimal cost impact on agriculture and a manageable impact on hydro-electricity generation.[10] To put this in some sort of perspective, the Commonwealth Government's economic body, the Australian Bureau of Agricultural and Resource Economics (ABARE), estimated that if 25 per cent of the flow was returned to the Snowy simply by opening a valve at Jindabyne, then irrigators would lose 3 per cent of their water and power generation would fail by 7.5 per cent.

The Snowy Alliance, using Sam Lake's science, claimed a 15 per cent increase in flow would make little difference. The next 18 months would witness some of the most extraordinary interstate politics seen in Australia.

Lad, you'd better stop away

The man at the centre of the political storm that blew up was Craig Ingram, a 34-year-old abalone diver, president of Native Fish Aus-tralia, and a force in the Snowy Alliance. Craig Ingram's family had lived in the Snowy catchment for five generations. The deep river where his grandfather had swum as a boy was now a shallow stream.

When the Alliance was fobbed off by the National Party, Craig decided to contest his local seat of Gippsland East (which had been a National Party stronghold for some 140 years), at the next election on the Snowy issue. He talked the talk that locals wanted to hear: 'People of the Murray say their river is in a bad state but they've had a finan-cial benefit from it in terms of irrigation and development and they've got prosperous towns.'

The Nationals' reaction was to lampoon him. Posters appeared. showing a small boy not able to reach the door handle of the premier's office, captioned with the words 'Don't send a boy to do a man's job.' The similarity with A.B. Paterson's poem was not lost on the Alliance. The election result? A staggering 24 per cent swing to Craig Ingram.

For Premier Jeff Kennett in the 'unlosable' election of 1999 the unthinkable had happened. The Labor Party, which all sides had believed was still in the wilderness, had won enough seats for three Independents to hold the balance of power. One of them was Craig Ingram, and the horse trading began.

During this time, Craig continued to campaign vigorously for the Snowy. He led a real horseback rally to Parliament House in Sydney. The 'men from Snowy River' rode down Macquarie Street in Sydney decked out in Akubras and oil skins.

In Melbourne, once it became clear that at least one of the other two seats would fall Labor's way, the race was on to woo the 'stripling'. Craig Ingram had power, and he needed it, because he would have to get both the Victorian and the NSW Governments to agree to the Snowy decision. New South Wales, under Bob Carr, was Labor.

Then Jeff Kennett made a fatal error. It was not that he had misread the importance of the Snowy issue, everyone had done that. Craig Ingram remembered, 'Jeff had leverage because there was a drought and NSW rice growers wanted more water from the Snowy. I had a discussion with Jeff in which I said, "You have a bargaining chip. Get New South Wales to agree to Snowy environmental flows in return for the extra water."' The Victorian Government signed off on the water anyway, due to the Federal and State political pressures, according to Craig.

Here was Labor leader Steve Bracks' opportunity. Labor convinced Craig Ingram that it would be much easier for Bracks to draw Carr into a deal. What tipped it? 'Steve Bracks was committed to it and said, "You and I both know Craig, I'm the only one that can get New South Wales to come on board."'

On 19 October 1999 Jeff Kennett resigned. In the end, a trickle had effectively brought down one of the toughest and most successful governments in Victoria.

Flying flint stones

An accord between New South Wales and Victoria was well and good, but the Snowy Scheme had three, not two masters. Enter the Federal Government, with ministers both pro and anti the return of Snowy water, reflecting the opposing interests that Murray water inevitably brings to the table.

At the bottom of the Murray, South Australia's Premier, John

Olsen, was furious about what impact a diversion of any Snowy water would have on the State's water supply. 'It's short term, Steve Bracks is propping up government policy because of one Independent. That is the tail wagging the dog,' he exclaimed.

Federal National Party leader John Anderson was prepared to accept 15 per cent for the Snowy, but in Adelaide politics grew juicier. Adelaide-based Nick Minchin was Federal Minister for Industry, and corporatisation of the Snowy Scheme, which had been delayed twice already, was his baby. His prime agenda was to get the deal through. But Environment Minister, Robert Hill, also Adelaide based, worried about the Murray, was opposed to any diversions and wanted a rethink. On 20 December 1999, Robert Hill explained, 'So far the debate has been unbalanced. I'm saying it has been dominated by romantic connotations rather than sound environmental or economic logic.' A day later, Victoria's Steve Bracks leaped on the comments as parochial. 'What I say to Senator Hill is negotiate and work with the whole of the country, not only South Australia. There are two major governments, the two biggest population centres who are pushing this. The onus is on him to be there for the whole of Australia and not just for South Australia.'

Halting for a moment

Over Christmas, Senator Hill's office worked overtime. The full implications of the Murray–Darling Salinity Audit, delivered late in 1999, were beginning to sink in. Taking water out of the Murray River was unthinkable. There is, however, one tool available to politicians in such situations. The stall.

In January 2000, Robert Hill commissioned a new environmental impact statement (EIS) for the corporatisation of the Snowy Hydro Authority, justifying it on the basis that 'the question must be asked, if savings are to be achieved in the Murray system, is it in the national interest that they be converted into a new flow down the Snowy or should they be better used to provide greater flows in the future in the Murray system?'[11] The move was immediately written off by the Australian Conservation Foundation (ACF) as an expensive and time-wasting stunt which was no substitute for leadership on river health. 'The EIS was a waste of taxpayers' money, telling us the bleeding obvious. There's not even a sensitivity analysis,' complained Sam Lake, referring to the standard methods of comparing various scenarios under different assumptions. What the EIS did do was to postpone the June 2000 deadline for corporatisation, for the third time in as many years.

If corporatisation had gone ahead, hydro property rights could have been entrenched and environmental flows to the Snowy would have been more unlikely.

Fancy riding

Victoria did not let Canberra or the 15 per cent figure distract it. In October 2000, the State reached a momentous decision with New South Wales, later endorsed by the Commonwealth. Twenty-eight per cent would be restored to the Snowy, with 21 per cent back within the next decade.

The media lapped up the announcement. 'Champagne flowed on the dry river bed' was the caption beside pictures of Bob Carr, Steve Bracks and ACF President Peter Garrett below the Jindabyne dam. Peter Garrett hailed the decision as 'a symbolic step forward for all Australian rivers'. Tim Fisher rejoiced: 'We're undoing four decades of environmental damage and community anguish over the impact of building the scheme. There were some people in tears yesterday when they were being briefed on the agreement.'

The beauty of the decision, according to its makers, was that no one would lose. Water was to come from efficiency savings from irrigators on the Murray and Murrumbidgee. The States would put $300 million into an independent body which would oversee the purchase of 65 gigs of water-efficiency savings. Victoria and New South Wales would share the cost of reduction in power generation, but interestingly, the cost of buying the water will be split fifty-fifty, notwithstanding the original 75 per cent/25 per cent split in water usage. 'It probably cost the Victorian Government more than it would have liked,' admitted Craig Ingram. 'Basically it reflects the commitment of Victoria to resolve the issue.'

It might well make the boldest hold their breath

After all the agony there is something almost too perfect about the result. There is one nagging question. In the months leading up to the decision, the idea of water from efficiency savings was dismissed by many as unachievable. Now, suddenly, the public line was that savings were the solution.

In case anyone was in doubt, Nick Minchin reminded them that 'flow levels in the River Murray will not decrease. These savings would need to be found first, before the environmental releases to the Snowy River are made.'[12] The 21 per cent claw-back in the first ten years should be the easier bit, although it will take State governments, the Murray–Darling Basin Commission and irrigation companies, which all have infrastructure responsibilities, to work together.

The other 7 per cent will have to come largely from increased farm efficiency, and while part of the $300 million set aside by Victoria and New South Wales could be used to pay for improvements in irrigation equipment, whether irrigators will accept necessary reductions in allocations is another matter. Murray Irrigation's George Warne for one thinks it highly unlikely. 'We're very cynical about the ability to actually save the water, and when I read the agreed documents between the governments, they're much more inclined to go and buy the water, and $300 million at the current market will buy the water. It will buy the water.'

But, according to George Warne, what buying water means is that farms will have to go. 'Does it mean that a farmer with 1,000 megs sells 100 and puts a new system on his farm that uses water more efficiently? That's a really nice thought. Or does it mean a farmer with 500 megs sells the lot and retires, because they're the highest bidder?'

South Australian Premier John Olsen doubts the 28 per cent target will be reached. 'And what's more,' he said, 'I object to the fact that they say to us, "Well, we'll have the pristine Snowy water flow down the Snowy River and to make up for the pristine Snowy that you don't get, you can have some Murray–Darling water", which is not what you'd call pristine before it starts its flow down the river system, so it ain't like with like, thanks very much.'

Already there are mutters about the Murray suffering from the Snowy re-routing. Peter Garrett has warned, 'It is important the Snowy is not held hostage to government failings on allocations, because it would be a tragedy if the beginnings of this unique decision on one of Australia's iconic rivers was derailed.' To the Greens, though, whether 28 per cent is achievable is now a secondary issue. The decision, dubbed 'the single biggest environmental windfall in living memory', sets a moral benchmark on environmental flows. The Nature Conservation Council's Kathy Ridge says most water management committees in New South Wales have seen 6 per cent restoration as enough.

Irrigators are bemused by the decision. 'If you had $300 million to fix up the Snowy, would you allocate it all for purchasing water for environmental flow?' asked Bill O'Kane, from the Goulburn Broken Catchment, who points to the inappropriate clearing and land use, poor riparian management and erosion down the Snowy.

It was Robert Hill who made the most astute observation, shortly after the decision was announced.

If Victoria and New South Wales are going to fund $300 million to purchase water to send down the Snowy, so be it. But from a national perspective we should understand that you buy the cheaper water now and if further water is going to be necessary for the Murray in the future, it will be more expensive.

In his environmental impact statement, released at the end of last year, the Minister raised concerns about how corporatisation would lock in the water set up for the Murray and Snowy for 75 years with a further option of 50 years. Even Craig Ingram has some sympathy. 'It will be very hard to change after this as you'll need to compensate the corporatised Snowy Hydro for loss of capacity. That's why we worked so hard to get the changes we did.'

Deputy Prime Minister John Anderson is no more comfortable than his Environment Minister. 'I've got very serious reservations on Snowy. The Commonwealth accepted it, but I wonder whether we've got the environmental equation right. There's an awful lot of romanticism here. I worry about the economics of it and I'm not sure that in the long term there won't be reduced flow to the west.'

For its part, the Commonwealth is expected to spend $70 million of its $900 million windfall from corporatisation on further irrigation efficiency savings to release more water down the Murray, hardly generous. Corporatisation should finally be achieved during 2001, the exact timing now driven by negotiation of the small print. National Party President Helen Dickie worries, 'I always was against corporatisation of the scheme, because while government had ownership, they had some obligation to irrigators.'

For the Snowy, revenge should be sweet. One of the loudest messages from its victory is the power of public opinion. Other rivers which have been 'brutalised' in this country go unnoticed. The romance of the Snowy has been its saviour, as Victorian Premier Steve Bracks demonstrated. 'To leave behind a legacy of the Snowy River running again for future generations is something I am very proud of.'

John Olsen despairs about the preoccupation with the Snowy, whose uppers reaches are a trickle, yes, but with other tributaries entering lower down, becomes a respectable river. 'The historic flow down the Snowy is 60 per cent, the historic flow down the Murray is 20 per cent. Let's get our priorities right. And the Man from Snowy River and the Old Regret horse and the Museum is not on the Snowy, it's on the River Murray, Corryong, yet this folklore that's developed on the eastern seaboard ignores the Murray.'

Yet the Man will always be remembered as being from Snowy River. And the tale is not over. The next ten years will be most interesting. As for Ben Chifley, the odds are that he would make the same decision again, even knowing what was to come. Perhaps he'd be a bit disappointed in the lack of Federal leadership to try and fix the problem for the long term.

9

Water Politics

Whisky is for drinking, but water is for fighting over.
Mark Twain

Boys and water pistols

John Olsen leans back in his premier's chair. 'May I say a few words before you ask me anything? Good.' He sits forward. 'I'm a water skier and as a result of that learned to bare foot and the River Murray and its calm waters has a great deal of personal appeal to me. My father used to take me to a place called Cobdogla and a caravan park at Barmera. We skied during Easter each year and out of that I developed a love of the river—would be the way you'd describe it—and ended up buying a shack there.'

The Murray River has done for South Australian Premier John Olsen what the Falklands did for Maggie Thatcher. 'I've mentioned this in speeches until my staff cringe: water is the gold of this century!' His adviser nods at both points.

The golden opportunity to champion the Murray came with the release of the 1999 Salinity Audit which officially warned of Adelaide's deteriorating water. John Scanlon, the former head of the State's Department of Environment and Natural Resources, remembers well. 'It didn't *actually* say the water would become undrinkable within 20 years, but, hey, it was close enough. Olsen needed an issue to run with and this one was a winner. The *Advertiser* ran a campaign, "Save the Murray". Other papers followed: the *Australian*, *Sunday Mail*. People were soon saying two in five days, you would not be able to drink the water.'

'We championed the cause, we raised the profile' said John Olsen. 'We fortuitously have a Federal Minister of Environment that is from South Australia and understands these issues. We have the ear of a prime minister and three cabinet ministers from South Australia. I let no opportunity go by that wouldn't champion the cause. The PM, to his absolute credit, recognised that this is an important issue. We said that we wanted it on the COAG agenda. He put it there. He put it as item number one. It's never been there before!'

There's no doubt that over the last three years the leverage Premier Olsen enjoys through his Federal mates from South Australia is extremely helpful, made even stronger now that Victoria, New South Wales and Queensland are all Labor Governments.

The 'State of Recalcitrance'

Water in Australia attracts plenty of politics, just not a lot of leadership. And the most colourful spats have been between the top and the bottom of the river.

Just a week after the 2001 election that put Peter Beattie back in power, he's cut the gangrene out of the party without anaesthetic and now looks invincible. A 10 per cent swing to Labor shocked everyone. 'I'm as stunned as all of you,' admitted the Premier on the night, 'but with a great result comes great responsibility.' The signature Cheshire cat grin from ear to ear looks bigger than ever.

Queensland's position on water is best understood by one graph, the one on page 90. The State takes just over 5 per cent of the water in the Murray–Darling system compared with 56 per cent by New South Wales. Queensland's take is also less than South Australia's. Peter Beattie is adamant that Queensland deserves a fair share.

The snag, of course, is that the same graph shows that 80 per cent of the water in the system is already being removed. For John Olsen down the bottom of the river, Queensland is 50 years too late. '[Peter Beattie] said at the Premiers' Conference, "Well most of my farmers would want me to put a bloody dam on the border and not let *any* water flow down to you." And that's when I said, "We are Australian, aren't we? We all live in one country surely, and in a developed country. How is it that anybody would have such a disregard for another fellow Australian that they would do that?"'

The problem for John Olsen is that Queensland's impact on his State is not nearly as bad as he makes out. According to the Basin Salinity Management Strategy, Queensland will contribute just 1 EC increase in salinity at Morgan in South Australia, as compared with the total predicted increase of 88 ECs by 2015. Indeed, about 24 ECs will come from the Mallee region of South Australia! As for the water impact on South Australia of Queensland's development thus far, it is also mild. The real impact is close to the Queensland border, in northern New South Wales. So from South Australia's standpoint, there's a touch of not letting the truth get in the way of a good story.

More practically, Peter Beattie now has substantial power as premier, with the opposition in tatters. Not so John Olsen. One wonders whether Peter Beattie really cares about what John Olsen has to say.

Peter Beattie's relations with the Federal Environment Minister are no cosier. Relations were soured long before the water fights, over land clearing. One hundred and fifty years of history has shown that clearing is not good for land or rivers, yet Queenslanders are going at it faster than ever. The last set of decent figures[1] showed that of the almost half a million hectares of trees felled each year, 425,000 or 85 per cent were in Queensland and, even more disturbing, over half of Queensland's clearing is in the Murray–Darling Basin.

What is tragic and unacceptable is that the Queensland State Government has flagged that tough laws on clearing may be just around the corner. This has meant that trees have been dragged out faster than ever, something John Olsen links directly to poor river health.

Peter Beattie has a dilemma because, even for a Labor premier, in Queensland farmers count. Some, as we'll see in chapter 10, are very powerful, as are farming bodies like Agforce. Farmers have been getting very ticked off about plans to halt clearing, particularly of what they call scrubland. It's the Brazilian argument (you've cut your trees down, what right have you to stop me cutting down mine?), and Peter Beattie makes no bones about demanding roughly $100 million dollars from the Commonwealth as compensation for protecting 30 per cent of Queensland bushland.

All this makes John Olsen furious. 'Look what Queensland's been doing and Peter Beattie said, "I want compensation." Excuse me, Peter! For 15 years we have had constraints on people clearing native vegetation in this State and we have paid $80 million compensation for farmers not to clear. I said to the PM, and I said to all the other premiers including Peter, "Your new found quest for keeping trees in the ground is admirable, but who's going to compensate us, because we've been doing it for 15 years?"'

At this point, John Olsen looked again at his adviser, unusually, a Queenslander!

Holding the purse strings is Federal Environment Minister Robert Hill. There have been numerous spats between the Minister and Peter Beattie, often played out over the airwaves. The day of our interview, though, Robert Hill is conciliatory. 'Other States have said they've adopted legislation to restrict clearing, but they've also already cleared much more than Queensland. In Victoria I think it's about 90 per cent. I think perhaps there is an argument for support in Queensland. The question is how it would work in practice.'

This then is the stand-off. Robert Hill wants areas of concern properly protected and is not moving on the clearing issue, calling on Queensland to clean up its act in the fight for good water quality.[2] But dryland salinity is not yet a big issue up north and trees continue to fall.

Queensland has also been dragging the chain on water. Much to the annoyance of Robert Hill, Queensland refused to agree to the Murray–Darling cap in 1995 and implemented its own water management review, through 12 water allocation management plans, or 'WAMPs' in different river catchments. The Federal Minister believes Queensland is simply not valuing water as it should. 'It's not a challenge in South Australia. It's already a high-profile issue. But in places upstream it's much harder, and even higher up, they're in denial.'

'To the contrary,' responded Peter Beattie, pointing out that his WAMP processes are not only inclusive of the community but the research behind them has received accolades from top scientists like Peter Cullen at the Centre for Freshwater Ecology. 'This research has developed the first model that proves there is a conclusive link between changes in river flows and the ecological condition of a river.'

'Look at the graph, again. Why should Queensland join on the basis of no further development?' Peter Noonan, who now runs the State's water management company, pointed to the State water use graph. He was instrumental in negotiating Queensland's conditional entry into the Murray–Darling Agreement.

Queensland has been rapped hard over the knuckles in the course of the water reform process. A furious Robert Hill let it be known at the World Water Congress in Melbourne in March 2000 that the State was doing its irrigators no favours.[3] 'Queensland still has not accepted a cap, and growth in water diversions, in particular floodplain harvesting continues to rise. Queensland indicated that a cap would be in place once its WAMP process has been finalised. However, that process is now more than two years overdue.' Don Blackmore of the Murray–Darling Basin Commission is equally frustrated. 'They've got to get closure here because quite frankly the other States are getting bloody annoyed with them.'

Fifteen million dollars of the second tranche of payments by the Commonwealth to Queensland for water reform (see chapter 7) was held back for six months, to the delight of ACF's Tim Fisher, who said that act virtually brought progress in new water infrastructure projects to a halt. It did. In mid-2000, the Queensland Government finally put a moratorium on dam building down the Condamine–Balonne

Rivers which flow into the Darling, until it had completed that area's WAMP. Once again, however, as Peter Noonan admitted, the State took its own sweet time, which meant a rush of development happened as panic set in with the irrigation community fearful they would be targeted. 'To me, where we've gone wrong, while we've known five years ago that the WAMP was coming in, the WAMPs took too long, time for opportunists to come in.'

Queensland missed the 1 July 2001 deadline to fall in under the Cap. This was expected to be finally agreed in early August.

It's an old joke that when a new dam is rumoured, you can feel an election coming on. Nowhere is this more the case than in Queensland. From a Green standpoint, however, the saving grace is that Peter Beattie was re-elected this year over the National Party's Rob Borbidge. 'That National Party made the extraordinary promise that it would build 16 dams if re-elected,' said Premier Beattie. 'Labor, on the other hand, promised a continuation of the proper planning processes through the WAMP.'

The Federation albatross

The spats between Queensland, South Australia and the Federal Government will continue. The Northern Territory and the West must be forever grateful of their distance from the Murray–Darling because the interstate bickering points to only one conclusion: there was a big mistake made in 1901.

Before Federation, New South Wales, Victoria and South Australia were all very keen to share the Murray. The question was: what role would the Commonwealth play? Unfortunately, while South Australia was well aware of the power of the States upriver, what it really cared about was river trading. The win for the State was that navigational powers went to the Commonwealth. The disaster with hindsight was that the Constitution left control of water in the hands of the States, providing in section 100 that 'the Commonwealth shall not, by any law or regulation of trade or commerce, abridge the right of a State or of the residents therein to the reasonable use of the waters of rivers for conservation or irrigation'. Justice Peter McClellan (of Sydney crypto inquiry fame) agrees this was a big mistake. 'What we didn't do and should have done at Federation in 1901 was give control of water to the Commonwealth. That was the genesis of the Murray–Darling Basin problem.'

The error soon became obvious. Post-Federation, South Australia was keen for the Commonwealth to support it in disputes with the other States, but the Feds weren't interested. Not until 1914 was the

River Murray Waters Agreement between the three States finally signed. Under the deal, New South Wales and Victoria shared the water measured at Albury, and South Australia was given a set amount (see chapter 5).

The situation was far from perfect, however. Over the years, South Australia became increasingly concerned about irrigation developments beyond its borders. The State erected numerous weirs below Mildura to ensure minimum depth in the river for barges. But there was also the matter of the quality of water by the time it got to Adelaide, which had not been specified in the Agreement. By the 1970s, the city's water had gained its notorious reputation. The State decided to actively intervene in the process of the allocation of licences and management of streams in New South Wales and use whatever legal means there were to do it.

Hired to commence the numerous actions during the 1980s was Peter McClellan. The cases were all stories in themselves but in the end they failed, and what is more sparked an angry political response from New South Wales to the intervention. 'As if to confirm the inadequacy of a Federation to deal with a national problem, the New South Wales Act was amended so that South Australia could no longer be a party to any proceedings involving an application for a water licence in New South Wales.'[4]

South Australia has continued to have problems, most recently with the Swire group's cotton expansions on the Barwon-Darling at Bourke (see chapter 7). Adelaide was livid but had to retreat from its support of a court case against the Bourke developers, because the licences activated were sleepers and perfectly legal.

In September 2000, a Parliamentary select committee in Adelaide raised the possibility of South Australia prosecuting upstream irrigators, particularly 'the New South Wales blokes and the western region'.[5] State water laws can have extraterritorial application, but a direct link needs to be demonstrated between the perpetrator and the damaged party. With a problem like salinity this is very difficult to establish. John Scanlon, former head of the SA Department of Environment and Natural Resources, explained the harsh reality to the committee. 'You would have to ask, "What does South Australia have to gain and what does it have to lose?" It could start to fracture the cooperative relationship that exists between all jurisdictions. The day we have to rely upon the courts to resolve this issue is the day we are all completely done for.'

Tensions can only increase as water is clawed back for the system. John Scanlon's answer is for the Feds to play a stronger role.

Federal impotence

As we have established, the challenge for the Federal Government is that the States control water in this country. It is not that water hasn't been important to the Feds. There are a number of Liberals who put Malcolm Fraser's loss in 1983 down to a decision to call an election during the drought. Suffering in hospital at the time, perhaps his judgment was clouded, but the skies were not. He called the election in March thinking the drought could only get worse but in fact it allowed Bob Hawke to enjoy an extended honeymoon while everyone loosened the belt a bit.

Realistically, however, water is not yet a mainstream issue like law and order, tax or health. Furthermore, like hospitals, the Federal Government can largely wash its hands of this awesome responsibility, because water is within the power and budget of the States. Water is, nevertheless, creeping up the scale of importance and as it does so it becomes more obvious that what is needed, desperately needed, is Federal intervention.

A big mistake was that in its eagerness to push through water reform the Commonwealth did not ensure that any of the $5.5 billion in tranche payments given to the States to help them with the reform process was linked to spending on the water industry. Instead, the cash has gone straight into State coffers, and despite the warnings from the designers of water reform that money would be needed for 'structural adjustment', the dollars are nowhere to be seen.

This is a matter that infuriates Deputy Prime Minister and National Party leader John Anderson, a long-time supporter of reform. 'There's no doubt that it was a mistake not to tie the tranches payments to the States into some form of adjustment package. No doubt in my mind.' The question now is whether the Commonwealth will do anything about it. 'I don't know whether we have much leverage,' John Anderson continued. 'Peter Costello and I met with irrigators recently to discuss water reform. We are very sympathetic and I've now asked the Prime Minister to meet with irrigators. New South Wales and Queensland are the real culprits and have got to be stopped. Unfortunately, I worry that the next set of tranche payments are not sufficiently large enough to provide the leverage we need.'

Federal intervention to assist the Green cause has also had little impact. Robert Hill has the good fortune to be Federal Minister of the Environment at a time when public awareness has given the portfolio a new importance. 'It's made it both easier and some other things more challenging,' said the Minister. 'The community has a better understanding. It wants the Government to act more responsibly. But the more challenging side is that the public is better informed and therefore wants more achieved.'

The Commonwealth had an opportunity to stick its oar into water last year. The *Environment Protection and Biodiversity Conservation Act* (EPBC) passed in July 2000 gives Robert Hill new powers to approve or refuse actions that might harm areas of national environmental significance. Unfortunately, while wetlands come under the six targeted areas, neither water use nor land clearing is included. If only it was, Robert Hill could walk quietly and carry a big stick.

So would the Minister consider it? 'The PM's National Action Plan is really the last attempt at a cooperative model. If this does not work, then whether it's the EPBC or new Commonwealth legislation, or constitutional reform, although that is the hardest change, people will expect us to act unilaterally.' At the moment, Robert Hill's words are more of a soft threat than reality.

Ecologist Peter Cullen for one is sceptical. 'I'm not confident that this Government is likely to take this sort of action. There will be a furore from the States.' Quite right. Peter Beattie's response was that 'Senator Hill seems to be suggesting that this Federal program will fail. It is the uncertainty and lack of vision by the Federal Government that is causing such concern amongst both rural communities and environmentalists.'

The States have already given up some sovereignty over the Murray–Darling Cap. 'John Olsen went as far as to suggest that the Commonwealth should have stronger powers but when questioned seemed to back off a bit,' Robert Hill pointed out. For all the benefits that intervention might bring South Australia, and for all the ties that exist, any major intervention still rattles the nerves on States' rights.

An area that has not yet been tested is the constitution. 'Section 100 gives the States the right to "reasonable use",' said Dr Stuart Blanch from the Australian Conservation Foundation (ACF). 'It's still unclear whether South Australia could claim that States upstream are being unreasonable.' States' rights are messy, however.

There has been one Federal manoeuvre of note. In January 2000, a high level cabinet task force of ministers consisting of John Anderson, Robert Hill, Peter Costello and Warren Truss was set up to 'prioritise' natural resource management with particular focus on the Murray–Darling. The rash of demands to the 'gang of four' came thick and fast: more money to do up infrastructure, more duckwater, stop land clearing, change farming practices, plant trees, recycle . . .

It was the task force that led to the Government's National Action Plan, the $1.4 billion for 20 catchments on the Commonwealth watch list around Australia. But even this initiative has been appallingly handled through the miscommunication with the (Commonwealth) Murray–Darling Basin Commission and its Strategy Management Plan of 21 Basin catchments, John Scanlon observed:

From what I have read, the National Action Plan appears to have been developed in isolation from the Murray–Darling Basin Minister- ial Council and Commission. This is reflected in the absence of any serious role for the Council or Commission in implementing the National Action Plan. One must ask the question—given the long history, success, and high recognition of the Murray–Darling Basin Ini- tiative, why is it being worked around rather than being actively worked with?[6]

As we saw in chapter 5, this bureaucratic *mélange* has created unbelievable confusion over duplication of effort and exactly who is in charge of the process and what the goals should be.

Of course the other key to the success of the National Action Plan will be how the money is handled and on recent experience the Com- monwealth Government has got to pull its socks up.

Where has all the money gone?

Telstra's partial sale in the 1990s was opposed by many. The big carrot put up by the Commonwealth Government was that $1.5 billion of the money raised would go to the Natural Heritage Trust, to be spent on mending Australia's environment. It was billed as a responsible transfer in investment between two long-term assets. At the time of the mid-year review in August 2000, Robert Hill was able to say that the Trust was 'achieving its principle objectives'. Yet even he agrees spending the dollars responsibly was hard work. Questions have been asked about where and how the money was spent and whether the Natural Heritage Trust had become more of a Natural Heritage Trough. But the real criticism was not about pork barrelling. The ACF, for example, is scathing about where the money targeted to Murray–Darling issues ended up. 'For rivers, the NHT funding has been next to hopeless,' said Tim Fisher. 'The Murray–Darling 2001 Initiative put hardly any money into environmental flow. Instead large chunks went to irrigation drainage programs, which in our view are part of the problem. Another chunk went to the repair of Hume Dam, which under the new full cost recovery rules should have been paid for by users, not the NHT. And even if some good was done, you'd never know, the monitoring and reporting was pretty non-exis- tent. In any other area of Government expenditure that would just not be acceptable.'

'Tragically, it's been neither strategically directed, nor issue focussed,' agreed Peter Beattie, 'and it was only four years of funding so projects could only be funded for a maximum of three years. Many natural resource issues can't be dealt with in two or three years, they need a long-term commitment.'

At the grass roots, the message is the same. Take Mayor Dianne Thorley who said dollars were handed over without planning and nothing has been allocated for maintenance. 'If money went to repair of riverbanks and it's not maintained, it's useless, isn't it?'

'The NHT process was harder to deliver than expected at the outset,' admitted Robert Hill. 'The idea was that effective outcomes would need integrated bids supported by regional bodies, but we rarely got them.'

All this is supposed to change with the new stash of $1.4 billion National Action Plan money. 'The National Action Plan sets integrated bids as a prerequisite, whereas the NHT said it would be a goal but continued to fund and to accredit schemes,' explained Robert Hill, adding that the system is now much better placed to be effective. 'If we had had the National Action Plan approach five years ago, the money would never have got to 10,000 schemes, it would have taken too long.'

The challenge is whether we are smart enough to organise, administer and deliver the funding and stop it from being caught up in a war between States or regions. 'I'm frightened about the implementation of it,' Robert Hill confessed.

The Federal Environment Minister seems unfazed by the calls from Peter Garrett and Ian Donges for $65 billion. 'In this area, you can just think of a number and double it. We find it's better to start at the other end and see what can be done.' To be fair, coaxing Green dollars out of a Treasurer and a Treasury which no doubt see the environment as a black hole can't be easy. And there would have been precious little without an asset sale. 'Costello hasn't been so involved,' confirmed Robert Hill, 'but the PM, whenever now you ask him to list the four or five top issues, he always includes salinity in that list.'

Some insight into the Treasurer's attitude comes from Jenny Hawkins, a rice grower from Berrigan: 'Compensation is not a word he knows. We had a women's advisory council meeting with him last year in Canberra. Costello said to us, "Don't think that we're handing you any money. You've got to find your own way out of your salinity, from your own self-funding process." He seemed to have a gilded view that there was some untapped resource in the country, a lifestyle factor. Well Jenny Russell from Blackall said to him, "You don't live in Blackall Mr Costello for lifestyle."'

When the National Action Plan was announced in October 2000, the Prime Minister also flagged funding for an 'NHT mark 2'. What everyone wants to know, however, is where the money will come from, especially if Telstra is not for sale.

The plumber

The Feds do one more thing. Along with the States, they fund the Murray–Darling Basin Commission. A commission to build dams and share water in the Basin has been around since 1917 (the River Murray Commission), and this remained unchanged until 1985.

From 1985 the politicians got involved in the belief that cooperation rather than conflict was the best way to cool the tension between South Australia and upstream States, particularly New South Wales. Up to three ministers from each State and the Commonwealth make up the Murray–Darling Basin Ministerial Council to which the Commission (now named the Murray–Darling Basin Commission) reports. At the World Water Congress in 2000, Robert Hill proudly announced that the 'Commission has been recognised internationally as an example of world's best practice in securing cooperation between different State jurisdictions.'[7]

Credit where credit is due: the Agreement in 1992 which signed the Commonwealth and three southern States up to the 'Initiative', and later Queensland and the ACT, to plan for sustainable water use was an achievement. It was this cooperation that led to the real breakthrough, the Cap on water allocations.

The Chief Executive of the Commission, Don Blackmore, fits the job well. He's an impressive speaker, suave almost. But his main job is plumber for the system, and he has little time for long-term environmental solutions. At the Canberra headquarters, Don Blackmore sits poised to explain his problems. The suave image soon falls away. 'I don't think we have the luxury of talking about the environment out there as if it's going to be self-managed. Somebody else has taken away Australia's right to rainfall. They've done it through greenhouse, northern hemisphere activities. So we don't own our clouds anymore. And I find that to be offensive. So if you're going to tell me that we're going to let the environment alone. No. Bullshit. What we've got to do now—we're into managed outcomes. We have to now be much more intelligent, determine and manage what we want as an outcome, and that's where [Peter] Cullen and others are playing a leading role. What they're challenging us with is whether we are intelligent in making these choices.'

Don Blackmore isn't exactly Mr Popular with irrigators. He promoted the Cap, after all. 'He's a faceless bureaucrat who's undermining the viability of the whole irrigation industry,' is how Robert Caldwell from the Lachlan described him—impossible to pin down, the consummate bureaucrat.

Aside from irrigator antipathy, the Commission has plenty of other weaknesses. It really doesn't have any teeth; it is beholden to bickering State ministers and no one State is responsible for progressing any issue. Moreover, the make-up of the Commission is not a panel of experts, as might be expected, but civil servants, heads of all the State Government water, land and environment departments who are employed on short-term contracts and must get tangled up playing politics.

This is how John Scanlon, a former Commission member himself, explained the set-up to politicians at a State Parliamentary select committee meeting last year: 'If you are head of a department, you are not a politician and not politically aligned, but you cannot help but be aware of the politics of your jurisdiction. If you come back with a decision that flies in the face of where your State wants to go, your minister will say, "What's going on?".'[8] John Scanlon wants to see experts in finance, business management, science and technology, law, engineering, conservation and management of natural resources as well as government on the Commission.[9]

Don Blackmore defends the status quo. 'It's made up of experts. You find me a head of department who isn't, 20 or 30 years, a person of great skill.' How about an expert without a political agenda? 'Ah, well, now you're talking about a different animal. Should there be an independent commission away from government? Well, that wouldn't do anything basically, because government if it didn't like anything would just cut it off.'

River ecologist Peter Cullen feels there's room for change. 'The problem with a Commission which is made up of State representatives is the States effectively have a veto power and it's lowest common denominator thinking. I would certainly like to see some people on that Commission because of their expertise. I think you've still got to take the States along, but I'd prefer some community groups, and I'd prefer some experts on it to widen the decision making and to make it a little less of a secret club so that we all know what goes on. Then people might be a bit more accountable for the way they vote rather than what seems to be remarkably slow movement.'

Luckily, the Ministerial Council does have its own Community Advisory Committee, run by Queensland grazier Leith Boully. John Scanlon reckons it's been rather more dynamic than the Commission, in no small part due to Leith's no-nonsense manner. 'They just say,

"This is it: this is what we reckon. If you don't like it, that's fine. That's what you're elected to do,"' John explained to his politicians.

By far the most serious issue for the Commission is whether the Federal Government will show some of its own 'initiative' and get rid of the Commission altogether. The confusion created by the National Action Plan bypassing the Commission is a veiled threat and Don Blackmore was clearly not impressed. 'The PM said we need a coordinated effort for Australia. My feeling is that it would be a tragedy. What's happening in the Basin is world's best practice. It might be not fast enough, but you're talking about a Commission that has an administrative budget of $5 million for an area as big as France and Spain and you get what you pay for basically in terms of integration. We can run harder —but I don't know that most of us can as it turns out. Look at me . . .'

He looks like anyone who had to confront eleven ministers and nine heads of department on a regular basis.

The bureaucracy

The finger is often pointed at the bureaucracy and, in the case of water issues, the bureaucracy often deserves it. But it should also be stressed that there are many, many men and women within these government agencies who do great work for Australia and receive little acknowledgment.

While the politicians have been slugging it out, a much more fundamental shift has occurred at the department level towards the green end of the spectrum and for many irrigators, it's far too sinister. Gung-ho developers have become eco-friendly civil servants. The intricacies of bureaucratic change, State by State, are avoided here, for fear of boring most well-adjusted people, but some changes are too big to ignore.

Cotton grower Keith Coulton can remember the days when the dams were going up on the Gwydir and Namoi Rivers, and men like Peter Millington from the NSW department were dishing out allocations to dam water like fete lamingtons, dams such as Splitrock, Copeton and Keepit. 'The Keepit Dam, I mean they nearly got on their hands and knees and begged people to take licences and use the bloody things,' recalled Keith.

In the last ten years, the Peter Millingtons of the Department of Land and Water Conservation are now doing other jobs like freelance engineering. 'We still had guys in the department who lived in the communities. It was driven by social and economic policy. Now it's driven by environmental issues,' said Berrigan rice grower Ian Mason.

In the last ten years, the situation has changed radically. 'It was a conscious decision to take out the development management people in the Department of Land and Water Conservation and turn it into a green department,' barked a frustrated Ted Morgan, former chair of the State body, NSW Irrigators. 'All good people were sacked. There is just no balance now whatsoever.'

For environmentalists the cultural change in New South Wales has been a massive victory and long overdue. It has been driven from the top. Premier Bob Carr's campaign to lock up native forests swept the community along with him, on a greening misison.

To NSW irrigators, the water department is now dysfunctional, because it is responsible for both managing and regulating water, answerable to an anti-development government. Ted Morgan, now of Jemalong Irrigation on the Lachlan River, says the department often takes three days to return calls and has diverted people into non- core business activities. 'While the DLWC is making no decisions, they've no interest in maximising the use of water. In fact they've been told to minimise use of resources, leave as much water as possible for the ducks. They should set up a new policy arm, leave water people to run the water.'

Interestingly, this is precisely what Queensland has done. Like New South Wales, the people who 'set the tone' at Queensland's Department of Natural Resources were water engineers and dam builders, some still hanging around from the Bjelke-Petersen days. Last year, the Queensland Government placed department veteran Peter Noonan at the head of Sun Water, the State's new water manager and developer.

One woman not at all impressed by the changes is Jenifer Simpson, who when she's not recycling, is Queensland's vociferous anti-dam campaigner. Jenifer has become somewhat legendary in Sir Humphrey circles, as one admiring grazier confirmed. 'Jenifer loves to attack DNR's dam builders and she gives them heaps, and of course she has such a beautiful accent. You know when an Australian swears at a bureaucrat, it usually means that the bureaucrat's won, but when Jenifer swears at them, she really pulls them in.'

'There's been a big reshuffle at the Department of Natural Resources. Everybody's gone up in the air,' explained the lady herself over lunch in her leafy Keel Mountain home. 'But they come down with different hats on but they're still the same people. They're the old Water Resources Commission and they used to go around with jackboots on kicking everybody out of the way, saying, "Get out of the way, we're going to build a dam here!".'

The rural vote and 'That Woman'

Ask any farmer to name their top politician and the popular answer is sure to be Black Jack McEwan. This extraordinary man was a soldier-settler turned politician for 37 years, including 23 days as Prime Minister following Harold Holt's death. As Country Party leader during the 1960s in coalition with the Liberals, he empowered rural Australia.

How times have changed. And the water issue with them. 'Without a doubt farming has lost power,' said National Farmers Federation President Ian Donges. 'We've been too slow as farmers at coming together at a national level to address national Government's involvement in water.' The fact that Federal agendas now impact the Murray–Darling Basin makes the situation all the more serious. 'The problem is that there are so many disparate groups out there that are representing river systems and various irrigator organisations and there's just no coordinated approach to really tackle the problem.'

National Party President Helen Dickie, based in rice country at Berrigan in New South Wales, agrees. 'It's death to give a local politician the water issue, because you can't win, you can never please people and farmers are the hardest people to please.'

There is a national body, Australian National Council of Irrigation and Drainage (ANCID), which aims to further the interests of water users, but if anything, reform has set rice grower against grape grower. A new lobby group formed in March 2001, the Council of Irrigators, is trying to rally users with the message that what holds them together is far more important than what divides them—their fundamental property right to water. President Laurie Arthur explained, 'Not so long ago, irrigators were soldiers, there was great pride in what we were doing. You read a children's essay these days on irrigation and it says it all.' He admits irrigators have been a bit of a rabble but, he believes it's time to close the doors, spill a little blood on the floor and come out united to capture the sort of rear-guard action created when the Cap was first introduced.

It is fair to say that the person who has done more for rural Australia than almost anyone is Pauline Hanson. Terrifying as she may be for most journalists and many of those who consider themselves well-educated and well-read city folk, 'That Woman' achieved something that Ian Donges and John Anderson could not. When almost one in four Queenslanders voted for Pauline Hanson in the June 1998 election, John Howard went bush.

The Queensland results were a shock for southern critics but it was, after all, Queensland, and stranger things have happened up there. But when One Nation achieved a Federal presence, self-imploded, was deregistered and written off by the media and then came back three years later winning seats in both Western Australia and Queensland, the same critics were dumbfounded. The question before politicians has been 'To preference or not to preference?'

'Pauline Hanson has no policies!' cries the media. It's the skirt vote, or more often, it's just a protest vote, one that must put the wind up John Howard. Really? Or is it just that she's not as sophisticated at articulating policy as some of the old boys? Naturally, the media can't wait to have One Nation policy articulated so they can shoot it down in flames as impractical . . . 'Pauline, have you thought about the ramifications of this . . . well, let's just follow this idea through to its logical conclusion . . .'

Yet at the same time a new undercurrent in global politics is developing. After the anti-free trade protests in Seattle and later Melbourne, Seattle Man has joined Pauline in a bid to take back ownership of his life. Ask yourself how many pure economic rationalists there are these days compared to a decade ago. Earlier in 2001, National leader John Anderson called for a stronger public interest test to be introduced to counter the sometimes brutal effects that competition policy and the free trade push is having in rural Australia. This could be written off as a knee-jerk reaction to the rural protest vote, but there is a growing body of opinion now that questions the value of being a 'loss leader' in areas like free trade—and argues that we could perhaps be a little more selective about free trade, competition policy and economic rationalism. Europe certainly is doing so, as is America.

Pauline Hanson is delighted with the new scepticism. 'They all criticised me when I was talking about how national competition policy and reform and free trade were destroying Australia. Look at what Mr Mahatir said in Malaysia. He said it would ruin his country and he wasn't going to have a part of it. Now I'm not saying we draw a big wall of total protectionism around the country, but we should be much better business people with the products we've got. I'm a small businesswoman, you know you don't sell your product short.'

The man who hears populism every day is Alan Jones. 'The only surprise to me about Pauline Hanson is that the figures don't demonstrate the level of support that she's got. They're sick of it, west of the Great Dividing Range. If they hived off west, she'd win everything. And they say she doesn't have a policy, so that's the way to put her down. Her policy is simply to wind back existing policy.' Transpose that idea to water in Australia and you'd win a lot of votes with irrigators.

So why hasn't Pauline Hanson thrown herself into water? 'I'm not certain she would be so capable of exploiting the water issue,' argued John Anderson. 'A lot of water users have a very sophisticated under-standing of their problems. So what does Pauline say? "I'll give all the farmers back their water".'

'It probably hasn't been brought to her attention yet,' said Ian Donges. 'Ultimately it would be the sort of issue—a simple line or two in there would, of course, appeal to a lot of people who live in those regional electorates, no doubt about that.' But try these for some simple lines from Pauline given to me on the road not far from Townsville. 'In north Queensland there's a company that owns all the groundwater around here and farmers are not allowed to sink bores anymore. They're starting to control the water. There are already reports that are being made under satellite surveillance. I don't want to say too much on specific water issues until I've been down and listened to people on the Murray–Darling, but I am very concerned about water. And look, I care about water and the environment, and I know cotton growing has destroyed a lot of the river, but I think the greenies have too much of a hold, the pendulum has already swung too far.'

Leith Boully, the Murray–Darling Council's Community Advisory Committee chair, believes Pauline is just the tip of the iceberg. 'The question of how best to deal with people who feel dispossessed and socially, economically but not politically marginalised by a changing Australia will become an increasingly difficult challenge for govern-ments: a challenge that will continue long after Pauline Hanson and One Nation have become a distant memory.'[10] Add to that the July 2001 defection of Bob Katter from the National Party and the ongoing pressure that Nats Leader John Anderson is living with in his own seat of Gwydir.

This pressure on leaders is to come up with something more than just listening, as a cornered National Competition Council chief Graeme Samuel warned when he weighed into the brouhaha in Feb-ruary 2001. 'Listening is now viewed as code for doing nothing and affected communities want to see some action beyond flapping ears.'[11] According to Graeme Samuel, politicians should do more to sell the benefits of competition policy. Either that or they should change the laws.

The green vote
'From when I started in the media,' mused Graham Morris, a former adviser to John Howard, 'the environment was a boutique little issue—interested some left-wing radicals and a few people in Sydney

who had nothing better to do. Nowadays the environment is a main-stream issue. It crosses all ages. It goes from kids right through to grandmas and the environment is here to stay as an issue. It is not going to go away and you are going to have to live with it.'[12]

Australia is getting greener. Almost all polls and surveys since the 1980s have seen a sharp increase in concern for the environment—fuelled by the media, rallies and no doubt a more comfortable standard of living and education, which gives us the luxury to think longer term and be more altruistic. By 1990, Labor and the Democrats were including green issues as part of their election platform, and today all parties claim to be keen on Green.

Certainly, the Federal environmental portfolio these days is not what it was ten years ago. It now matters. Robert Hill is well in control and seemingly not the least bit unsettled by those outraged by his policies on uranium and greenhouse. The Natural Heritage Trust money, despite its questionable impact, was a coup for any Environment Minister, but true Greens, like Bob Brown, accuse the Government of greenwashing. 'Robert Hill, he's a great parliamentary performer, perfect for Government. People think he's saying good things about the environment but there's the logging issue and he's put a bandaid on salinity. Yet he has had enormous success in promoting the $1.5 billion of NHT money when the impact of the dollars spent has seen no change in the major environmental indices. Greenwash is the policy of the major parties.'

Senator Bob Brown is Australia's lone Federal Green politician, the man catapulted from prison to politics after the Franklin blockade. He talks with a healthy swig of cynicism. 'Big parties are not going to change, because they can't exist without corporate largesse. As the famous German Green, Petra Kelly, said, "The major parties put on their green spots when in opposition and shed them in government."'

To Bob Brown, what is true of water can be said about all green concerns. He very seriously puts down the country's high suicide rate to the helplessness young people feel today about the environment and not being able to do anything about the planet. 'We believe the environment is supreme. The economy can't exist without the environment, but things can exist the other way around. Yet the dollar reigns supreme and decisions are made in the short term without the next generation in mind.'

As of May 2001 there were in all ten Greens holding seats in parliaments around Australia. Not many really. The truth is that for all the polls, Australians are not passionate yet about being green. David

Farley, former chief at Colly Cotton, believes at some point there will be a vote over the Murray. 'I still remember the Franklin. It was surprising how many people supported it. River health will become part of someone's campaign, you know, "Why should so few be making so much money?" Bob Carr did a similar classic campaign on the national forests.' Until then, though, the real power of green politics is in the ACF, the Wilderness Society, Greenpeace and State conservation groups.

The power of the Rainbow Serpent

To celebrate Federation as 2001 arrived, the Rainbow Serpent lit up on the Sydney Harbour Bridge. In July 2000 perhaps half a million Australians walked across the same bridge in the name of reconciliation. For many Indigenous people this was a worthy exercise, but the harsh facts are that it did little to change the way that some of the big indigenous issues are travelling. One of the most important is land rights, for which also read water rights.

Establishing exactly whether Aboriginal Native title rights exist on land and associated inland water and precisely what these rights are is essential, because it establishes the ground rules for negotiation between indigenous people and others such as developers or miners. So far, two major cases involving inland water rights have been played out in the courts and the findings form a critical precedent for the future. But first, for the 99 per cent of people for whom the current position on Native title is highly confusing, a brief history.

Until Eddie Mabo's watershed case was ruled on in 1992,[13] the law had not recognised that Aboriginal people had existed in Australia before white settlement. This was the principle of *terra nullius*, the empty land, belonging to no one. The judgment went in favour of Mabo's Meriam people, for the first time acknowledging property rights for indigenous people. 'The antecedent rights and interests of the indigenous inhabitant constitute a burden on the title of the Crown.' The Mabo decision was catapulted into the political arena and led to the (Commonwealth) *Native Title Act 1993* which recognises Native title and also provides mechanisms to compensate Aboriginal people where those rights are changed.

The other critical case for the issue of Native title was Wik in 1996,[14] which focussed on the thousands of pastoral leases in Australia and whether pastoralists had exclusive possession. The court said no. On these leases, Native title can coexist with rights under the lease. Only when there's conflict between the two rights would the pastoralist rights prevail.

Such was the resulting confusion that the Coalition won government in March 1996 saying it would 'amend' the *Native Title Act* to make it workable. After much tussling in a Parliament that depended on the vote of Tasmanian independent Brian Harradine to carry the Government, in July 1998 the *Native Title Amendment Ac*t went through with a new schedule, the Prime Minister's Ten Point Plan, attached, significantly threatening the strength of Aboriginal rights.

Point eight of the Plan read, 'The ability of governments to regulate and manage surface and subsurface water, off-shore resources and airspace, and the rights of those with interests under any such regulatory or management regime would be put beyond doubt.' What is happening now is a series of crucial test cases before the courts.

The challenge for indigenous people is to be able to demonstrate Native title still exists on the land in the first place. This has been the big problem for the Yorta Yorta people of the Murray–Darling.[15] These Kooris were the first to lodge a land rights claim following Mabo, in February 1994, in this case over land along the river in the Murray-Goulburn region, where they had lived until the 1840s and within a generation were sent off by white settlers.

According to Dreamtime legend, Biami, the Yorta Yorta creator being, was worried about his wife who had been sent off in search of food. He asked the Rainbow Serpent to look for her and it was the serpent's writhing body which created the river and water holes. After that time, explained spokesperson Monica Morgan, the Yorta Yorta became strong on the bounty of the river and its wetlands.[16]

It seems that as early as 1860, the Yorta Yorta were asking for a share of the river's resources. The Victorian Protection Board noted the intention of one of the clansmen to approach the Governor 'to request him to impose a tax of £10 on each steamer passing up and down the Murray, to be expended in supplying food to the Yorta Yorta people in lieu of fish which have been driven away'.

On 19 December 1998, after a court sitting of 114 days, 201 witnesses and over 11,000 pages of transcript, Justice Olney knocked back the claim on the basis that by the end of the nineteenth century the traditional forebears of today's Yorta Yorta people had been lost to European settlement. 'The tide of history has indeed washed away any real acknowledgment of their traditional laws and any real observance of their traditional customs.'[17]

The decision was appealed, critics outraged that the trial judge had used a 'frozen in time' approach and had failed to recognise key historical evidence of a connection. But in February 2001, Justice Olney's judgment was upheld in the Federal Court.[18]

'We wanted to see which water was available for environmental flows,' explained a bitterly disappointed Monica Morgan. 'That's what we were after, flows and quality of the river. It's a concept the white-fellas find it hard to believe that we don't want to use the water for financial control. We were looking for sales water, but have been told that in no part of the river is there water available. The long-term hope is that in 100 years from now, the river will still be alive and our descendants will be able to drink the water.'

Not surprisingly, Monica Morgan is pretty cynical about the process. 'Any move from the status quo means changes to people who have been consuming water for their own economic gain. Look, the Murray–Darling Basin Commission is not about looking at water from the environment's point of view. It's about managing water and carving it up for commercial use.'

The Yorta Yorta people will probably be going to the High Court, according to Monica Morgan. But even if they could get the ruling overturned, she fears it will be a hollow victory in terms of the water. 'The *Native Title Act* [amended] basically now extinguishes any rights we have to water.' It's one reason why so many today favour a new treaty which would give an opportunity for negotiated outcomes. So far, this too is falling on deaf ears.

The second Native title claim has done a little better. Lodged by the indigenous Miriuwung Gajerrong people of the East Kimberley, Western Australia, in 1994, there was no trouble proving traditional links with the land over which they had claims. This land included some of the most valuable areas of the northwest area including the massive Lake Argyle, the Ord Irrigation Scheme area and land in the Northern Territory.

In November 1998, the Federal Court put the wind up developers and irrigators. Justice Lee decided that Native title existed over more than 7,500 square kilometres and that the traditional owners had rights to 'possess, occupy, use and enjoy' the land and potentially to royalties from resources.

The huge victory for the Miriuwung Gajerrong was short-lived, however. In March 2000, the Full Federal Court agreed that Native title rights existed, but the majority on the bench ruled that these rights had been severely pared back because of big capital projects. Justice Beaumont said, 'The majority holds that Native title has been wholly extinguished in respect of the areas covered by the Ord Irrigation Project and by the Argyle Diamond Project.'

Further, it held that rights to minerals and petroleum were also extinguished. Very significantly, for the first time the Federal Court ruled that Native title could even be extinguished on pastoral leases where land had been improved and enclosed so that Aborigines no longer had access.

It is this new idea of Native title as a 'bundle of rights' which can therefore be partially extinguished that concerned Indigenous leaders. Norman Fry, head of the Northern Lands Council, described this as 'bucket loads of extinguishment by the back door', leaving Aboriginal people with a bundle of hunting and fishing rights on Crown land and a need to go back to the High Court to reinterpret Mabo and Wik so that indigenous people could negotiate commercial agreements with pastoralists and developers. The question lawyers were working overtime on was: are Native title rights now a bundle of rights which can be knocked off one by one, or are they a 'root title', where the tree can be pruned back but the title remains underneath whether pastoral leases come and go?

Needless to say, the case went to the High Court in March 2001. The decision is not expected until late in 2001.

A third and interesting case is happening not far from the Miriuwung Gajerrong, just south of Broome, which is a test case on Aboriginal rights to groundwater in the largest undeveloped sedimentary basin in the country, the Canning Basin. The Karajarri people have made a claim over groundwater in the Great Sandy Desert, living water, or *kurnangkul* as they call it, but the same water is also in the sights of a cotton enterprise, Western Agricultural Industries. Traditional owners understand all too well how the 'on top water' that seeps up into wetlands depends on the 'bottom water' and they fear the consequences of development by the cotton mob.

The company wants to use up to 90 per cent of the groundwater and, later, water from the Fitzroy River, an ambition which has already raised concerns from the Greens about wetlands at Roebuck Bay and Eighty Mile Beach. The West Australian Water and Rivers Commission has to issue licences for cotton, and indigenous lobbyists managed to convince it to investigate the cultural values of the Karajarri, the first work of its kind in the State. Anthropologist Sarah Yu (wife of former Kimberley Land Council CEO Peter Yu) prepared the report, which stresses that while the Karajarri were not anti-development, that 'for the traditional owners living waters are physical evidence of the continuity between the *pukarrikarra* (Dreamtime) and the present, and are so fundamental to the conceptualisation of the country that it is said: without our living water, our country has no meaning.'[19]

Tribal elder John Dudu Nangkiriny has said:

What happens if this goes? What's going to happen underneath? What happens to the roots underneath? To the pukarrikarra underneath? He put everything on top—Nangarrangu [people] and every living thing—but pukarrikarra is underneath. So we worry about underneath. We feel that a big wind is gonna come soon, as the underneath go wrong. This is the place here that belongs to Aboriginal people. We are only talking about Karajarri country. But this is one law from the pukarrikarra that goes right through. One law for Aboriginal people in the Kimberley.[20]

The report recommends the Commission take a bicultural approach to the development. Long-time lobbyist for Indigenous rights Peter Garrett believes the Karajarri have an uphill struggle ahead of them. 'I think they're going to have to arm wrestle with lawyers every step of the way to get an outcome. It will be a long, hard struggle. The Kimberley is different from Victoria, because the land is unimpeded and there are clearly a number of areas where Native title should apply. Their prospects are much better than the south where there is now so much entrenched development.'

The issue of Native title is one of the more complex areas of Australian law, law which is in the process of change. What developments in the Native title arena have done, is to create a climate of uncertainty. This has major ramifications for the nation's water. It is a developer's nightmare, particularly in the northern parts of Australia such as the Kimberley. For the Greens, however, the confusion from Native title provides a welcome breathing space for the environment.

'I took the strongest position on indigenous rights. We're the strongest critics of the disempowerment of Aboriginal people,' said a passionate Bob Brown. 'The whole idea of looking people in the eye about future generations comes from the indigenous people.'

One thing is certain. Future generations can only become greener.

10

Water Wars

You may think you know what you're dealing with, but believe me you don't.
Noah Cross in Roman Polanski's Chinatown

Cotton tales

There is no crop so controversial as cotton. With the water it uses, the chemicals it needs and the money it can make for growers (and the country), cotton has become a political hot potato.

Cotton has also turned farmer against grazier and farmer against farmer, as we got a taste of in chapters 4 and 6. When cotton was first introduced in Australia, it marked a social divide between the waning squattocracy and the crop grower, the nouveau riche. This is something many on the land will not talk about.

It should be no surprise that two of Australia's biggest water fights have involved cotton. The first, which took place in New South Wales, was a bitter fight which started in the 1970s, led to a full government inquiry and turned two close families into lifelong enemies. The second in Queensland is the story of a powerful fiefdom which transformed two towns and is still one of the most vilified agricultural businesses in the country. What both tales provide is an extraordinary insight into just how far people are prepared to go when water gets short, and why government and the bureaucracy should be ultimately responsible. It was Cotton Australia's former chief executive, Gary Punch, who observed:

> Water disputes are also certainly not about readily understood issues like pay and conditions. They involve the need for high science but are often characterised by a dearth of the same, or worse, where limited data on river flows, water quality and economic effects is held by the bureaucracy, it is treated with a level of secrecy the likes of which I have only ever seen once—when I

was the Minister in charge of Australia's Defence, Science and Technology.[1]

The Whalan Creek inquiry

Whalan Creek is not much of a creek at all. Lying just south of the Queensland—New South Wales border, it is more a string of lagoons that join up if the Macintyre River swells enough to spill water into it. Quite how much it used to flood is part of a long running dispute. Yet this tiny creek occupied Sir Laurence Street and his State inquiry for several years. In fact, Whalan has been a 30-year saga.

The two men at the centre of the inquiry were Alec Holcombe, a long-time grazier, and Keith Coulton, the region's pioneer in cotton. Both are old men now, but neither has recovered from the events of Whalan Creek.

Alec Holcombe's family from Polatingle, north of Warialda, has been in the district for 110 years. 'My grandfather came up in 1890 and selected 2,560 acres, very small selection on the Whalan water course, and he built that up to 20,000 acres by the time he died in 1921. He left everything in credit in 1921 with three daughters. And my mother inherited the homestead and got what is still 6,500 acres. Now that country is a bit like Topsy, when it is good, it's very, very good and when it is bad, it's horrid—because it relies on floods very much. My eldest son Hamish lives there now.'

Alec is charming, a traditional gentleman grazier. He insisted that his mother leave the land to his sisters and bought a bit of rough country near Yetman, and managed to make enough of a go to buy back the land from his sisters and from outside the family.

Keith Coulton's father, Alfred E. Coulton, bought land at North Star, not far from the Holcombes in 1924, naming the property Getta Getta. He died when Keith was just 17. In 1949, at a time when grazing sheep and cattle was the only livelihood any white farmer had ever had in the region, the young Keith planted 300 acres of wheat. By 1953, he was the outright owner of Getta Getta.

Keith Coulton is a battler in every sense of the word, with a larrikin charm. His interest in water started in the mid-1950s and he built his first bore on the property with an old auger and about 70 feet of piping, slowly working his way down into the ground. When the locals passed by he used to pretend to be picking up sticks, 'because I reckon they'd put me in a lunatic asylum. But I found water, good water.'

The old days were fixed in old ways. For Keith Coulton there were obstacles at every turn. As he reminisces, it's clear that it was the pluck and cheek of a new kid on the block that kept him going. 'When

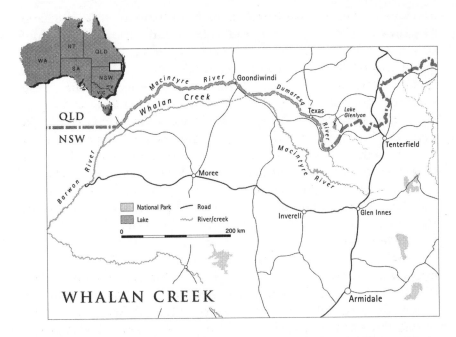

I wanted to start wheat farming—my bank head-office was in Sydney—I went down to see them. They told me that if I wanted to increase my sheep and put up fences for rabbits, they'd be quite happy to finance me, but they couldn't see their way clear to finance me into wheat farming. Then when I wanted to irrigate, the banks told me the same thing. So I used to borrow money from the banks to put up a fence, and then I'd bloody well get some irrigation equipment. That's what I had to do. And they used to come down and inspect, and I used to show them the same piece of bloody fence every time they came down!' He chuckles and looks across at the family birds from the deck of Getta Getta. Coulton's training was to set him up well for the battles ahead.

Keith's irrigation soon took off. By 1965 there were 800 acres of hay, which drought-proofed the property, and in 1977 he planted the first cotton crop on the Macintyre River, of 300 acres. By the 1970s, the water authorities had caught up with Keith Coulton and were encouraging other farmers to get licences for irrigation. Keith had bought more property and had formed a syndicate with four other farmers. They built about 9 kilometres of channels to water the five properties.

To make water available, in 1969 the NSW Government built the Pindari Dam, on the Severn River and the Macintyre system, a 37,000 meg storage. A bigger dam, Glenlyon on the Border River (running along the State border) to be shared by New South Wales and Queensland irrigators was to be built between 1972 and 1976. The Severn River flows from Pindari Dam into the Macintyre which runs to the New South Wales–Queensland border, where it is joined by the Dumaresq River just upstream from Boggabilla. The Whalan Creek is an anabranch of the Macintyre. Together, the Dumaresq, Macintyre and Barwon are known as the Border rivers.

With the prospect of these dams coming on stream, irrigators like Keith Coulton spent up big, developing their land. Unfortunately, it turned out that Glenlyon failed to deliver anything like the amount of water that irrigators had been promised. Keith reckoned it to be just 20 per cent of their allocation.

There are many conflicting views in this story, but the first is Alec Holcombe's view that Keith Coulton was never entitled to Pindari water. Keith adamantly disagrees. He founded the first water users' organisation which covered both sides of the border. 'The NSW Commission asked me to set up a council that could handle the NSW side of a regulated stream from Pindari, Glenlyon down to Mungindi. So I formed this and called it the NSW Border Rivers Council.'

'Keith's licence was set from Glenlyon dam only,' insisted Alec. 'The conditions were that he would not try to get water from Pindari. Pindari was a separate NSW [dam], and nothing to do with the Border rivers. But Glenlyon was showing that it was not going to have the supply that they thought it would. It was down to billyo. This frightened Keith tremendously and he had it set on getting the Pindari water into the Border rivers.'

It's likely that both men are right. The culprit was the then NSW Water Resources Commission, first for vastly overestimating Glenlyon, and secondly, at a time when Holcombe and other graziers were not yet irrigating, the Commission allowed Pindari water to go to the Border rivers irrigators, which slowly became the norm.

By the mid-1960s, it had become clear to graziers down the Whalan Creek who had stayed away from crop growing that irrigation was where the money was being made.

Keith Coulton described what happened next: 'A friend of mine, Alec Holcombe—when I say a friend, a good friend; we'd go to their place, and they'd come here, we'd drink more than we should have done. He had country down the Whalan Creek and he wanted to get into irrigation. And this was quite early in the piece. And even though he used to call me a Chinese gardener, I went in and had a look at what they proposed to do with a group of people.'

Alec Holcombe too, describes Keith as a friend, but the friendship was even stronger at a family level. 'It can't be underrated, the relationship that there was between the two families,' says Alec's son Nick, 'because we had four boys in each family, and a girl each and the boys, my mates were Dave and Mark, Charles' mates were Ben, and there were nights at home with Keith playing the piano and Mum singing, or that sort of thing. There was great camaraderie, two great families, great friends. Dave was my best mate at school, we made a pact we'd be best mates for life and then this sort of thing comes along and it makes it pretty hard to be best mates.'

By the 1970s, the graziers had decided they wanted a share in the water. The problem was they had no water licences. The Whalan Water Users Association, which had been formed in 1968, finally sprang into action and in 1979 Alec Holcombe was elected president.

The Whalan mob were after licences for water to be pumped from the Macintyre, before it reached the Border River, into Whalan Creek, both to irrigate and to offset the loss of flood water that the river used to bring before the dams went in.

'I knew nothing about irrigation and neither did the secretary,' admitted Alec Holcombe. 'So I was talking to one of the very early irrigators on the Gwydir and he advised me to go and see Alan McLaughlin at the Commission . . . And I said, well the point is, these dams are taking our gentle floods, even though we get the disastrous ones. He said, "That's a story we hear all over the place." That's from the executive head of the Comission.'

A meeting was held with at least four top jocks at the Commission, including the licencing officer, who actually filled in the Whalan Users application for a licence from Pindari Dam on behalf of 56 people. A licence entitled the holder to 976 megs, so when the subject of the amount of water came up, according to Alec Holcombe, 'Commissioner Mead said, "Oh you better put down 50,000 megs a year, that's what they'd need and we've got that available." And that was our application.'

The Whalan application wasn't the only one, however. With Glenlyon woefully overallocated, Keith Coulton and some of his irrigator friends had lodged 14 applications for licences to Pindari water at about the same time. The Commission had already advertised them in the Goondiwindi *Argus*, which was the due process, to see if there were any objections. If anyone does object, the issue is put before the local land board for arbitration.

The Whalan application had also been lodged, but had yet to be advertised. Slightly worried, Alec Holcombe decided the best plan was to object formally to the 14 applications, and he spoke to Keith to explain why. 'I rang him and said I've lodged an objection and I want to tell you we won't go through with it. I was objecting to the Commission not advertising ours.'

However, when the Holcombe Whalan Users application was eventually advertised for 50,000 megs, it was far too much for Keith Coulton. Keith was by then chairing a council that represented water users right the way down to Mungindi. Allocations for him and the other users had already been wound back significantly. Licences had also been changed from area licences of 400 acres to volumetric licences of 972 megs a year which he reckoned worked out at around 6 megs to the hectare, very tight for growing cotton. Keith objected. Increasingly worried, Alec again rang Keith. 'I said, "Keith, I see you've objected to ours." He said, "That's right." I said, "Well I trust that you'll withdraw it. I did for yours." He said, "I might." There was no assurance. I rang up later on. I said, "Look Keith, I think we better have a talk about this."'

The two men met with a few others and Alec Holcombe told Keith Coulton he was expecting a report from the Commission showing figures on the Pindari Dam that would support his application. According to Alec, he later rang Keith to confirm the report was supportive. 'His remark was, "Don't you try to intimidate me." I said, "Oh." So I rang up another one, quite a nice bloke and he said, "Don't you try to intimidate me," and slammed the receiver down. Then I rang another bloke and he said, "Don't you try to intimidate me." Well, it was that obvious and Keith was the one who had said to say it.'

These are not Keith Coulton's recollections. 'Anyway we objected to it,' he agrees, 'and of course that was the end of the budding friendship, friendship that I had had with Alec Holcombe for many years, and his kids and all the bloody rest of it. I don't bear him anything. This is the first time I've even discussed it.'

The two sides were scheduled to go before the land board to decide upon whether the Whalan Water Users application should proceed, but a critical event happened before the hearing. The Commission completed a stocktake on Pindari Dam water and again it became clear that there wouldn't be enough to go around. As a result, the NSW Government in August 1982 announced a blanket embargo on all new allocations on the Macintyre and its sister river the Severn. Notwithstanding this, the Commission's attitude remained that 50,000 megs a year might be allocated to the Whalan scheme! By the end of the year, it had reportedly revised this to 26,000 megs a year.[2] Deputy Premier at the time and local member Wal Murray in support of the Whalan Water Users, emphasised that 'on 5 May 1981, the Commission made it absolutely clear that the Whalan Water Users Association was a beneficiary of Pindari Dam and there was sufficient water in the dam to meet the 26,000 megs allocation'.[3]

With 972 megs the allocation for one licence, and only 26,000 megs available, in the end, the group had to shrink from 56 to 29 applicants. But Alec Holcombe has always maintained it was a community project, family farmers.

The Commission was in no position to make promises, as Keith Coulton points out. 'I do think that the Department at that stage thought they were Jesus Christ. I think they probably said, "We'll go along with 26,000." They could never officially say that.'

To this day, Keith Coulton maintains that Alec Holcombe brought the water problems on himself. 'Twenty-six thousand megs is a hell of a lot of water in one hit. Now if Alec had of got in, and this was the truth of it all, if Alec had of got off his butt, and got the bloody job done, he'd have had his 26,000 megs of water. And I would have not quibbled about it, and would have been arguing about the Munroes, or someone else, you understand. He could see that everybody else was going mad on licences. Why didn't he get it done? He didn't. He just lost his place in the queue, and he must have known surely to Christ that it was getting tough out there. The only reason I opposed the Whalan was because I was the chairman of the NSW Border Rivers Council. It was my job to turn around and say, "Hey, that's enough."'

It is now that a familiar figure should be introduced. Even today, players in the water game are relatively few. The young man who would represent Keith Coulton at the land board hearing was none other than now Justice Peter McClellan, who was later the QC to run the Sydney crypto inquiry. The Whalan case still fascinates him. 'The Whalan inquiry is about water, yes, but the real story is a social story. Keith Coulton, as you can imagine, was an energetic, driven young person. The reason he's deaf today, by the way, is that he used to drive bulldozers without ear plugs. But he was what you would call "new money" and determined to make a dollar. Coulton was not a member of the Moree club, you know, didn't wear tweed jackets to the polo, didn't send his kids to Kings. Coulton borrowed money from the bank, a lot of money. It's risky, you know about cotton, the money's all up front, but Coulton and his like developed the land and made money. The Holcombes of this world said, "We're graziers, we don't take risks." Everyone had the same opportunity, but it became evident that those who were taking it up became very successful.'

Alec's son Nick strongly disagrees with Peter McClellan. 'Keith sent his children to the prestigious schools of NEGs and TAS. I think

Keith went to TAS too. Keith was a polo player who rubbed shoulders with royalty complete with tweed coats, he was a member of the same club as Alec in Goondiwindi. To say Alec doesn't take risks is untrue, Alec at 83 is still borrowing many millions to expand his land base and would have borrowed whatever was required had we had water from our creek to apply to and develop our land.'

Nevertheless, a divide of some sort is in both men's memories.

According to Alec, 'Keith, very capable in so many ways, wasn't a nuts and bolts man. He'd go over the top, he wouldn't open his mail, that's just the way he was. He was one of the first to start irrigating out of bores. He was innovative, but politically he got great support from Labor, he was on advisory panels for Wran.'

For Keith, 'This was sheep country and it wasn't a social thing to do to be a farmer. And then all of a sudden as time went on, you'd go into town and there'd be the graziers down one end of the bar and the wheat cockies would be up the other end of the bar, and eventually the graziers thought, well, we'd better go and find out what these fellas are doing, so they'd say, "My son wants to muck around and do a bit of farming." And then all of a sudden it became the social thing to do to be a farmer.'

The first fisticuffs

The first week of the land board meeting finally commenced in October 1983. Emotions were clearly tense. Of his meeting with Keith, Alec Holcombe said, 'He came across to shake hands and I declined. He claims that I was the one that ended things, and I think he had done it before that.'

Interestingly, during the hearing, Keith remembers that the Whalan group's lawyer cut his farmers to ribbons. 'He was saying, well they only need half the water because they can rotate crops. He wanted to know how much money they were making. Now there's nothing wrong with making a profit, it wasn't an offence. And of course what they were getting at is that they should have a share in all this water, because we were making money and they should be getting something.'

This was the first time the concept of fairness and equity was raised.

Alex agrees with Keith's assessment of his QC: 'I thought he was hopeless, terribly self-important. And when we went to Moree, he antagonised quite useful witnesses. I mean, he treated them as though they were on murder charges—people there that I knew.'

The second week of the land board meeting commenced at Warialda on 20 February 1984. Alec Holcombe shook his head thinking back. 'Even in that week, Coulton's people beat us to book out the motel at Warialda. We had to stay in Moree, they beat us on that!' It was a sign of what was to come.

Peter McClellan recalled a critical conversation from the weekend before: 'On the Saturday night at a barbecue, I worked out the point we'd make with my junior. On the Sunday, I said to Coulton we can run with this point and if the other side are clever, they'll match it, but if not, we'll win.'

What happened was a disaster for Alec Holcombe's lot. It turned out that the Whalan group had filled in the wrong form in its application for a licence, the mistake being that a Section 10 licence application (for water supply to one person) had been filled in, instead of a Section 20 (a joint water supply application). So even though the application had gone in before the embargo on new licences, it was actually invalid.

The meeting lasted less than a morning. Alec Holcombe's lawyers weren't clever enough to find a way around the problem. The magistrate said he was not in a position to be able to do anything, and the case was referred to the Supreme Court.[4]

'I remember the CWA and everyone at Warialda,' said Keith, 'They were there in droves and they had tables all laid out and cups and bloody scones and God knows what, for this week-long bloody marathon and we finished by lunchtime. I don't think the CWA ever forgave me.'

The report from the Commission that Alec Holcombe had pinned his hopes on and mentioned to Keith in the earlier phone call, was never presented at the land board. 'It was never properly investigated by the people that were supposed to do it,' he said.

Off to court

The farce continued. In the Supreme Court, the judge took the view that the application failed because the Whalan group had only worked out that there was a problem with the form *after* the embargo. The Commission had organised for the Whalan users to sign an amended licence, after the embargo, assuring Alec Holcombe that the application would be treated as pre-dating the embargo. The judge disagreed and saw the amendment as a separate licence to the embargo.[5]

By the time the appeal arrived Alec Holcombe had replaced his counsel with Bob Ellicott, later attorney-general in the Fraser Government. At this point, Alec Holcombe's legal team caught up with Peter McClellan. For he had known that the actual embargo on licences was also technically invalid! Under State law, the embargo was to be enforced only when the total amount of water allocated under irrigators' licences exceeded a set level. The mistake made was that the calculation which took the water allocation over the limit, included the Whalan Water Users 26,000 megs. And that 26,000 megs wasn't licenced at all—it was just an application and the limit was not breached.

'The embargo was triggered by our application, as though it had been granted!' exclaimed Alec. 'In law, they could impose an embargo if any more applications would exceed the supply in the dam. Our application was taken into consideration and the embargo itself was illegal. That was a winning argument!'

Quickly, the Whalan team tried to have the invalidity of the embargo raised when the case was appealed in the Court of Appeal. The court said it was prepared to consider the issue, but because this was deemed new evidence, the decision was then taken to the High Court by the Coulton team to determine whether it could be considered at all.

According to Keith Coulton, he approached a couple of distinguished barristers to take the case up to the High Court.[6] In the end Roger Giles took on the case. The High Court set about pondering whether the validity of the embargo could be pursued in the Appellate Court. In 1986, the High Court said no. Keith Coulton had won.

Interestingly, the dissenting judgment came from William Deane, later Governor-General, who, like the Court of Appeal, attempted to look at the fairness (or lack thereof) of the situation.

'He [Sir William] said it should never have come to the High Court. His judgment was about 14 pages, much bigger than the majority,' Holcombe stressed. Justice Deane also pointed out that the embargo issue could be challenged in another court. This duly happened in 1990.[7] A final irony was that the Judge in this case found that the embargo on new water licences was indeed invalid, but unluckily for Alec Holcombe and the Whalan Creek Users, he also found that the *1986 Water (Amendment) Act* had given the embargo retrospective validity! 'It was Catch 22,' said Alec.

The years of litigation did a great deal for lawyers' pockets, but nothing to resolve whether or not the Whalan Creek boys should get their 26,000 megs. Alec Holcombe and his mates set about lobbying hard within the Coalition parties to get new legislation through—Ian Causley, John Fahey, Wal Murray. 'And then Wal said, "Look we've got something coming up, you'll be surprised"—that was the announcement of the inquiry.'

The Street report
Veteran mediator Sir Laurence Street, former chief justice of New South Wales, was brought in to head the inquiry which sought a solution to the Whalan Creek dispute 'that is fair and just to all parties and that gives full weight to the consideration of public interest'.[8] It began in December 1993. The media tagged along with the inquiry, travelling to Goondiwindi, the nearest large town to the Whalan Creek, and back again to Sydney. Sir Laurence created a task force of his own advisors, bureaucrats, and two members from each opposing side: Keith Coulton and NSW Irrigators chief Gary Donovan, and Alec Holcombe and local grazier Ian Uebergang.

'It was going to be finished in a fortnight,' said Keith. 'He came up here in January when all the judges were away and all the lawyers went up to Noosa. I think it was nearly three years before it finished. He had a full staff at the Premier's Department, the Department of Water Resources [formerly the Commission] virtually did little else, they were all shit scared, and you take all the records and all the stuff they had to get out, it cost us the best part of three-quarters of a million dollars. It cost the opposition three-quarters of a million dollars. I suppose all in all you could say it probably cost ten million dollars.'

Both teams worked hard. The inquiry was wide-ranging, even examining efficiencies. The Holcombes had a visit from the National Parks and Wildlife's Dr Richard Kingsford, who they say treated them with 'validity and credence' because of the Whalan Creek Users plans to only put around 15 per cent of their land under irrigation. But they still had to fight the Border River group's claim that pumping water down the creek was in itself wasteful.

One of the crucial arguments concerned what the original flooding patterns were down the Whalan and what had been lost to the graziers. Alec Holcombe's son Nick, who happened to be an engineer, was brought in and with the help of his old professor, hydrology expert Ian Cordery, he analysed flows using data collected since 1890 to refute the view of the Department of Water Resources that there was never very much water down the Whalan anyway. 'The Department didn't accept it, but Sir Laurence did,' said Nick. 'Sir Laurence said that we were probably right, he would err on the side of the Whalan in terms of the denial of flows and we won that battle but lost the war.'

It took until March 1996 for Sir Laurence Street to report. But by then, as Peter McClellan pointed out, Whalan Creek and the whole of the Macintyre had changed dramatically.[9] Since the 1980s, water licences have been predominantly used for cotton growing. The years had shown allocations were serious overestimates, on-farm storage of water was growing and water trading had picked up any slack. Most importantly, the Greens were now demanding their share for the river itself. Laurence Street had apparently asked Gary Donovan, 'You're a mediator, what do you think?' The response was, 'I don't think there's any water.'

That was the only conclusion Sir Laurence could draw. To add to everything, the decision was delivered at a time of drought. Alec Holcombe and his family are bitter, but particularly about what they believe drove the decision, which was an eleventh-hour letter from the Director-General of the Environment Protection Authority. 'In my view he was worried that there wasn't enough to get down the Darling and the Murray,' said Alec. Sir Laurence found that the process of litigation had

> . . . brought about a result which can be characterised as manifestly unfair to the users. The train has now left the station and, whereas others were granted licences virtually for the asking ten or fifteen years ago, there now exists a situation in which the combined operation of the perceived gross over-licensing of the Border rivers system and the incontrovertible needs of the environment leaves no room for establishing a new irrigation scheme such as a Whalan user envisages.

In the end, both Alec Holcombe and Keith Coulton were paid their full costs for the inquiry, but the Whalan Water Users lost out. There was no compensation—after all, how do you compensate people for what they might or might not have made out of irrigation? Some go broke. The decision was a watershed. New South Wales, a State which had been pro-development for over a century, had ended a dispute without allocating water. Even Peter McClellan admitted, 'The case was won on a technicality. I understand the great sense of injustice rightly held by the Holcombe people.'

Dr Geoff Syme from CSIRO's Research Centre for Water in Society has done more analysis of the ideas of fairness in water use than perhaps anyone in the country.

> Despite uncertainty and the quality of the decision making at the time, the reversal of 'first come' or 'first used' precedence is hard to contemplate, regardless of the 'fairness' of this situation. But where does this leave the recent and almost arbitrarily defined 'missed outs'? Are fairness studies and justice criteria too esoteric? Should we 'pay' only those who take the financial risk to drink first from the trough, even if the size is uncertain?[10]

To his friends and fellow irrigators, Keith Coulton is a bit of a legend who fought hard for water for the district: the go-getta from Getta Getta who never takes no for an answer. In 1979, he received an OMBE for service to the community.

Alec Holcombe too, as his son Nick describes him, was 'a really constructive member of the community, fought for Australia, loves Australia, without being too emotional, has been a tremendous pillar in society, done the right thing by his family, doing all the wonderful old-fashioned things to develop the country in a friendly way and the country gave him a great big kick in the face.'

Ten years later, Keith Coulton drove a tenfold expansion at Pindari Dam for irrigators in the district. It was a joint venture between irrigators and government, but this time, he made sure that the contract with government was literally 'watertight'. 'It stopped any more licences being issued until they could show that the reliability factor gets past 70 per cent, and that's going to be a bloody long time, because it hasn't all come up to expectations, and it would be, probably be, around 50 to 60 per cent reliability, which is a bloody lot better than 20 per cent just the same.'

But even this new expansion was controversial in who it helped. Keith Coulton said his wife had suggested to Mrs Holcombe that Alec put some money into a larger Pindari Dam, but that Alec Holcombe had refused. Alec and his family insist they were never approached, and yet again missed out. 'The Coultons said after they managed to get the water out of the first bit of the Pindari Dam, "there's a big enlargement coming",' explained Nick Holcombe. 'This was the conversation around the district, therefore when it goes from 37 to 370,000 there'll be more than enough for everybody, so don't get too upset about these court cases, that water will be yours obviously. And then of course when the next lot of water came on line, they were just as vehement they were going to have that as the first instance.'

Wal Murray, still a Whalan supporter, attacked the watertight Pindari contract as a disgrace, precisely because it gave security to irrigators—security at the expense of others who missed out. He labelled the Department and the Government involved as 'at best unconscionable.'[11] Keith Coulton maintains that Wal Murray was present at the sod turning ceremony for the expanded Pindari Dam. The bitterness continues.

'We don't talk now,' said Keith. Alec still has practically no water.

Today, the Coulton family controls over 30,000 acres of land in the Boggabilla, Yalamar, North Star and Moree districts, employing between 50 and 100 people depending on the time of year. Harvard

University has ranked it in the world's top 20 privately operated cotton producers, although a foray into organic cotton practically broke the business, costing them over $4 million.

The Holcombe family properties have two licences but Alec says development of the land to use the licence is difficult in the area. 'The ultimate irony,' said Nick, 'is that my brother lives on the country we were hoping to irrigate, and is now surrounded by other irrigators who are using the water which was initially earmarked for the Whalan scheme to irrigate country.' The irrigators come from the Border River. According to Nick, the Department justified providing these irrigators with licences because it believed they would be irrigating on already developed country and not in the virgin Whalan area.

Keith Coulton does not stand for talk of inequities. Perhaps he feels privately he deserves the last laugh after all the ridicule of the early years, or that he has done enough for many irrigators in the district and a Johnny-come-lately cannot be helped. 'Whatever you did had to be good for everybody—that's what we've done, tried to do,' he said, and when pushed on the unfairness of Whalan replied, 'It's like saying, "Well, look I haven't got nearly as much country as you. Why can't I have some of your country?" Really, we're talking virtual bloody communism.'

Whatever Keith has done, he has ingrained the value of water into his boys. 'Our forefathers split up the land—and we split up the water,' explains Ben, one of the younger sons. Sam, the eldest, is a chip off the old block, although tougher. 'We are going to fight for that water until I die. We're not going to give it away, it's our lifeblood, and I expect the graziers will say that as well.'

Cotton Queensland style

The downsizing of rural towns has been given much coverage both by the media and at the parliamentary dispatch box. But this has done little to stop the rot of bank and post office closures. Many outback towns are now teetering between 'blink and you miss it' and gone altogether. No so, two towns next door to each other in southwest Queensland (next door being 100 kilometres away). St George and Dirranbandi have been booming, and the reason is cotton.

This second cotton tale is about the controversial growth of these two towns, and the very powerful cotton growers behind them. And it's about a politically charged ten-year war over water—not just between cotton growers, but between government, graziers and the Greens as well. A war which continues today.

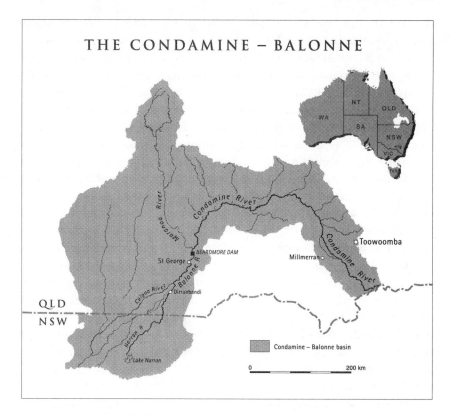

THE CONDAMINE – BALONNE

Importantly, the location is Queensland, the State that is being dragged kicking and screaming to the table on Murray–Darling Basin water restrictions. According to Federal Environment Minister Robert Hill, 'Queensland has achieved extractions from that system in the last ten years that has taken the rest of Australia 200 years in the lower end of the Murray–Darling Basin and it's just not good enough.' The events in St George and Dirranbandi play a big part in influencing the State's behaviour.

Unluckily for Queensland, it has one main river system that forms the beginning of the Darling River. The Condamine–Balonne runs from Toowoomba, just west of Brisbane, 1,350 kilometres southwest through the State and across the NSW border. And as the State's head water bureaucrat, Peter Noonan will admit, in terms of responsible development in the last five years, Queensland has gone from being the best to the worst in the Basin.

The little people

The St George–Dirranbandi country lies about six hours' drive inland from the Queensland coast, just north of the New South Wales border.

For the first half of last century it was sheep grazing country, but as the wool boom faded, so did the towns.

Between the 1950s and the 1970s, in an attempt to diversify, the Queensland Government sold land plots of between 100 and 250 hectares to about 40 farmers to grow cotton, but the blocks did not have river frontage and lacked water. The pro-development National Party Government built the Beardmore Dam on the Balonne River, just north of St George, with the promise of a water supply of 88 per cent reliability, delivered through a system of channels for the land-locked farmers who began producing cotton. At around 80,000 megs, the Government storage wasn't big, but during the 1980s it was adequate, holding water from the floods of February and March to be used in August as pre-crop irrigation for cotton planting, which took place in the first week of October.

Over the last ten years, however, life for the channel growers, as they are called, has gone from manageable to catastrophic. As with the Holcombe–Coulton story, the performance of the bureaucracy played a major part. In 1989 the Department of Natural Resources in its wisdom decided that there was enough water for extra entitlements to be issued to extract water from the dam itself and the river system. Since then, experts have reassessed the capacity of the dam at just 80 per cent of its design capacity. But worse was to come for the channel growers.

By the early 1990s, the Department came up with a new rule: anyone whose property had access to the river was allowed to apply for new rights to harvest water from it during floods. Lots of water. The idea was that this would not interfere with the land-locked channel growers, because the harvesters were required to put up their own private dams and could not take water from the public storage. But in reality, harvesters were allowed by the Department to 'park' some of their water in the Beardmore Dam, taking up valuable storage space. Overall, the channel growers say their reliability has sunk to as little as 15 per cent of their licence entitlement—the equivalent of one useless watering for the year.

The bitter battle between these two sets of predominantly cotton growers has been going on for almost ten years, dividing a community and driving channel growers off their land. Ken Pierce, a former Department employee hired by the channel growers, is 'PNG' amongst the harvesters. He spends his life trying to get the plight of the growers noticed by mainstream media, but despite the inequities, events have overtaken him.

For the harvesters, the battle is ancient history. Possession is nine-tenths of the law and they have won. They point to the fact that several of the channel growers now also have harvesting licences anyway (and no doubt they paid highly for them). But there is a much bigger battle these harvesters face which makes the channel grower dispute look like a playpen squabble.

The water harvesters

The average annual flow of the Balonne through St George is around 1.1 million megs. In some years, the flow can be 8.5 million megs with 250,000 megs a day passing St George during a flood, and yet in others, it can get down to just 85,000 megs in a whole year. (That low level one year led to the creation of the channel system, according to Peter Noonan, who spent many years working at the Department of National Resources in St George.)

It was this incredible variability that led a very pro-development Queensland Government in the 1980s and 1990s to support whole-heartedly dramatic development of private dams all along the lower Balonne River. At St George and Dirranbandi, monster storages cover around 40,000 hectares, more than half built in the last three years, all sucking water from the lower Balonne River, at the top of the Murray–Darling. Fifteen years ago, farmers talked about 6-inch pumps, now they have 26-inch pumps capable of moving megalitres at great speed.

In central-south Queensland, dams are a lifesaver. Farmers can go three or more years without a decent dump of rain. But to make the most of the licences given to them, a couple of harvesters have built dams so big that they cater for the flood that comes once in ten years. In other words, these dams are only full one in ten years, but can take a massive amount of floodwater. Critics of the harvesters say the dams were never, ever, meant to be built as big and it was never intended that these water licences would be developed to this degree. In Keith Coulton cotton country on the Macintyre, for example, dams are built to take a one-in-three-year flood.

So how does water harvesting work? The idea behind these licences is that more are activated as the river flows gets bigger. Actual numbers help in understanding.

Clearly what is sitting in the Beardmore Dam is largely for the channel growers. However, once the rains have come and water starts flowing over the dam, this flow is carved up into licences. The first 750 megs a day to flow down the river is for stock and domestic purposes. When the flows reach 1,200 megs a day, harvester licences start being activated, giving the holder the right to suck out 86 megs a day to store in their own dams. As the flow levels get higher, more licences come into operation. A typical water harvester might have bought one of the few licences at quite low flow levels, say 2,000 megs a day and a few more for when the river reaches 8,000 megs. What is critical to remember here is that there is no volumetric restriction: once the river is over the flow set out by his licence, a harvester can pump his 86 megs a day for as long as that flow is going past. There is also no limit to the number of licences one holder can have. In terms of water use, it makes off-allocation water from New South Wales and off-quota water from Victoria look extremely mild!

Fruits of harvest

Irrigators have transformed the Balonne region, which now produces an estimated \$190 million a year in cotton and horticulture like melons and table grapes. Ninety-five per cent of the cotton is exported.

Water transformed Dirran and St George beyond recognition, as if a new goldrush had begun. The change was most dramatic in 'Dirran', the smaller and more isolated of the two, where the population grew from about 450 to 1,100. House block prices soared from \$150 to between \$5,000 and \$7,000, more than 3,000 per cent! In fact it became hard to get a block in Dirran. The local Aboriginal community had employment and three new motels appeared, plus a new hospital with a maternity wing—at one time there were a reported 28 Dirran women pregnant.

For the Greens, however, such a thriving community has come at an enormous cost. Remember in Queensland as recently as the early 1990s, no one was saying anything about 'environmental flow'. There was one prime target on which the Greens could set their sights, the Big Daddy of cotton properties, Cubbie Station.

The rise and rise of Cubbie Station

It's not easy as a journalist to get onto Cubbie Station. Its owners are sufficiently ticked off with the way the press has painted them as an environmental pariah that very few get to meet and greet. Luckily for the journos, a quick chopper flight over Cubbie provides all they need to make the point. More than the eye can take in, really.

Cubbie is the largest privately owned irrigated property in Australia. Treeless flat plains of cotton on dark soil stretch as far as the eye can see. Cubbie has amassed enough harvesting licences to capture more water when it floods than there is in the whole of Sydney Harbour—500,000 megs of water. It brings in \$50 million a year, but compare this with the entire South Australian agricultural product grown under a self-imposed cap of 700,000 megs per year, which brings in billions of dollars a year!

General manager of Cubbie, John Grabbe, has been quick to point out that 'it is only when we have a flood and the Caribous are out dropping fodder to stock that they [dams] would all be full—this is something the environmentalists just don't understand.'[12]

It was telling that when ANCID, the national irrigators council, organised a tour of Cubbie Station as part of its annual conference last year, most of the irrigators from other States returned from the trip aghast. From the air, 20,000 acres under water certainly makes an impression. Nearly 30 kilometres of continuous 5-metre-deep dams (2 metres of which are lost to evaporation during the year) dwarf the river from whose floodwaters they are drawn. The picture is quite ghostly, with drowned trees sticking out of the water. It suddenly becomes clear why the layman suggestion that we just cover dams to stop evaporation is most unhelpful.

Dianne Thorley, Toowoomba's mayor, who made the same trip as the irrigators was flabbergasted. 'I thought I'd start taking Valium after the flight up there. I don't think anyone could have explained to me. And I don't believe anyone driving past could have any comprehension of what's happening. Now, I understand we've got to have farming, because it's what we live and die by here in Toowoomba, but at the end of the day, it was beyond anything I could have imagined. It just seemed to take forever to fly over, there were just acres and acre-feet of water.'

On the ground while being shown round Cubbie, despite the help of the CB radio, the guide simply 'lost' the tractor we were looking for. In 1,000-acre fields or cells, this is not hard.

From an engineering standpoint, Cubbie is magnificent. The homestead is built at the top end of the flood plain, and with the help of an imperceptible gradient, each of the 1,000-acre cells runs slightly downhill, minimising the need for pumping.

The great man

Cubbie was the vision of one man, Des Stevenson. Stevenson died in the middle of 2000, a shock for all who knew him (and he was a very popular man). He went into Royal Brisbane for a hernia operation and never came out.

Like Keith Coulton, Des Stevenson was a pioneer, but far more entrepreneurial. He ran the first refrigerated trucks from Brisbane to Darwin and then he had coalmining leases near Ipswich in Queensland. By the time he was thinking of getting into cotton, as luck would have it, the National Party in Queensland was falling over itself to promote development. In the hot seat was Natural Resources Minister Howard Hobbs, whose seat also happened to be at St George.

Dam development on the scale of Cubbie needed big dozers, precisely the D10s that Des Stevenson had for his coal business. He could do a job that looked impossible because his cost base was very cheap. But the other critical input required some work. That was water.

Just how Des Stevenson managed to acquire enough water harvesting licences to fill an inland Sydney Harbour has been the subject of much media speculation, and it keeps the channel grower lobbyist Ken Pierce awake at night. Some licences were obtained when Cubbie bought out other graziers as its cotton operations expanded over thousands of hectares. But Des Stevenson's stroke of genius was to hire one of only a couple of men who knew a great deal about the licensing process, John Grabbe. At the time, John Grabbe was working for the Department under Peter Noonan and spending much of his time designing irrigation lay-outs for farmers during the pro-development era. From then on Cubbie steadily built up the access to high flows in the river.

At the same time there is evidence that at the very least, governments were incompetent in their handling of Cubbie. And there is no doubt that Des Stevenson and John Grabbe had friends in high places. Witness a bid by Des Stevenson in the mid 1990s to build a dam 10 metres high, which was knocked back in the courts. When Cubbie decided to go ahead with a new dam less than 5 metres high, the Director-General of Water Resources whose job it was to gazette all dams and provide opportunities for objections, reportedly did none of this. Instead, Minister Hobbs decided to ignore a Department recommendation and do away with gazetting dams on the basis that the cost was too much for growers. When asked why the Director-General was not making the decision on gazetting, as should be the case, Hobbs reportedly responded, 'I guess they were just covering their backsides.'[13]

'Free' water

Irrigators across Australia might wonder at what Cubbie Station actually pays for all this water. The answer is $3,700 a year. That's right. For up to 500,000 megs of water and over 50 licences. Once again, the rules were changed for Cubbie so that instead of paying for individual licences, they were all rolled into one and the charge of $3,700 was set by the Director-General. Channel farmer Ray Kidd pays $30,000 a year for just 1,000 megs, whether he gets the water or not.

Like all irrigators, water harvesters get pretty annoyed when the accusation of 'free water' is thrown around. The vast dams they have built to catch the water are all constructed at personal expense. As one harvester explained, 'I know what we pay now is minimal, but when

it's on a catch-as-catch-can basis, we're not supplied from any government and we've got no expectation of that happening and we can
go for three years with only seven days of pumping in three years—
yes maybe we can pay a bit more, but it couldn't possibly be as much
as a charge of water coming out of a government store.'

As the price of water increases, however, a return on the high up-
front costs begins to look very attractive. A man with the vision of Des
Stevenson surely would realise that over time he was an outright
winner. It is another reason why even other cotton growers do not
think Cubbie is a good showpiece for the industry.

Cubbie's fiefdom

Des Stevenson had another talent. His workers adored him. 'The only
employer around here who would pay you within three days,' commented local welder Bill Wuth, giving me the guided tour of Cubbie.
'Beautiful, isn't it? Ah, I love this farm!'

Cubbie is a twenty-first century fiefdom, which has developed
under a commercially tough but benevolent landlord, who has had
his workers' livelihoods in the balance. In less than 20 years, the
land held by Cubbie, which used to support a handful of graziers and
hundreds of cattle, now serves hundreds of people, but has only one
corporate owner. Des Stevenson's son-in-law runs another vast
irrigation property.

'Look over there.' Bill Wuth pointed to a vast expanse of flat land,
looking severe as planting was just beginning. 'That piece of Cubbie
is called Urandle, 7,000 acres. There would have been 150 to 200 of us
working on that project at any time, from scrub to cotton in seven
months!'

Except for the really big bulldozing, everything is done under contract: water management, sowing, harvesting, stick picking, you name
it. And far from being the nasty corporation that operates the fly in
and fly out of employees, most contractors live in town. Des Stevenson also insisted that his workers shopped in town.

The owner's largesse is the stuff of legend in the community. Des
Stevenson lent his machinery to build a protective levy bank round
Dirran and made his chopper available for free in emergencies when
the floods came. It may have been magnanimous, but it was also good
PR and good business.

A hard man to Grabbe

Today, John Grabbe manages Cubbie Station, although it would not
be surprising if he had a reasonable stake in the business. A number

of folk in the area say Des Stevenson treated John like a son.

It took a long time to meet John Grabbe. For two years, Cubbie Station's reputation as an ogre developer had grown in the media. Finally, in September 2000 Cubbie agreed that I could do an ABC *Landline* piece about St George and Dirran.

After a rushed day of filming, John Grabbe and about 20 irrigators from around town gathered at a new Dirranbandi motel. That afternoon, an article entitled 'Sold Down the River' had been published in the *Canberra Times*, referring to the Cubbie approach to natural resource management as 'use, grab or destroy', and to the area as having had 'a two year dam building orgy'. The irrigators were busy absorbing the accusations. I was hastily informed that John Grabbe would say hello at the motel, but would not be doing any formal interviews. 'You media lot, you're all the same,' he later said to me at the motel with a casual airiness. His charm faded measurably over a question on how Department decisions have led to problems with over-allocation in the river (still over a beer). 'There is no overallocation on the Condamine–Balonne,' he said flatly. This was indeed a surprise. Especially as this man, presumably more than anyone, knew exactly how allocations in the river worked and the real concerns of over-allocation. The Department admitted it, as well. When I mentioned that at least two irrigators that evening had confirmed that the river was overallocated, he stood up. 'All right you blokes,' he shouted across the pub. 'Which one of you says the river is overallocated? Come on, which one?' This did not seem to be a joke.

Green pressure

It is an ongoing frustration for Ken Pierce that his constant barrage of accusations against Cubbie and water harvesters are so often ignored. The harvesters write him off as obsessive. Greener critics, however, have made more of an impact.

Today, only 20 per cent of the river flow gets through the system. Between St George and Dirranbandi, the Balonne River bifurcates into a number of smaller rivers. Unfortunately for the harvesters, one of these, the Narran River, runs into some rather special Ramsar-listed lakes. Stuffing up the river now has much bigger consequences.

So what, it might be asked, is all this damming and cotton growing doing to the environment? Enter Dr Duck. Dr Richard Kingsford from National Parks and Wildlife has spent many hours on the question and his views are sobering.

Richard Kingsford believes Cubbie and company have changed the land forever. Because while flood harvesters insist they don't make much impact on big floods, they have very successfully got rid of the gentle floods, as Alec Holcombe describes them, which keep the system ticking over. 'What we do understand about our inland rivers systems,' explained Richard, 'they're really boom and bust systems, so they're looking for the big boom period where, you know, the fish go mad and waterbirds do their thing and all that sort of business goes on, but in between, they don't have a human life cycle. They've only got 10, 15 years at the most to live. They need to be kept ticking along to be ready for those big boom periods, where they can really get breeding.'

We saw in chapter 6 how important Australia's wetland sanctuaries are for so much of our inland water fauna. According to Richard Kingsford, in the early 1990s the Condamine–Balonne system had 20 per cent of all wetlands in the whole of the Murray–Darling Basin, 1.3 million hectares of floodplains, of which most are in New South Wales. Half of that, the coolabah and redgum country, is expected to go.

Narran Lakes is a wetland gem treasured not just by the Greens, but by Aborigines. Ted Fields, custodian of Narran, who has lived many decades in the area, told me the Dreamtime story of how Narran came to be.

> Byaami, our creator, left his two wives to go hunting. But when he got back, his wives were not there. He saw tracks of Garriya, the crocodile, cutting across to the waterholes. He followed the tracks, every waterhole Garriya had emptied, followed him all the way down the Narran. When he got to where the lake is now, he speared the crocodile. And it thrashed around, knocked all the trees down, made a great big hole in the ground. And he got his wives out and cut Garriya open, but when he got him open, all this water came out, filled the lake up.

Ted Fields explained that a rocky outcrop with indents next to the lake is where Byaami knelt down to drink water from the lake. He has also seen the changes. In the last ten years, 75 per cent of Narran's water has been snatched by irrigation. 'The river is not a river anymore, it's just a bunch of water holes.'

Narran is one of the largest breeding grounds for the straw-necked ibis (the farmer's friend because of the way it feeds on pests like locusts), rare species like the freckled duck and other migratory species from Siberia and China. According to Richard Kingsford, as a direct result of irrigation, the wetland may have lost 100,000 breeding pairs.

As to Narran, irrigators point to the wheat that's grown close to the Narran Lakes as floodwaters retreat and accuse the Greens of double standards, but neither Ted Fields nor Richard Kingsford sees wheat (not irrigated) as the culprit. Cubbie, too, is quick to point out that it takes water out below the Narran turn-off, from the Culgoa River, and couldn't possibly affect the wetland. But critics of Cubbie argue that its diversion is so big, that even though it's downstream, the water gets sucked away from the Narran River into the Culgoa, as if it's heading for a plughole.

Cubbie also points to the richness of birdlife on its dams, something which is certainly in evidence. Dr Duck, however, is unconvinced. 'Well you do see that for the first couple of years, but as the flooded trees die out and [as the] water is deep, it becomes less and less fertile. The thing about wetlands is that a lot of them are very shallow, to cope with invertebrates and frogs that fish and birds feed on.'

Beef pressure

Pastoralists have had their gripe with Cubbie too. The issue of social divide and a new break up of 'haves and have nots' is very similar to Whalan Creek. The naturally flooded pasture is going and with it, so are the fat cattle. It's a very good reason for graziers and Greens to get together.

John Hill, a fourth-generation grazier unfortunate enough to be just downriver from Cubbie Station, says the Culgoa River used to run nine months a year. 'We could come down three o'clock of an afternoon, catch enough fish in two hours to probably feed ourselves for a month and you'd be lucky now to throw in a line and catch a fish of any talkable size. I just feel the whole development of management plans from the DNR [Department of Natural Resources] has just failed something dramatic.' John Hill has been trying to get a licence since 1989.

Like Keith Coulton, John Grabbe argues that the licences were there to take up and if John Hill and other graziers had had a bit of foresight, they wouldn't be in this position.

Leith Boully is another grazier who moved to the area about 15 years ago, when Cubbie was just kicking off. She's also in the powerful position of being chair of the Murray–Darling Basin Community Consultative Committee and has been one of the most strident critics of the big water harvesters. Dianne Thorley wonders at three women being the most outspoken on the Condamine–Balonne—herself, Leith Boully and Bobby Brazil, chair of the Condamine Catchment. But as she said, 'I'm in the happy position where my electorate isn't going to kill me with what I say about the Murray–Darling. Now those girls have to live out there among the farmers.'

Interestingly, Leith now has a cotton joint venture with John Grabbe. She is still a critic of big irrigators, but possibly this has dented her credibility among the Greens. She is not alone, however. One of the signs of the times in the St George region is that many of the graziers who used to be very strong advocates of the environment are changing their tune. As making a living from grazing gets harder, many now want not environmental flow, but equity: a share in the water to grow crops like cotton as well.

WAMPer stomper

In the late 1990s, development was so manic around Dirranbandi and St George that it was quite difficult to get hold of a tractor. Chains were dragged over bush, dozers moved in and new dams were going up every week or so, ready for the next flood event. Even the bank was said to be gung-ho, offering much higher borrowing limits to Queensland growers than to cotton growers further south. Then in July 2000, something struck developers mid-track. It was a WAMP, a water allocation management plan.

As we noted in chapter 9, WAMPs are the Queensland Government's means of deciding how to divvy up water for users and ducks. Queensland has been very unwilling to tackle this issue, but under political pressure from southern States already suffering under the Cap and other restrictions, it could hang back no longer. Out came 12 WAMP documents for major catchments in the State with their signature 'traffic light' diagrams showing irrigators what development was doing to rivers and presenting cutback scenarios.

In some areas the recommendations were good for growers, like the new dam proposed for the Burnett catchment. But the Condamine–Balonne was different. Different because it formed the start of the Murray–Darling Basin and different because, for the last five years, development had been almost feral. When the draft WAMP finally came down, it was a lot tougher than irrigators expected.

A moratorium on any further dam building was put in place on 14 August 2000, effectively ending the bull run on the lower Balonne. Then there were the water restrictions: the middle WAMP scenario, which was anticipated by most with interests in the river, as the one Government would demand, cut back water harvesters by 15 per cent and channel farmers by 10 per cent. Not nearly enough for the Greens, but harvesters were outraged.

It turned out that water restrictions varied dramatically, depending on which part of the river the farm accessed. A 15 per cent cut-back was not a 15 per cent cutback at all. When irrigators went to the modelling at the back of the WAMP to find out how their own case was affected, the cutbacks looked more like 40 per cent or 50 per cent.

The government was at best economic with the truth in the WAMP. It forgot to mention sleepers and dozers upfront; in the modelling, the WAMP assumes that all sleeper and dozer licences are activated to the full extent. It's a fair assumption, because that is exactly what is happening. The result for water harvesters is that the flow drops and so does the amount of time that water harvesters can pump. And just in case growers want to appeal, that right has gone. The old land board, a similar body to that in New South Wales, is to disappear.

The shit hits the fan

Harvesters were never going to take the WAMP lying down. Spearheaded by John Grabbe, they hired a PR firm in Brisbane and commissioned their own scientific report to fight the WAMP's damning ecological survey, and a socio-economic one to look at the damage to the community, something they say should have been done by the Government. A sexy-looking 'Smartrivers' website hit the reader with the line, 'Think you have a right to water? So did we.'

The irrigators got smart as well. With the tall poppy reputation of Cubbie, it was decided that the best defence was to move the focus away from the big irrigators and onto the WAMP's impact on the rural community. Bill Wuth took over as spokesman for the newly formed Balonne Community Advancement Committee. Bill was a contractor to Cubbie, but as a welder, represented the community as compared to the 'big irrigators'.

There's no doubt that St George and particularly Dirranbandi are not what they were. Those less sympathetic have suggested that the cotton slump had as much to do with a drought and, more controversially, that development was maturing anyway. But the WAMP clearly knocked business confidence. For Sale signs peppered the streets. Dirran, which had swelled to 1,100 people at the height of development, slumped back to about 700. A spanking new motel expecting 70 per cent occupancy was lucky to get 5 per cent. The young owner had invested close to a million dollars. Contractors who had borrowed to buy massive equipment such as harvestors were in debt up to their eyeballs. The PR machine worked overtime on the sensitive topic of rural towns to sway State opinion.

Price Waterhouse Coopers found that three-quarters of businesses and social and government bodies in the area rely directly on irrigated farming to survive.[14] The Smartrivers website asked, 'How would you feel if your children, wives, mothers, fathers, brothers and sisters were all acutely affected by a socially devastating Government Plan?'

The real row, however, was over duckwater. The community's scientific report was interpreted by the Balonne Community Advancement Committee as slamming the WAMP as 'flawed science'.[15] A general lack of data, inappropriate benchmarks, unjustified extrapolation from the data and little reference to the 'natural' conditions of the rivers. Far from being severely degraded, the expert, Dr Lee Benson, went so far as to say, 'The regional picture is one of favourable similarity with nearby catchments, some of which are relatively unaffected by water resource developments.'[16] In particular, results on water quality and fauna were comparable with the Paroo, a totally unregulated and almost untouched river.

The truth is that, despite all the money paid to experts, neither side can prove what damage is being done. Irrigation may be worth almost a billion dollars to the region every five years, but who is to know what the future cost will be to the environment? That's reason enough, according to Richard Kingsford, to take precautions. 'We can't actually treat every new river system as an experiment and say, "Oh, well, we're going to learn from this once we've collected the data." We need to apply the lessons that we've learned from other river systems, and the Murrumbidgee, the Macquarie, the Gwydir, the Namoi—all of those systems have the same level of water diversions upstream—and you're looking at virtual loss of large areas of wetland and floodplain eucalypts and so on downstream.'

As it is, Richard believes even the harshest cutbacks under the WAMP will only deliver a flood event every 14 years. The ibis needs one every five years. The birds only live for eight.

Irrigators are disbelieving. They point to Queensland's paltry 5 per cent take out of the Murray–Darling, compared to 56 per cent by New South Wales. If they let duckwater through, they guarantee it will simply be sucked up by 'those greedy cotton bastards in Bourke'.

The river which the kids in Dirranbandi (below the cotton) swim in all summer looks fine. As irrigator Henry Crowthers said, 'Go talk to some of the local fishermen in town, they're catching catfish here they haven't caught for 15 years. It's just not true. I would never be silly enough to say we don't have some effect on the river system. Obviously we do, but until they can come up with some evidence that's believable, well then I don't see how I can do anything but fight.'

Interestingly, most of the lobbying by harvesters during the WAMP process was targeted at the State Treasurer and the Premier, not the Primary Industries Minister. It reflects a major shift in Government focus during the last decade, whereby decision drivers on river management issues appear to have moved from the Primary Industries Ministry under the National Party (a time of development), to the Environment Ministry, under Labor.

The State Government missed its 1 July 2001 deadline to impose specific cutbacks in allocations under the Cap. The final outcome for those on the Condamine–Balonne was expected in early August 2001.

That damned bureaucracy

People have become pretty emotional around St George. Some irrigators are in trouble, having borrowed big on the potential of their irrigation only to be caught by the moratorium on new dams. It is hard not to feel sorry for those who started damming late, either because the money wasn't there before or because they were a young couple just starting out.

But how much sympathy do most of the water harvesters deserve? Shouldn't they have known something like this was going to happen, particularly the sophisticated ones with those great contacts? All the signs are that they did know, and this is precisely why development in the St George–Dirranbandi area has been at warp speed in the last few years. Leith Boully, for one, agrees that the water harvesters should have seen the cutbacks coming.

In defence of the irrigators, the Department and the Queensland Government have a lot to answer for. Not only has the bureaucracy done little to stop things, until very recently they continued to encourage development. This is inexcusable. Licences were still being issued to irrigators almost up until the WAMP came in. Laid out in black and white, each licence requires installation of infrastructure within two years and demands that farmers use the water or risk losing the licence.[17] According to John Grabbe, if the river is being destroyed, it is being destroyed because irrigators are doing what they were told to do.

Added to this, growers complain that the Government was endorsing development by handing out interest incentives on loans. That is cheeky, given that it was some of the more powerful irrigators who sat on the Committee which agreed to the incentive with the Government, but does not remove the Government from responsibility.

Leith Boully said she warned Environment Minister Rod Welford of the scale of water storage soon after he'd been elected. 'I did the sums on an envelope at dinner with him one night and I said, "This is what it is, Rod," and he said, "Oh, no, the department doesn't tell me that. You've got to be wrong Leith". I said to him, "I'm not wrong Rod," and months later he came and said, "Well, you're right but what do we do?" And I said, "Don't ask me!"'

It is now very apparent that the Department screwed up and issued far too many water harvesting licences, particularly at levels of low river flooding. According to Leith, Rod Welford is not enamoured of the Department. 'He hates them, he just hates them. Because what they're doing, and this is typical of agencies across the country, they're actually playing a political game. They try and second guess what the Minister of the day wants.'

Government is also squirming about any idea of compensation. Peter Noonan asked quizzically, 'What security did Government provide in the first place?' Whatever was promised during the Howard Hobbs era, Government is washing its hands. It says the State simply offered an opportunity, but that irrigators took the development risk. And no, blue sky was never promised to irrigators.

The WAMP battle is far from over, but the odds are that Cubbie Station will do better than most. The aim has been for licensees to be treated equally, but *some* licences are more equal than others. Under the WAMP scenarios, water users on the Culgoa River, of which Cubbie is one, frequently do better than those on the Narran or the Balonne, particularly in the big wet years.

Even before he died, Des Stevenson was still pressing to increase dam height from 5 to 8 metres. He apparently pitched it as a way to be more efficient with water, because deeper dams mean less evaporation. What this also leads to is a bigger use of water. It is true that when the floods come, dams will only hold as much as they can, but if Cubbie could expand its cropping area, it can put more water on the land, which then frees up space in the dams for more floodwater. Des Stevenson clearly had a premonition because under the *Water 2000 Act*, dams can now be up to 8 metres high before they are referable in the State.

It seems that Queensland will remain a doubting Thomas about the long-term damage that massive irrigation projects can do to the land. The WAMP says nothing about floodplain management. Levy banks run for miles. In Queensland the rules are now that floodplain management is a community issue unless there is a dispute. One wonders how long that will take.

11

Out of the Ord

Water is the true wealth in a dry land; without it, land is worthless or nearly so. And if you control water, you control the land that depends on it.

Wallace Stegner, 1954

November

November is not the time to go to the town of Kununurra. Here in the top left-hand corner of Australia, eastern gateway to the Kimberley, it is the so-called suicide season, towards the end of a long nine-month dry spell when days get hotter, and hotter, and still no rain. Temperatures hover around 39°C.

Very few white folk stick around in Kununurra to old age. 'One thing unique to the Ord', explained local Tim Croot, 'it's not old farming families, it's new generation families. There may be some old families in the future, but I don't think there will be that many of them, because it's a very tough climate and people still move up here for a while and then move away.'

What has brought Tim Croot and others like him to the area over the last 50 years is the promise of lucrative farming, for which growth is astronomical and the water never runs out. Come flood time, the Ord River is one of Australia's most powerful, now captured by one of the biggest dams in the country.

When first mooted in the early 1940s, the Ord irrigation area was billed as the great white hope for high-production agriculture, with an abundant supply of reasonably priced land which could be transformed by just adding water. Most farmers in Australia will know that this turned out to be overly optimistic. But now developers are planning a much bigger expansion in Ord Stage 2. The plans have divided opinion into those who believe that the Ord will finally pay its way and those who fear that, despite the lesson of the Murray–Darling, we are on our way to yet another ecological disaster.

Aside from the heat, what hits the visitor to Kununurra is that the great Ord River irrigation area thus far is really quite small. About 100 farmers work a patchwork of neat fields across the Packsaddle and Ivanhoe plains; over half of them rely on off-farm income. To the west lies the cattle country of Ivanhoe and Carlton Hill stations, both owned by Kerry Packer's Consolidated Pastoral.

'Ord 1 is a piddle in the ocean,' exclaimed irrigator Robert Boschammer, 'except for the fact that it's worth $50 to $60 million a year, but look at the size of businesses. Ord 2, they're talking about $150 million worth of sugarcane. That would give an impulse for the area to develop other things. I believe that mangoes and bananas could go ahead and be worth more than sugar. Sugar would be the backbone.'

The scheme

The Ord River Irrigation Scheme was a dream of Kimberley Durack, of the pioneering family, which settled the East Kimberley ranges almost a century ago and were made famous in Mary Durack's book of 1959, *Kings in Grass Castles*. Kim Durack firmly believed in irrigation and started the first experimental plots.

In 1941, the West Australian Government set up the first research centre on Ivanhoe Cattle Station, next to the Ord River, and engineers began searching for dam sites upstream. It took until the late 1950s for the State Government to commit to a major irrigation project, but in 1958 it agreed to build a small diversion dam and help develop the Kununurra township nearby. By 1966, about 30 farms were being irrigated.

What changed the land forever was the construction of the main Ord River Dam, built over three dry seasons and finished in 1971, creating the awesome Lake Argyle. Whereas the low level diversion dam backed the Ord's water up to create small Lake Kununurra, it was built to open fully when the floods arrived. About 50 kilometres upstream, the Ord River Dam blocked the entire river, and all the waters from a catchment area of over 46,000 square kilometres. It was an engineer's delight and would never be built as it is today.

At normal storage levels, Lake Argyle extends 980 square kilometres, holding 10.7 million megs (about 20 Sydney Harbours). But when in flood, the area expands to 2,072 square kilometres and holds over 34 million megs, or 60 Sydney Harbours. A spillway at the northern end of the lake, quite some way from the dam wall, allows for flood overflows.[1]

Unfortunately, such was the success of the dam that it flooded out

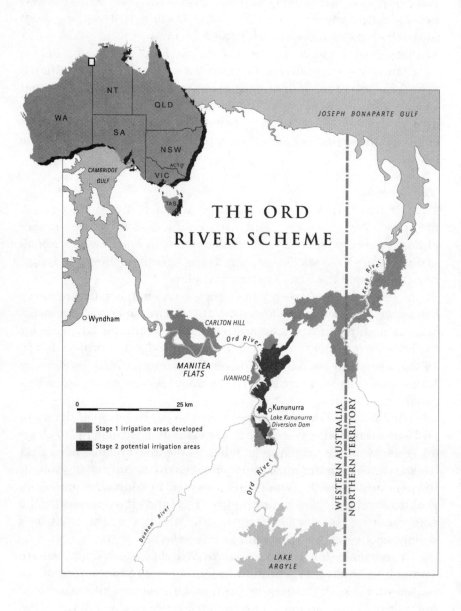

THE ORD
RIVER SCHEME

JOSEPH BONAPARTE GULF

CAMBRIDGE
GULF

○ Wyndham

CARLTON HILL

Ord River

MANITEA
FLATS

IVANHOE

○ Kununurra
Lake Kununurra
Diversion Dam

0   25 km

Stage 1 irrigation areas developed

Stage 2 potential irrigation areas

Ord River

Dunham River

LAKE
ARGYLE

WESTERN AUSTRALIA
NORTHERN TERRITORY

Keep River

NT
QLD
WA
SA
NSW
ACT
VIC
TAS

the Duracks. Victorian water watcher Don Rowe recalled, 'Everybody in Western Australia was as sad as all hell when Mary Durack and the Durack homestead went under water. But in fact it was the idea of the Duracks. The reason that they wanted a dam put in there was to water pasture so that cattle could be fattened all year round. You may recall back in ancient history there was a huge abattoir at Wyndham and the Air beef scheme at Glenroy Station. They actually had chiller DC3s taking the beef straight from the abattoir at Glenroy, and to this day there are still crocs all around Wyndham.'

In the last ten years, the lake has been raised 6 metres by a concrete plug to the spillway. Today, the lake provides about 310 gigs for irrigators with a 98 per cent reliability (a level inconceivable for irrigators in the Murray–Darling).

Cotton flop
With the dam in place in the early 1970s, the West Australian Government enthusiastically encouraged farmers to move into the Ord and take up cotton, which had been so successful in the east. About 7,000 hectares were planted, but the dream was to turn to a nightmare of biblical proportions.

'Very few people, virtually no active farmers are here who were cotton growers. They all went broke,' said Robert Boschammer. The big mistake was growing cotton through the wet season. Job's plagues arrived. It was a bug field day.

One of the few people still around in Kununurra who grew cotton in the early days is Peter McCosker. 'The suite of insects on cotton when the crop started,' mused Peter McCosker, 'we had rough bollworm, pink bollworm, *Spodoptera litura* and punctigera, but mainly roughies and spods. Through the 1960s gradually punctigera became the issue because of the spray resistance, and through the seventies it was armigera.' No amount of chemicals seemed to fix the problem. Armigera soon developed resistance and while farmers tried natural pests like parasitic wasps to deal with the bugs, the situation remained hopeless.

Peter recalled the huge costs farmers incurred fighting the bugs, costs most could ill-afford. But real disaster came in the 1973–74 wet season, with massive rains. 'We were growing over the wet season, we bogged a lot when we were planting it. And at the time we were planting, cotton prices were very high but by the time we were harvesting, the prices were pushing historical lows. The industry made a decision to stop growing for at least a season to try and break the cycle.'

It was over. No one had the capital to handle the pain and the

Government had pulled out, leaving farmers bitter and broke. The area under farming shrank from the 7,000 hectares planted in the 1970s to less than 3,500 hectares in the early 1980s. It was thus that the Ord earned the label of white elephant.

The comeback

Today farming in the Ord survives, but not yet on cotton. Growers learned the lessons and now plant crops through the dry season when the risk of plagues and bogged machinery is less. Sugar does well, and a range of high-value horticultural crops has been introduced, including bananas, pumpkin, melons, mangoes, chickpeas and leuceana pasture for cattle fattening. Supporting the industry is Kununurra's Frank Wise Research Station (named after the State premier who as early as 1928 tested crops in the Kimberley), with a plethora of plants being tested for commercial use.

It's not an easy life, though, the summer is still hot and nature keeps growers on their toes in a frontier mentality, vigilantly guarding against diseases like sugar smut or assessing the constant threat of fires from electrical storms. 'The Ord will always be an area I believe that doesn't tolerate fools very easily,' said Robert Boschammer. 'If you make mistakes, you go down quick. It's an expensive area to make mistakes, but it's rewarding to do the right thing.'

The Ord makes a different breed of farmer. Tim Croot arrived as a disillusioned grower from the struggling West Australian wheat belt. He'd got well into grower politics and even organised the big truck demonstrations in Perth, but the Ord has been quite different. 'Here, there's no political power except that you say "piss off", and they bowl you over coming back,' said Tim, hopping down off his tractor and chuckling at the beautifully graded dirt roads in the valley recently put in by the Government. 'About $11 million of roads for 20 farms. That's bullshit! I mean they wouldn't do it anywhere else, but really what the Ord growers said is "No, you can't help us, we want to put the sugar mill [in] ourselves. Go away! Because last time you came here, you got people growing cotton and then said, 'No we don't want to do it any more' and sent everybody broke." But Government couldn't help themselves, "We'll bloody help you bastards," and they did.'

Farming is also different because the remoteness means almost every input into the business is more expensive than elsewhere in Australia. They even pay through the nose for power, despite electricity coming from a 30 megawatt station built at the dam in 1995. There's one exception, however—water.

South Australian wine growers would be wide eyed at the set-up. In a region where there is 800 millilitres rainfall per year, but almost 3 metres also evaporates every year, about 360 kilometres of open channels and drains service Ord irrigators, who now farm around 13,000 hectares. Flood irrigation is most common, although some horticulture is sprayed. Charges for water late last year (2000) were about $70 per hectare for the larger farms and $700 to $800 per hectare on the river banks for the small horticulturalists. Ord 1 irrigators use around 310 megs a year. The fact is until 2001, water had been cheap and plentiful with no consumption-based charges. And you can still catch barramundi the size of wombats in the Ord. Too good to be true?

First signs

Already in the Ord there are signs that irrigation is changing the land. Forty years of watering and the water table in places is well under 2 metres from the surface. Ord Stage 1 has few salinity problems. The major threat is waterlogging. Pressure on the underground water table is caused by the filling of Lake Kununurra and leaks in the main irrigation channel, M1. As a result, water collects underground in ancient gravel riverbeds of palaeochannels formed by the River Ord many thousands of years ago, that used to run north. These riverbeds run under the 60 per cent of ground where water now collects. De-watering bores keep the water table at bay, but the issue clearly has to be managed.

The soils don't help. Most common is Cununurra cracking clay (with a 'C'). The alkaline version in particular can be gluggy. 'More what on the Darling Downs they call Sunday soils, too wet on Saturday and too dry on Monday. Less forgiving,' expounded Rob Boschammer.

As head of the soon to be privatised irrigation company for the area, these are problems that Elaine Gardiner now has to face. She and her husband, who run the growers cooperative, have a farm of bananas and sugarcane on the Packsaddle Plain. From the ute, she pointed across to a field of thriving bananas in plastic jackets. 'All this is spray irrigated and our sugar is flood irrigated,' she explained in one of those pleasing Scottish accents that never leaves a Scot, 'and these are our drains, all draining off into Packsaddle Creek, which runs into the Dunham River, which then runs into the Ord, just here, which is not far downstream from the bridge.'

At the time the drains were put in, it was best farm practice—'good drainage'. Today, draining farm run-off into rivers is just about worst farm practice. Unfortunately, the Dunham River is a slow, winding affair, accentuating any problem with fertilisers or chemicals that might arrive. And they do. Cununurra clay is not as good as the black soil of the east. Add fertiliser, however, and melon, sugar and pumpkin do very well. Several of the crops also demand a fair whack of chemical. 'Horticulture—rockmelons—are quite demanding in some years,' Elaine admitted frankly. 'Nineteen ninety-eight was a really bad year for bugs and people were basically tearing their hair out trying to work out what to do with them. We did get a few hits in the river of endosulphan, which kills fish, and so straight away, of course, we were on the front page of the *West Australian*, "300 Fish Dead!"'

The situation is serious enough to raise concern at the West Australian Water and Rivers Commission, which commented that farm practices were

> . . . of a major concern to the Commission. Inefficient use is causing extensive groundwater recharge throughout the Stage 1 area. Excessive irrigation return flows are transporting harmful chemicals and nutrients to the Lower Ord River. These represent a threat to health of the Lower Ord aquatic ecosystem and this threat will increase as more water is diverted from the river in the future. [2]

For the first time, the pressure is now on growers on the Ord to cut back on water and recycle it, by collecting run-off containing nutrients and chemicals in tailing dams. 'We've been told by Water and Rivers Commission that Stage 1 has to reduce its tail water by 50 per cent in five years, so we have to become 50 per cent more efficient,' explained Elaine Gardiner as we both retired to the welcome cool of her small air-conditioned office. While achievable, according to Elaine, tail-water recycling is a hot issue, not least because the system was designed to drain a lot of clean or 'float' water through it during the flood brought on by the rainy season. If tailings dams are to work effectively, such drainage would be impossible.

Much of the pressure is coming from improved practices in the Murray–Darling, but here's where Elaine Gardiner gets cynical. 'Recycling hasn't come in across the eastern States because it was an environmental requirement. It came in because there was no water, and realistically until it is commercially right, environmentally it's great, but somebody somewhere has to come up with the money to do all these things.' According to Elaine, there's a real fear that cheap tailings dams will exacerbate the problem, acting as holding ponds that leak into the ground.

In a pre-emptive strike, farmers have prepared their own five-year water management plan, with Chair Tim Croot urging 'a gentleness and respect towards our use of the natural area—we cannot continue to "bang them about".' The plan's targets are to keep groundwater below 2 metres, and within five years to achieve on-farm irrigation efficiency of 65 per cent, a 40 per cent reduction of chemicals and nutrients in tailings, and 50 per cent reduction in terms of what ends up in the Dunham River.

If a bad bug year hits, it's hard to see how these targets will be achieved. Elaine Gardiner is looking at various methods to slow water flow over irrigated land and reduce run-off, such as slower siphons, connecting fields, and more laser levelling of the ground, technology which can ensure scrapers create almost perfectly flat landscapes that prevent run-off. But cutting back on water in the Ord is not the easiest of messages to get through to growers. When floods peaked in March 2000, the flow over the spillway was 900 cubic metres per second! 'Two days' supply over the spillway, and that's just the excess water out of the lake this year, [it] was our allocation for the whole year. That wasn't the rest of it going down the river, just over the top of the spillway weir. Now to tell a farmer he has to be efficient with his water when there's untold water going down the river, flooding, is pretty hard,' said Elaine.

A further challenge is to manage the transition of the Ord district from an irrigators co-op to a privatised irrigation company, following in the footsteps of irrigators in the Murray–Darling States.

Come the new privatised irrigation company, Elaine plans to keep water as cheap as possible for farmers, and Peter McCosker has warned that pricing high water users (like sugarcane growers) out of the market is bad for business. 'My personal view would be to structure an environmental levy that's based on consumption, to encourage efficient use of water on the high-demand crops,' he said.

The biggest challenge to a real breakthrough on environmental improvement in Ord Stage 1 is money. Farmers are either not able or not willing to spend the capital. And as the community's plan says, 'For improvement to actually happen, it will be necessary for all those involved in using land and water resources to change what they are currently doing.'[3] Ord Stage 1 never had an environmental impact statement and its critics believe the 'she'll be right' attitude remains today. Meanwhile, most green eyes are on the next, much bigger development, Ord Stage 2. The temptation is to forget about Ord Stage 1 altogether.

Ord Stage 2—sugar

In February 1994, the West Australian Government endorsed further development in what has become known as Ord 2. Peter McCosker's job is as project manager for the Ord 2 development and his involvement now spans over a decade. It's rather a sore point, as will become obvious shortly.

The project dwarfs the old irrigation scheme. According to McCosker, 76,000 hectares of cracking clay and surrounding areas have been earmarked, running in a northeasterly direction from Kununurra, with about half the land in the Northern Territory, making the development a two-State affair. The Territory's head of Resource Development, Howard Dengate, sees the project as a fantastic opportunity to attract investment. 'People say, "Gee, if they're prepared to spend $600 million, let's have a look at this area."'

Under the new Scheme, a new (M2) channel would form the main artery supplying Lake Argyle water to about 30,500 hectares allocated for farming. Wesfarmers and Japanese trading house Marubeni won the mandate to develop Ord Stage 2 and plan to expand the sugar industry in the northwest, beating the rival cotton proposal from Colly Cotton. There are plans for 400,000 tonnes of raw sugar to be grown on 29,000 hectares, which would represent up to 10 per cent of the country's sugar. A new sugar mill and port development at Wyndham caps off the development.

Economically, the project would be a shot in the arm for Kununurra. 'The labour component is projected at about 550 full-time jobs and about 300 indirect jobs, and it has a construction component that will peak at around 650,' said Peter McCosker. Wesfarmers expects to generate $170 million in annual revenue.

Plans are for up to 3,000 hectares of the new land to be put aside for independent farmers to grow whatever they want. In addition, once the land is developed, Wesfarmers plans to sell down its holding gradually over 20 years. The bulk of the land required for the West Australian part of the Ord 2 project (both for agriculture and conservation areas) will come from Kerry Packer's Carlton Hill and Ivanhoe Station pastoral leases.

There is a separate development also on the cards: to the northwest of Kununurra lie the Mantinea Flats, rich loam country with soils deposited along the levee banks of the lower Ord River, perfect for horticulture.

Ord 2, the developers say proudly, will be eco-friendly. 'Ord 1 is flood irrigation which drains out the other end, there's no tail-water return, it's not a recycling system. Stage 2 is entirely designed to be

that sort of system,' enthused Howard Dengate. Water supply efficiency will be 85 per cent compared to 65 per cent in Ord 1, and
instead of the old drainage problems, leakage is expected to be less
than 2 millimetres a day, world's best practice. West Australian Water
Corporation will be responsible for the channels and drainage system.

Things have not been going quite as smoothly as the developers
would have liked, however. In fact, the project has been ongoing for
some five years. And not everyone agrees that Ord 2 is eco-friendly.
There are serious tensions, because like many big projects, the economics require the development to be built in a hurry, over three to
four years, rather than incrementally, during which the impact on the
surrounds could be assessed. In a national climate which is increasingly anti-irrigation and pro-duckwater, the West Australian Waters
and Rivers Commission has had the wobbles on allocation for many
months. 'I'm glad I'm not in Peter McCosker's shoes,' commented
Elaine Gardiner. 'It's pretty hard for them to make the big decisions
when the Commission can't make a decision as to what their true
water allocation in normal years will be.'

Wobbly bureaucrats

The developers' first problem is Ramsar. In 1990, Lakes Argyle and
Kununurra and the lower Ord floodplain, including Parry Lagoons,
were singled out as critical dry-season refuges for ibises, ducks and
geese. Up to 200,000 waterbirds have been counted around the lakes,
including the only major presence of magpie geese in the west.

But Peter McCosker is not impressed. 'The interpretation here is
that doing anything here screws up a Ramsar-listed area. It's got to be
pristine. That's not the Ramsar agreement at all. The other tactic is
to try to expand the bloody area that's supposed to be Ramsar.'
He's right. These are exactly the plans ecologist Peter Cullen has
for Ramsar wetlands. Mantinea Flats, in particular, seem to be the
focus, now the subject of an environmental impact statement over
surface hydrology.

Federal Minister Robert Hill is taking a keen interest in Ord 2.
'Under the *Environment Protection and Biodiversity Conservation Act*, the
legislation is limited to endangered species, but there are other issues
not listed that concern us,' he says, carefully. 'One of them is future
salinity and the Prime Minister has included the Ord as one of the 20
catchments in the Action Plan. Now I know there have been objections
from that region, but the issue is what problems they are creating in the
future. I mean if the Condamine–Balonne had been managed appropriately ten years ago, we would not be in the position we are now.'

Robert Hill could make life difficult, but the ongoing dramas have been State based. The West Australian Water and Rivers Commission has been ruminating over how Ord 2 should go ahead, most importantly, what level of duckwater should be allowed to flow down the lower Ord. This has turned out to be a $64 million question, or indeed even more. The vexed situation faced by the Commission was that no one had bothered to record what the Ord was like before it was ceremoniously plugged up by the Argyle Dam. What everyone agrees upon is that the area below the dam after 30 years is dramatically changed.

Instead of the flash floods that occurred as frequently as one in two years, the lower Ord is predominantly a steady flow. Flooding is now dependent on the Dunham River and is an event that happens perhaps one in 10 or one in 20 years. Equally, the Ord during the dry months used to stop completely, reduced to a series of pools. Today, it runs constantly through the dry season not least because of the demand for hydro-electricity. Post-dam, the river course is covered with melaleucas and pandannus palm vegetation that could not have survived the extremes of the past. A panel set up by the Commission looked at the impact of returning the area to a more natural state by trying to mimic high and low season flows. It was a rushed process, a one-day workshop and a week of site inspections in Kununurra. Nevertheless, the Commission's conclusion was that while the return of such flows would do little harm to birds and some invertebrates, crocs, fish and water quality would all suffer.[4]

Crocs and fish weren't the only consideration. About 100,000 visitors pass through Kununurra each year,[5] most after a taste of Croc Dundee country. Regular flows in the lower Ord are one of the attractions of a tourism industry bringing in a very respectable $50 million a year compared with some $65 million from Ord 1 agriculture. Seasonal flows would be the last change the fishing safaris would want.

Faced with these issues, the Commission released a draft interim plan in May 1999 for public review, and in December the State Environment Protection Authority threw in its views, stating that 'as the riverine environments downstream of the existing dams on the Ord River are already substantially modified, there may not be value in trying to maintain a downstream river flow which mimics pre-dam flows'. Instead the EPA favoured 'protecting environmental values which are sustainable under post-dam flows and so preserve the riverine ecosystem which has adapted to these changes.'[6]

Tim Fisher at the Australian Conservation Foundation is outraged at this logic—keeping flows regulated because of damage to a man-made environment. If the Argyle Dam is not to come down, the very least that can be done is to mimic the old environment. 'It's absurd!' he exclaimed. 'So what if the flows change things? Monsoonal river systems soon recover.'

However, the reality is there is little Green presence in the Ord. Despite the land clearing and further diversion that Ord 2 brings, it's all a long way from HQ in Melbourne or Sydney and an expensive protest to maintain. The State's Conservation and Land Management (CALM) Office has cried foul on occasion, especially when it was suggested that roadways in the M2 area of Ord 2 doubled as biodiversity corridors! But generally CALM is pretty toothless. The aim is for all irrigation and first flush stormwater to be retained on-farm. However, unlike most of Ord 1, salinity is a threat to the Ord 2 area, albeit in the longer term, and even Peter McCosker admits some groundwater management will be needed within ten years. Surface water management would also aim to reduce the amount of fine clay sediment that may be carried with irrigation water, the problem here being that pesticides attach themselves to the clay particles.

Federal Forestry Minister Wilson Tuckey took up the Green slack when he visited the district in 2000 and called for all channels in Ord 1 and Ord 2 to be converted to pipes to check leakage, evaporation (nearly 3 metres a year) and salinity, which in the longer term is a threat in the Ord 2 area. He was laughed out of the district, the economics unworkable.

The bureaucrats at the Commission have been left to make the tough decision on duckwater. And all at a time when their counterparts in eastern States are dealing with the consequences of over-allocating water in earlier years.

By late 2000, the Commission had honed down its ideas to a few scenarios (assuming Ord 1 irrigators continue to get their 310 gigs a year). It looked at the effect of various levels of minimum duckwater flow downstream of the Kununurra diversion dam, and compared them with the current water levels of about 50 channel cross-sections along the river. While admitting to a lack of science the Commission settled on a 'rule of thumb': changes in the wetted perimeter of over 20 per cent were of 'considerable concern'.[7]

Under the first scenario of a minimum duckwater flow of 40 cubic metres per second, the Commission allocated 710 gigs a year for Ord 2 and 70 gigs for downriver horticulture development, such as Mantinea, delivered at 95 per cent reliability. The second scenario involved a higher level of duckwater, 45 cubic metres per second. It means the developers would need to live with 710 gigs a year at 87 per cent reliability, or 625 gigs at 95 per cent reliability. And there would be no water left over for downstream development.

Back at Wesfarmers' HQ in Perth, the man in charge of the Ord 2 project has been Andrew Hopkins. He was careful not to give away too much about how the economics of Ord 2 stack up. Clearly the company would have to have taken a view on long-term trends in sugar prices. But the fly in the ointment is not the sugar price. 'All I know is they're talking about environmental flows of 40 to 45 cubic metres in the worst part of the dry season. We're still going through the latest numbers. It's fair to say that we've had a nasty little surprise over the last couple of years.'

Black and Green

Not only do the environmental goalposts keep changing, but the whole Green approval process is tied up in the maze of Native title issues and as of July 2001, Native title negotiations were at a critical stage.

'Kununurra' is, as it sounds, an Aboriginal word. It means 'meeting of big waters'. In 1994, the indigenous Miriuwung Gajerrong people of the Kimberley lodged a claim for the return of their waters (see chapter 9). Law is in the making and the case will set a critical precedent in the water rights drama.

The claim was made over 7,500 square kilometres of land in the Kimberley, including Lake Argyle Dam. The court decision in November 1998 was to be the first successful determination of Native title on mainland Australia. All hell broke lose when Justice Lee gave extensive but fairly undefined rights to the water resources for local people.[8] Significantly, the judgment implied that Native title holders might be due royalties from the development of resources in the area.

'The argument since Mabo,' Peter McCosker pointed out, 'is about what is or is not extinguished by public works, and where in fact the extinguishment occurs. Our view is that the public work extinguished Native title.' The problem for the developers is that, unlike a mining lease where in 30 years' time the land (at least in theory) can be returned to the traditional owners, farmland would not be returned. Developers need complete extinguishment of Native title.

On appeal in 2000, Justice Lee's decision was shackled and the Full Federal Court swung in favour of the developers. While the ruling given was that Miriuwung Gajerrong people do have an ongoing connection with the land, Native title water rights were pared back. Critically, it was determined that development on all land within the Ord 1 irrigation area and Lake Argyle has extinguished Native title.

These determinations were put before the High Court in March 2001 and a decision is expected towards the end of the year. Peter McCosker for one believes a decision is unlikely before the end of 2001. Unfortunately for Wesfarmers-Marubeni, the project is a 'green field' or brand new development, and at the same time a critical test case for water rights.

Precisely what the local indigenous community gets from Ord 2 lies somewhere between the courts and the negotiating table. Elaine Gardiner has put the challenge for developers this way: 'How can they go ahead with this and not forget the community they're coming into, particularly the Aboriginal? Part of my work here is Aboriginal liaison. We sort of know who's involved, the Kimberley Land Council are the ones to talk to.'

In July 1998 a presentation was given by Wesfarmers to the traditional owners, and community consultation has been ongoing in Kununurra and Wyndham. Warren Ford, the company man based in Kununurra, seems to have real empathy with the local indigenous people but decisions are made from head office in Perth.

'Wesfarmers' favoured position is to have an indigenous land use agreement with the Miriuwung Gajerrong people over the entire project area,' Wesfarmers' Andrew Hopkins said generously. With about 40,000 hectares of the proposed area left undeveloped, Elaine Gardiner says the hope is that Aborigines can manage some of it as part of their country, but she agreed that some Aborigines have asked for an irrigated block. On this key question, Andrew Hopkins volunteered that Wesfarmers 'hadn't got to that stage yet'.

The natural instinct of a developer is to hang on to the best land. The area for development, we are reminded by the developers, is just run-down pastoral leases worth a mere $2 to $5 a hectare. Once developed, the land sells at $5,000 to $6,000 a hectare. This is a crucial point in negotiations with Indigenous groups. The developers argue that the cost of preparing the land with lasering and infrastructure makes the project almost marginal. 'If you take the value of the land developed and work back from that, in the terms of the cost involved, like roads, water supply and electricity, it's a negative value,' explained Peter McCosker. Surprise, surprise. It seems traditional owners, like their Green colleagues, have a rather different value system when it comes to land prices.

Ernie Bridge grew up in the Kimberley and is still Australia's only Indigenous Australian to hold a cabinet position, as a former Labor Minister for Water resources in the west. It's clear that he would like to see the indigenous community owning part of the action, that

means, irragable areas which they could farm. 'Australian governments have been absolutely remiss of one thing,' he stated emphati-
cally. 'They've never understood the real value of genuine stakeholding. You've got to say some of these people out there have a
genuine and a legitimate argument to talk about their equity, because
at the end of the day that's the only place they'll assume a genuine
stakeholding position. You can talk about compensation until the
cows come home. That's the basic fundamentals of good business you
know, and I believe that this Government has not yet figured out the
very basis of good business.'

Trying to establish exactly what indigenous groups are entitled to
in the Kimberley at present is impossible. Yet regardless of the court
outcome, some form of participation by the Aboriginal community
will happen, and negotiation is under way. Like all country towns at
the coal face of reconciliation, frustration exists on both sides. 'They
have to meet commercial type arrangements, they have to meet all
the competence requirements, and it can work,' Peter said firmly. 'But
there's no doubt here, there's been a real issue and a number of us
have spent many years trying to get Aboriginal community into the
mainstream economy, and one can get very cynical about it I can
assure you. But the opportunities are there for them. Now the old
women certainly very much want them to get into it. There's a bit of
a perception about what you get out of the young people, who everybody is trying to get into work, what they see as their criteria, priorities at the time, and that's a big problem.'

Elaine Gardiner agreed: 'There's very little Aboriginal employment, not because we don't want to employ them, because I've tried
and it just doesn't work. They just don't turn up.'

Other growers, like Robert Boschammer, are more cynical about
the water rights agenda. 'I believe it's promoted by a lobby based in the
ranks of the EPA that are very anti-farming, so they promote
Aboriginal land use, and it's a political agenda by, let's say, the Green
lobby, the Blackfella lobby and anyone else with a bit of get up and go.'

Rob Boschammer and his neighbour Tim Croot have had their
own frustration with red tape. As Tim explained it, 'If you go to a
public servant and say, "I want to build a wetland, and I'm going to
build it here, and I want to use the water out of the wetland to do all
this citrus and the blackfella's going to own a quarter of it," you get
everything together and there's another public servant somewhere
that's going to say, "Oh, no you don't. You haven't got permission off
me yet." So you have to be really careful to include as many as you
can. There's a political route, and it's about time we understand that

there's a commercial route and it's a lot quicker and more sensible.'

Ord 2 is as political as it is commercial and perhaps it is no bad thing. In Kununurra, bureaucrats move quickly in the air conditioning, most blackfellas move slowly in 40°C heat.

Ord on hold

If there is one thing developers cannot abide, however, it is uncertainty. Wesfarmers' Andrew Hopkins has been battling on Ord 2 for three years now, a few less than Peter McCosker but then, when he took it on, it was supposed to be a 12-month job. Adding to Wesfarmers' frustration over most of the project's life has been a Liberal Government that has changed its attitude somewhat, from giving the initial pro-development tick to the Ord by former Premier Richard Court (Dad did Ord 1, I'll do Ord 2) to a schizoid inertia.

'The Court Government through one of its ministers went out to industry in a gung-ho, let's develop the Ord, let's get industry in, let's drive [style]. That's when we came into it,' said Andrew Hopkins reviewing the situation. 'At the same time, other parts of Government, the EPA, have been less keen for the project to proceed as have lately the Waters and Rivers Commission. So we're talking about a project that's been in the planning stages for five years and yet we still have hurdles to jump and hoops of fire to launch ourselves through with a number of departments within that same government.'

Since my conversation with Andrew, Labor's victory in February 2001 has been another change to contend with, although government policy is to support development. For Wesfarmers and Marubeni, the nightmare is that indecision will rule. The Water and Rivers Commission, terrified that Lake Argyle might not be the water equivalent of the magic packet of Tim Tams, will insist on waiting for the outcome of new Murray–Darling Basin duckwater flows before it will decide on its own levels. 'Paranoia with a capital "P" is the best way to describe it,' said Andrew Hopkins. 'I think all these guys around here are waiting for [Murray–Darling flows] to get sorted out, for that to take the lead. To use precedents that have been used over there and then stand back and say, "Yes, that's the way things are done in Australia, just look at the Murray–Darling."'

Ord 2 will go ahead, said Howard Dengate. 'Oh yes, it's just a matter of when. The water's been sitting there for 28 years, waiting to be used.'

Peter McCosker agrees. 'Well certainly we won't turn earth in 2001, in my view. At this stage of the game, it's still certainly feasible that earth could be turning by the dry season of 2002.'

Impatience is building on the project. Wesfarmers has already had its mandate from Government rolled over twice and according to Andrew Hopkins, it is unlikely to want that to happen again. 'I think some time [2001], I couldn't say exactly when, Wesfarmers will come to the view: yes we will proceed with this project, or no we have been at this for so long it is so frustrating, we will just devote our resources to something else.'

For the Greens, a dejected Westfarmers-Marubeni is the best hope. The general manager of Carlton Hill Station, George Warriner, certainly isn't preparing for an imminent handover of property. 'Our cattle will still be feeding out on the plain right up until the night before the dozers move in.'

Cotton extra

Those from the east fearing a deluge of cotton on the Ord River would have been relieved at the news that sugar, not cotton, is to form the backbone of Ord 2, at least for the first decade or so. But cotton is far from dead around Kununurra.

Across the Packsaddle Plain around 900 hectares of BT genetically modified cotton was grown in 1999, carefully nurtured and monitored by the State's Frank Wise Research Station. Elaine Gardiner reports that the new varieties are good at resisting bugs but they don't like the cold weather and night-time temperatures of the dry season.

'The integrated pest management system being looked at now is very promising,' said veteran Peter McCosker, 'and that's from someone who was highly sceptical at the start.' Rob Boschammer feels cotton still has teething problems. 'From the end of the wet season to be able to get it in, to the end of the dry season to get it off, our climate is a bit limited at that time. Seed crops like cotton are also a lot harder to get going, so as you get into terrain that is slightly more difficult, it gives cane a better advantage.'

Whether the Ord becomes extra-ord in national farming terms remains to be seen. Geographically, the Kimberley couldn't be better placed for Asian exports. In a country where politicians go on about Australia's ideal geography, it's as well to remember that Sydney is nearly as far away from Thailand as Western Europe. In irrigation terms, the potential is there, but so is the ability to stuff up one of Australia's more pristine corners.

It's hard to find an irrigator in Australia who will predict disaster in our rivers. Tim Croot cannot see it happening to the Ord. 'The reason I don't think this will go the way of the Murray–Darling is that we haven't got that sort of conflicting land use or ideas on land and also we've got a flow off system. There's always going to be water flowing off this area and you're never going to want to prevent it.'

'Never say never,' say the Greens.

12

Dinosaur Water

As the drill is plugging downward at a thousand feet of level,
If the Lord won't send us water, oh, we'll get it from the devil;
Yes we'll get it from the devil deeper down.
            A.B. Paterson—'Song of Artesian Water'

Water that dripped off the dinosaur's back

Bottled water is on sale at every deli in the country. In one of the more ludicrous global developments, we happily ship in European bottled water and pay extravagantly for it. The last place anyone would think it came from is Australia's dry interior.

More than 900 kilometres west of Brisbane the small country town of Mitchell, 'Gateway to the Outback', is busy bottling local water. Not from above ground, though, the town gets just 525 millimetres a year. Mitchell's water comes from 90 metres under the ground. 'Great Artesian Water,' boast the attractive orange labels, 'from the water that dripped off the dinosaur's back.' A cheerful-looking dinosaur also graces the label.

The marketing idea came from Brian Arnold, the local business catalyst. 'I was actually in England and drinking some spring water there. They said it dated back to Caesar's time and I immediately thought of Romans walking around in their togas and maybe some water dripping off their shoulders onto the ground and seeping into the water table—and I was drinking it. So when I got back to Australia and the outback I thought, well, we've got Great Artesian Basin water and it dates back to the Jurassic period, so this could be the water that dripped off the dinosaur's back.'[1]

You can do more than just drink the bore water in Mitchell. You can 'sooth the savage beast within' as the literature advertises, by relaxing in the Great Artesian Spa. In fact, so luxurious is the spa and delicious the local wild boar sausages that tourists find themselves making the trek back the same way on their return.

The dinosaur water comes from a bore belonging to grazier John Douglas, who reckons the water is so good he drinks it in preference to water from the rain tank. And he certainly won't let his wife use any of it for her gorgeous garden.

It is this same dinosaur water that was the lifeblood for Australian settlers in the nineteenth century, a huge underground oasis that allowed graziers and their animals to survive inland Australia. Sadly, like the Murray–Darling system itself, we have abused as much as used this extraordinary asset. And we are only just coming to accept that something has to be done.

The making of Jurassic Park

The basin, or the 'GAB' as it has become known, is the largest artesian groundwater basin in the world. 'Artesian' means water from underground that is also under pressure, named after the old French province of Artois, where water from bore wells was found to flow freely. The GAB is enormous, 1.7 million square kilometres running under one-fifth of Australia, from the Great Dividing Range to Lake Eyre, from the Gulf of Carpentaria, in Queensland, through New South Wales, South Australia and the Northern Territory. It stores 8,700 million megs, or 17,000 Sydney Harbours, sitting between just a few metres and 3 kilometres under the ground. The quality of water varies, from Mitchell quality to quite saline.

As well as the massive pastoral industry, many of Australia's outback towns are dependent on the GAB. Take a shower in Cunnamulla and you'll soon notice the bad-egg aroma from the sulphurous fumes which make you wonder whether the body is cleaner after the process. In fact the fumes soon dissipate and if water is left to stand, it's quite pleasant. In some areas, showers get close to scalding (bore water ranges from 30 °C to over 100 °C in the deeper areas). The west Queensland town of Thargomindah was the first to convert the heat to electricity in 1893.[2]

If you look at a cross-section of the Australian continent, the GAB has been called a mummy bear bowl in a daddy bear bowl, like the diagram on page 262. It was formed when layers of sandstone capable of holding water became sandwiched in between harder clays. As noted, this aquifer can be up to 3 kilometres thick in places.

About 200 million years ago, at the beginning of the Jurassic period, the edges of the basin were pushed up and rain formed rivers which began creating floodplains of sand and gravel. When the sea rose early in the subsequent Cretaceous period, over 100 million years ago, it invaded much of the basin and dumped a seal of shale and

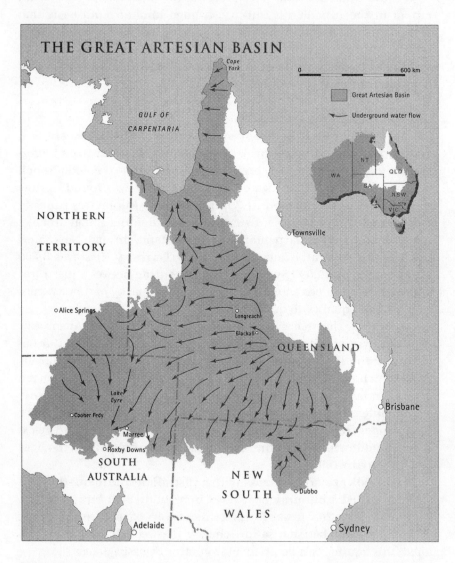

THE GREAT ARTESIAN BASIN

Source: Great Artesian Basin Consultative Council

mudstone. The final touch occurred about 30 million years ago, when the mountain ranges of the Great Divide were formed and exposed some of the permeable sandstone, which to this day allows water to seep back into the GAB on its eastern border. Dinosaurs, by the way, disappeared around 65 million years ago.

As a result, GAB water moves at between 1 and 5 metres a year, mostly from Queensland towards South Australia, but in the northern part, out towards the Gulf of Carpentaria. There is some water recharge in South Australia, but most water seeps into the Basin in the east, travels across three States and comes back to the surface as springs predominantly in the middle of the desert in South Australia.

The 'mound springs', as they are known, are almost biblical in appearance, bulrushes poke out of the desert, and up close water oozes from the ground. Some of the more spectacular ones, such as the Blubber and Blanche Cap, are true mounds and give the springs their name. The water, which has picked up salts on its long journey through Australia, deposits them around the spring creating earth mounds, some perfectly regular and surrounded by 'fluffy' soils from all the bicarb. By the time it gets to South Australia, GAB water is said to be 2 million years old!

The squatter's gift

The earliest discovery of GAB water came near Bourke in New South Wales in 1879, when a bore was sunk into an old mound spring called Wee Wattah. Queensland's government geologist tried to prove there was water under Blackall, but before he had found it, water was struck at 200 metres on a property near Barcaldine in 1886, and at 300 metres at Thurulgoona Station near Cunnamulla in 1887.

The biggest obstacle to squatting had been overcome. Water has filled Lake Eyre only nine times since white settlement. The arid inland which overcame Burke and Wills in 1861 was invaded. In the end, the explorers may have 'contributed more to European knowledge of the country by their failure than they would have in a successful race to the north coast and return. Expeditions sent out to find [Burke and Wills] explored vast areas of the Basin's surface and the pastoral industry soon followed.'[3] However, in many cases explorers found the horse tracks of pioneers ahead of them.

By the early twentieth century, pastoral lands had spread west from the Mitchell grass plains around Longreach, and floodplains of Moree, across miles of gidgee mulga, brigalow and spinifex grass, and right into the unforgiving country of the Oodnadatta Track and beyond.

In 1899, more than 500 wells were spilling up 1,000 megs a day, and by 1915 flow from the GAB was peaking at just over 2,000 megs per day with 1,500 flowing bores working right across the Basin. Some flowed at rates of over 10 megs per day—that's five Olympic swimming pools gushing out a day. Along with the bores came many thousands of open channels, dug across vast tracts and running the bore water vast distances to reach vulnerable stock. These bore drains or channels became permanent water troughs, sometimes 100 kilometres long. Bores were a lifesaver. As A.B. Paterson wrote:

It is flowing further down, to the tortured thirsty cattle, bringing gladness in its going; through the droughty days of summer it is flowing, ever flowing.[4]

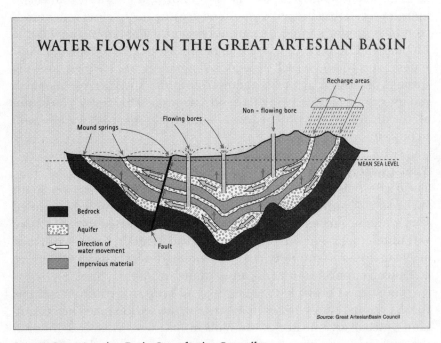

WATER FLOWS IN THE GREAT ARTESIAN BASIN

Recharge areas

Non - flowing bore

Flowing bores

Mound springs

MEAN SEA LEVEL

Bedrock

Aquifer

Direction of water movement

Fault

Impervious material

Source: Great ArtesianBasin Council

Source: Great Artesian Basin Consultative Council

Bored stupid

Until whites arrived, the GAB (like everywhere else in Australia), was in equilibrium. Such was the pressure in the aquifers of the GAB, that in the early days any bore intersecting water would gush up. Unfortunately, while the basin is big, only a limited amount could be taken without affecting this balance. And a total Basin recharge of only 300 megs a day in Queensland couldn't keep up with the squatters.

The change that bores brought to the landscape has been dramatic. The natural flows to the mound springs are estimated to have declined by 30 per cent over the last 100 years, and many have ceased to flow altogether. There is also growing evidence that the base flows in some of the rivers highly dependent on groundwater are also disappearing.[5] But there was much more widespread damage in grazing territory. In the words of biologist Ross Blick, 'If you want to kill a dry landscape, just add water.'

Ross Blick's study of about 1.6 million hectares around Thurulgoona Station makes fascinating reading. He based his study on the work of R.L. Heathcote, who in the 1960s described how the gradual increase in bores and bore drains in the period 1886 to 1914 had created permanent artificially watered country.[6] Using Heathcote's work, Ross plotted the availability of all surface waters in the area. Nowhere is over 10 kilometres from open bore drains or waterholes, presumably to spread grazing pressure. What he calls the 'unmeasured consequences'[7] of these bores is that pasture has been devastated.

Ross Blick's analysis goes like this. Before white settlement, the country was ungrazed or only very lightly grazed by animals, such as bilbies and rats, which needed very little water. 'Before the bores, roos could smell a storm maybe three days away and they'd start heading for it, and by the time they'd got there, the silt had spread across the land and there were new shoots for the eating.' The land was also very prone to fire, which kept the vegetation as savanna. As one by one the bores were sunk, water brought first many more kangaroos then cattle, which led to overgrazing and a reduction in fuel for big fires. Across the landscape, in came the mulga and tougher woody weeds, like brigalow and saltbush. In came the feral cats, dingoes and foxes that took out the tiny marsupials. Compare this, he suggests, with Charles Sturt's exclamation in 1828 of the grass in drought country around Moree in New South Wales, 'waving as it did higher than our horses' middles as we rode through it'.[8]

To make matters worse, the GAB water was adding salt. 'Sixty million megs of water has been taken out of the GAB,' said Ross, 'that's roughly 60 million tonnes of salt and a whole suite of salts, not just sodium chloride.' Bore drains channelled the salt down to salt-scalded wetlands at their terminus. Run-off increased and the earth-walled dams that graziers had built to reduce evaporation became highly effective silt traps, insidious earth robbers of the landscape.

It is a sad admission that as long ago as 1891, the penny had dropped with water users. The water pressure was falling because already too many bores had been sunk. Bores which had flowed freely stopped and became 'sub-artesian'. Still, their number increased, sinking ever deeper, some up to 2 kilometres. Between 1912 and 1939 there were no less than six interstate conferences on bore management, and a report was finally published in 1954, but little was done. By the mid-1980s water pressure had fallen and the level of water in bores sunk to 80 metres; about one-third of all bores had stopped flowing.[9] By 1995, flows from the GAB had fallen back from the 1915 height of 2,000 megs per day to 1,200 megs a day.

There are wonderful archival pictures of open bores, but they really have to be seen to be believed. A typical one on the Oodnadatta Track in South Australia put down many years ago and still running today gushes piping hot water 24 hours a day into a pond of heat-tolerant algae. It seems on a number of occasions, a distraught tourist's pet dog has accidentally boiled when it rushed out of the owners' car for a dip. From the pond, the water feeds down the bore channel which eventually runs out of sight (a long way on the flat land around Lake Eyre). But with channel seepage, 40°C heat and evaporation of 3 metres a year or more, just 2 per cent of the water is actually drunk. Thirty-four thousand kilometres of open bore drains are currently in use, that's enough to run to the other side of the world, with pipe to spare.

The bore fixer

Various weak efforts have been made to halt the water exodus—efforts that have not been helped by the usual problems of State differences. State laws from as early as 1910 required that all artesian bores be licensed, yet as recently as in 1998, head of the NSW Department of Land and Water Conservation estimated that unlicensed bores in the State could top 50,000.[10] Literally hundreds of uncontrolled bores are still running. Some were built with valves which controlled the release of water, but many of these are now corroded and useless. The steel casings that line the bore hole to prevent seepage into the surrounding rock are also corroded and now leak.

In 1989, State and Federal Governments launched a program to rehabilitate old bores, replacing the casings and adding new heads. With all the pressure and heat, this can make for highly dangerous work. The joint government plan offered to pay 80 per cent of the costs of fixing up bores if the grazier coughed up 20 per cent. Unfortunately with a drought hitting hard, farmers were closer to choking. However, one bore near Moree which needed immediate attention demonstrated quite successfully what benefits rehabilitation could bring, with pressure returning to the area months later.

Not far from where some of the first bores were sunk lives the man with the weighty task of stemming the flow. John Seccombe has a 26,000-hectare sheep and cattle property, Kenya, some 120 kilometres north of Longreach, where during the 1980s he designed his own reticulation scheme. Since then he has been slowly but surely drawn into the bigger picture and now chairs the GAB Consultative Council, a body set up by the Commonwealth in 1995 to rally all governments, business and communities into saving the GAB.

Late in 2000, after two and a half years of deliberation, the Council came up with a new plan for the Basin:[11] not just fixing up the bores, but replacing the open drains with polypropylene pipe. A similar scheme had been initiated in southwest Queensland from 1993, but at last this was a national effort to address water wastage. Of the 3,700 or so bores known to be still flowing, the Council's plan targets 880 of these for rehabilitation. It also aims to have 34,000 kilometres of bore drains replaced with poly. The Council's goal is for all bores to be controlled and piped in 15 years, with 30 per cent completed in the first five years, along with environmental work in areas like the mound springs.

Once again, graziers have been given a sweetener of matching State and Federal funding. Just to make life complicated, the contribution from farmers differs in each State: in Queensland and South Australia it's 60 per cent government, 40 per cent farmer; and in New South Wales the contribution varies from 60 per cent down to 20 per cent depending on how remote the property is. In addition, in all States farmers are able to contribute 20 per cent in kind with labour and machinery.

At the grassroots level, however, many farmers have been far from enthusiastic about the upgrade. Benefits do not kick in until months later with increased bore pressure and better grazing as ferals are reduced and cattle and sheep can be rotated by closing a bore in one area, and opening one in the next door grazing area.

Apart from the 'if it ain't broke' logic, 20 per cent of the changeover cost is no small beer even in good times. In New South Wales some farmers have now had three failed crops in a row, but governments aren't much good at understanding cycles, according to John Seccombe. 'This is worse than the dairy situation outside Victoria. Dairy farmers had to carry on with 30 per cent less income, but graziers are having to contribute anywhere between $20,000 and $60,000. This is why it's so hard to convince government that it is an equity issue, and not just one of private benefit to the farmer.'

What is even less appreciated in cities, thousands of kilometres from where these bores lie, is how the move will change farming. Old channels ran 24 hours a day, 365 days a year over hundreds of kilometres. Once piped, troughs have to be checked for water regularly, and contrary to popular belief, most farmers in arid Australia do not have a light aircraft available to cover the vast distances between troughs. As John Seccombe's number two in Adelaide, Lynn Brake recalled, 'I sat down with this bureaucrat and it made me angry. [I said] "You go and do his water runs when it's 45 degrees! You say you're paying 80 per cent, but we're changing this guy's lifestyle."'

The art of persuasion is left to the Seccombe team. To the farmers they represent the shackling process of government, and in town they battle to explain the farmers' position. Unobtrusive and patient, it's not hard to see why John Seccombe was the man for the job. Above all, he's a cattleman. Precisely the man needed to convince graziers to cap their bore heads and make the change from open channels to pipes and troughs. 'It's a human psychology thing and bureaucrats don't understand this. Farmers need someone there to give them confidence, not to stand over them saying, "You're going to do this and you're going to pay".'

In fact, the coaxing side has been going quite well. What John Seccombe didn't bargain on was being let down by governments. In the Council's plan, on the Net for all to see, was the promised funding of around $300 million over 15 years for the bores and drain replacement. The document was even signed by ministers from all governments involved. Unfortunately, the Federal Government did not lock the States into upfront matching funding. After two years of coaxing and planning, John Seccombe is dumbstruck. Inconceivable? Not where State money and elections are concerned. 'That's obviously why there were conditional tranche payments to the States with the COAG reforms,' John said bitterly, alluding to the $5.5 billion paid to the States for water reform.

Naturally, some States have been more cooperative than others. The most uncooperative has been Queensland, the State where most work is needed. 'I've spoken to Peter Beattie a couple of times,' continued John. 'He says he's committed to "environmental outcomes" but I don't know where his credentials are on this. In Queensland, they say, "Well we'd rather fix up Lang Park than put the money into a rural issue."'

Even if governments did come to the table, the job is much bigger than anyone wants to concede. John Seccombe for one admits that the 880 bores due for rehabilitation is really a fictitious figure. 'Out of the 3,700 bores there are, 880 are the only ones without a tap, but for most of the others, if you turned the tap off it would blow the bloody casing out of the ground. But government, they really don't want to know about it.'

Elsewhere there are questions about the economics of 'subsidising' graziers. Few people have heard much about the pillaging of the GAB and even fewer give a damn, which is infuriating for John Seccombe. 'Of all the water used in Australia, a third to a half is groundwater—that's in almost every State—yet where do you ever see groundwater mentioned? Government only responds to what you see. I get so wild I feel like suing them all in court sometimes.'

As Ross Blick sees it, GAB water is a sovereign asset and 'Government should fund the graziers out of the whole bloody lot. It should recognise that the Queen owns the land and the Queen owns the water. We decided that the other day.'

The GAB is one of those sleeping tragedies, a jewel greater than the Barrier Reef for Australia but we can't see it, so we can't see it. Fixing it up would help farming and bring tourism to the mound springs, and indigenous communities should be able to return to places where there was once easy access to water. There might even be room for new users (read irrigation and mining). Cheap at the price?

The bore war
Graziers are not the only beneficiaries of groundwater. About 11,000 megs a year are used by irrigators.

Groundwater presents us with yet another twist in the tangle of water rights that are confronted by growers and regulators on a daily basis. Water reform in the area of groundwater finally got going in 1996 but none of it was linked to the State tranche payments (see chapter 5) and COAG has been warned that if the rules aren't the same between ground and surface water, cutbacks in water allocations from rivers will mean that everyone will be sinking bores.[12]

If the task of deciding duckwater levels in rivers is hard, imagine how it is for underground reserves, deciding what the right extraction levels should be to keep the GAB resource sustainable. Nowhere is this a more sensitive issue than in the cotton-growing country of the Namoi Valley in New South Wales, almost all of which consists of family farms. When the NSW Government produced its white paper on water reform in early 2000 proposing new water extraction levels for users, growers found themselves looking down the barrel. Some faced cutbacks of up to 80 per cent of their bore water over the next ten years. 'Financial compensation as a result of adjustment to water allocation is ruled out,' the State Minister for Land and Water Conservation said on his trip to Gunnedah. That same day, 1,250 furious irrigators and locals rallied in the town. Even McDonald's shut for the occasion.

Deputy Prime Minister John Anderson took more than a passing interest in the Namoi irrigators. He was their local member. He set up a task force which recommended structural adjustment, the funding to be shared equally between the Feds, the State and the irrigators. In June 2001, Anderson offered $40 million Federal compensation for Namoi irrigators, but absolutely conditional on the State contribution. 'We'll put our money on the table, but they have to agree,' he said firmly. 'My real concern is the moment we allow a State government to make decisions which impact on farm communities in social or economic terms without having to face the issue of adjustment because the Commonwealth will step up, they will feel completely unencumbered and able to act without any responsibility.'

Regardless of the outcome, the adjustment package won't be anything like the estimated cutback of 170,000 megs annually, a value that Jim McDonald, chair of the local groundwater committee, put conservatively at well over $100 million. Now the task is to decide who should get the assistance (if and when it arrives), given that the licences cover a range of uses and include dozers amongst them. There is also a difference of opinion over what the value of permanent water in the valley is. It's a task Jim McDonald has not enjoyed.

Mining water

Aside from towns, graziers and farmers, there is another big user of GAB water: miners. About 11,000 megs a year, the same as for irrigators, is taken from the GAB by mining, but it's a figure still dwarfed by the pastoral industry's 500,000 megs.[13] And the largest individual taker is one of the most powerful blue chips in the country: Western Mining Corporation.

Back in 1975, Western Mining's first exploratory drill at Roxby Downs, 500 kilometres north of Adelaide, found copper. Hopes were dashed when a further eight holes came up blank. The driller was in the process of boring the tenth at Olympic Dam when, according to the company's community relations bloke, David Stokes, 'Hugh Morgan and Arvi Parbo flew into Roxby Downs homestead, decided to stay overnight and then come out and give Ted the bad news next day—that it's all over—and by the time they got there, he had found—at 350 metres—nearly 800 metres of copper, gold and silver, and so they changed their minds.'

Western Mining has never looked back. Olympic Dam now holds the world's largest known deposit of uranium, producing an estimated 4,300 tonnes of U_3O_8 a year, almost enough for the whole of the country's electricity requirements.[14] And it's the world's sixth largest copper deposit, with gold and silver to boot and a mine life of up to

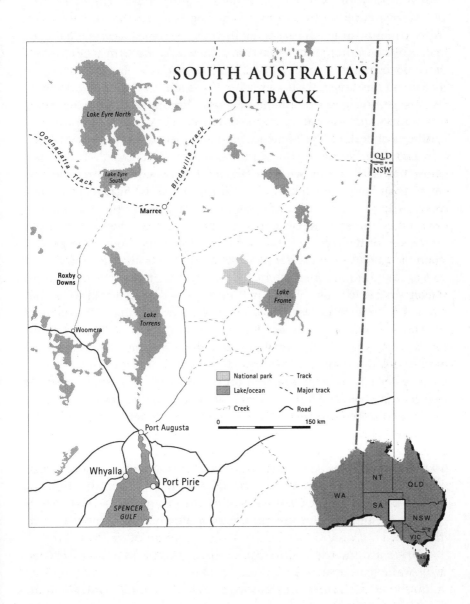

200 years. It is a massive operation, 200 kilometres of road underground—the ascent alone is 7 kilometres to the surface.

Like all mines, Olympic Dam is controversial and two points are worth a mention before a bit of DIY digging. First, it is one of the many hypocrisies of consumers and the media that there are only three times mining appears in mainstream news reports: when there's a death, when there's an industrial relations dispute, or when there's an environmental crisis. Yet mining allows us to have the cars we drive and the houses we live in. Equally, it is as well to remember that Western Mining is one of the best cost-cutters in the business and feeds the South Australian Government with royalties of hundreds of millions of dollars. The State has bent over backwards to help it along.

The chief executive of Western Mining, Hugh Morgan, is Australia's most powerful miner with top-drawer Liberal connections. He was more than happy to explain how mining shapes up against rural Australia when it comes to water efficiency. 'If you were to take out the amount of water per dollar of revenue exported or produced, we're right at the top of the tree. Just take Olympic Dam. Its exports are greater than the whole of the wine industry in South Australia. And you've also got to take into account the quality of the water that we are using. I mean, we'd love to be able to have our share of the fresh dam water that cities and irrigators' use. Most potable water is, what, 30 parts per million of salts. We vary across the company from 3,000 to 200,000 ppms! Discovering it, harvesting and then delivering it is a significant cost for us to support the mining activities and communities.'

Western Mining is as reliant on water as any grazier, in order to separate its bounty from its trash. Each day, thousands of tonnes of rock are crushed to a powder and water added to create a slurry. The precious metals float to the surface and are skimmed off; uranium sinks to the bottom. As luck would have it, sitting just 100 kilometres to the north of this vast natural (mining) resource lies another, also under the ground: the Great Artesian Basin. Western Mining now takes 34 megs a day, or 0.3 per cent of all GAB water used. Of this, 6 megs is used in the purpose-built town of Roxby Downs, not far from the mine.

Drive around the country at Olympic Dam and it is easy to see why water comes at a premium up here. My guide, David Stokes, admitted it's almost as precious as what the company is digging out. Rainfall at Olympic Dam is just 160 millimetres a year, often falling all at once.

The land is low, undulating plains of saltbush and the odd stunted mulga, interspersed with unwelcome sand dunes. This is the land of paper lakes and rivers, alive only on maps until the blue moon shines. It must be said that the actual dam at Olympic Dam, named at the time of the Melbourne Olympic Games, falls a long way short of expectations. As we sped past the dam in the pick-up, David volunteered, 'I know a lot of our detractors and people on the east coast, when they come over here to look at Olympic Dam, they expect something like the Yarralumla or the Hume Weir. I don't know where they get that idea from.' Well, the name.

The company baulks at the idea it gets its water for nothing. It *is* free, of course. The cost is in the delivery, at around $2.80 a kilolitre (about three or four times the average delivery price for Australian cities). And town water for Roxby costs a whopping $6.80 per kilolitre, partly to encourage thriftiness. But how Western Mining has obtained its water and is gathering more, even today, is fascinating.

During the 1980s, Western Mining gained State Government approval to take Basin water and nine wells were sunk 110 kilometres from the mine known as 'Bore Field A'. At that stage water was carted by truck while pipeline issues were being sorted. And in those days, sorting was pretty straightforward. 'It was pre-Native title,' explained David, 'It was also pre the requirement for heritage surveys. There's an awful lot of water that has run under the bridge Aboriginal-wise over the last 15 years.'

Water extraction from Bore Field A peaked about five years ago at 19 megs a day, but it soon became apparent that Bore Field A was not good for the Basin. It had been thought that a fault zone separated the bore field from the mound springs but this was not the case and by 1990 spring vents had ceased flowing altogether.[15]

The problem for Western Mining was that its operations were expanding, from 85,000 to 250,000 tonnes of copper a year. More water, not less, was urgently needed. By this time the South Australian Government had done its bit to fix up bores in the State, improving overall water pressure. It was well aware of the mine's economic benefits. Luckily, most were government-owned anyway, either on stock routes or railways. Rehabilitation (and improved efficiency) is expected to be complete by 2004. The company too has been working on improving water efficiency and over 20 years has managed to halve its per tonne usage. Yet this was clearly not going to be enough.

The solution? A costly one of $150 million. In 1996, Bore Field B was constructed, another 90 kilometres further from the mine than Bore Field A with wells sunk to 1,500 metres in the deepest part of the Basin in South Australia. As a result, 28 megs a day are taken from Bore Field B, with A cut back to 6 megs, with three of its wells closed.

A high-tech desalination plant treats about a third of the water required for the town and salt-sensitive operations at the plant. 'It's drinkable,' said David Stokes cheerfully of the bore water, 'but as Crocodile Dundee says, "You can live on it, but hey!" The stuff coming out of Bore Field B is better than A. The further north you get, the better the water gets. I don't think I'd want to try and live off A water.'

It wasn't long before Western Mining found itself needing still more water from Bore Field B. The constraint this time was not mound springs, but a rather clever government shackle.

What Western Mining has to work to is a drawdown restriction. At all times, the company has to keep the water table of the Basin within 5 metres of the surface at a certain perimeter around the bore field. This has proved rather tricky, because while perimeters on maps can be circular, the way underground water behaves when it is drawn upon is not regular. It depends on rock type and water pressure. Suddenly, Western Mining was finding that in some parts of the perimeter, it was dangerously close to breaching the limits. And yet Hugh Morgan has plans to expand to 300,000 tonnes of copper, a figure said at Roxby to need not 28, but 42 megs a day.

Something had to be done.

Wooing the graziers

The answer was just over the back fence. Graziers, as we now know, take far more water than any other sector and yet they actually use only a tiny portion. So if Western Mining could coax or coerce the pastoralists into capping their bores, pressure would go up and the company could then take more from its Bore Field B.

It's been estimated by the company that around 134 megs a day are used by South Australian pastoralists of which between 40 and 100 megs a day could be saved by capping. The company wasted no time trying to get the graziers on board. The job of wooing the graziers was given to my guide, David Stokes. The first steps were taken when Bore Field B was sunk in 1996 and seven properties with 21 bores around Olympic Dam were targeted. The carrot held out by Western Mining was to pay the cash contribution that Government was demanding from graziers to cap and pipe each bore under the GAB bore capping and piping program.

It will come as no surprise that the graziers were not a walkover.

Some of the pastoralists approached were cooperative—providing labour, machinery and fuel for installing the new technology and operating the bore. In one comic instance David Stokes recalled that Western Mining was quite wrong-footed. 'We saw one pastoralist. He had installed everything and he was operating, except that he pulled the bungs out of the bottom of his trough, so his troughs were still running back and still had bore drains. "Well you didn't tell me I wasn't allowed to do that."'

Like Lynn Brake and John Seccombe, David Stokes admits that life for graziers will not be the same once free-flowing bore water for hundreds of kilometres is taken off properties. 'Whereas in the past, a pastoralist and his family could go south for Christmas for two weeks knowing that there will always be water available, those days are over, and that is a massive change in lifestyle. And the compensation is that they're not having to pay for the infrastructure. But one of these days, if it's not WMC, it's going to be government—somebody's going to realise what's going on and the hammer is going to fall and there won't be WMC's money to help and you're going to have to do it yourself.'

David Stokes had his own problems with some of the graziers. He remembers one, Paul Broad at Etadunna Station, deciding to put down new bores to irrigate cattle feed. 'He was pulling close to 22 megs a day. That threw our planning right out. That was fine as far as he was concerned and he got ministerial approval to say so. So we were pretty jacked off about that, the fact that nobody bothered to tell us and it did affect our drawdown.'

Paul Broad's recollections are not quite the same. 'Look, I hate wasting water, water is the most valuable thing we have and capping is a brilliant thing, but I have to chuckle when Western Mining talk about water savings. Savings is the wrong word I reckon.' Paul Broad said he was taking more like 5 megs, not 22 megs, and he also has another use for the water which ran free down the bore drains. Paul is one of the blokes with real initiative in the area. Not only is he irrigating oats in winter and Sudan grass in summer for cattle food, but he's also experimenting with aquaculture. 'The yabbies are really successful with all the calcium in the water,' he said. Of course, this high-value use is directly in competition with Western Mining.

Paul Broad also maintains that contrary to David Stokes' story, it was the big miner who wrong-footed both him and other graziers. 'We drilled that bore a year before Bore Field B even came along, spent private money to do it, but the company knew they were going to sink bores right next to me all the time.'

Western Mining tried everything to get the graziers to cooperate, including offering to buy out Paul Broad. 'He declined to sell the property to us and he showed himself to be a very independent young man. He's a big bugger, ten foot tall and bullet proof. He's a black belt in karate and a shock of blond hair and a couple of teeth missing. If you bumped into him on a dark night, ahhh!' enthused David Stokes, adding in the true spirit of liaison, 'But he's a bloody nice bloke. We were embarrassed because there were people in the organisation who thought they could control Paul. Silly people.'

Paul Broad is amused at David Stokes' take on things and the 'generosity' of Western Mining's offer to pay the grazier's cash contribution towards bore capping. 'In their contract,' he explained, 'we have to apply for every new watering site, even if it's for cattle. I mean, we're basically handing over control, handing over our future. I offered to sell them water, but they refused because they don't think I have a right to that water. Yet they wanted to buy the property, so they obviously do realise I've got that right.'

To give an idea of just how important the water is to Western Mining, the company extended its 20 per cent offer to *all* pastoralists in South Australia using GAB water, costing the miner $1.2 million. For Western Mining, it's worth every cent. 'By the end of this year, we will have shut in enough bores to equal the amount of water we are presently using at Olympic Dam, 34 megs a year,' said Hugh Morgan. And does he believe that Western Mining's use is sustainable? 'Absolutely sustainable.'

Not everyone believes that Western Mining is helping the situation. According to hydrologist Dr Gavin Mudd, from the University of Queensland, if Bore Field B continues to go like the clappers, pressure will fall in the area and there won't be any recovery of springs in the long term. What he and some Greens are calling for is Bore Field C to replace both A and B to be built at least another 100 kilometres further to the north-east where the water belt deep down is much thicker. Naturally this would be another very costly venture for Western Mining.

Australian Conservation Foundation (ACF) President Peter Garrett accepts that mining can be environmentally sustainable, but only if the full cost of rehabilitation is met by industry at the beginning and during the mine life. As it is, he has warned, 'The drawdown of Western Mining out of the Basin is hugely excessive and it will lead to the long-term deterioration of the land and water resource.'

Coexistence

There is another very important group of people with long-held interest in the Basin: the Indigenous people. About 15 per cent of the GAB is Aboriginal homeland and another 20 per cent is under claim. In the Bore Field A and B areas, artefacts such as stone spearheads and stone tools, perhaps 5,000 years old, are still scattered across parts of the landscape. Not surprisingly, the mound springs have special Dream-time significance.

The nearest outback town to the bore fields is arguably the most fascinating in Australia. Marree in South Australia sits where the Birdsville and Oodnadatta Tracks come together, about 70 kilometres south of Lake Eyre. Most of its residents are a startling mix of Abori-ginal and Afghan. 'Marree used to be a rail head,' explained David Stokes, as we drew into the town on the Oodnadatta Track. His liaison duties make him Western Mining's contact with the local community. 'The Afghans used to run all the camel trains, caravans into Marree with supplies to the Alice back before the railway reached Alice Springs. The 1920s was about when they stopped caravans, and when the camels were no longer required a lot of them were shot but the people stayed in Marree and got jobs on the railway. Then the rail-ways closed down 25 years ago and people stayed.'

According to locals like Blanche Dodds, the shallow aquifer under the town that once used to supply good water is now soiled by diesel leaks from the railway days.

David Stokes explained that there are many tribes in the area—'the Arabunna, the Dieri, then there's the Kuyani, the Kokatha. It's a very, very divided community in a lot of respects. I like to say there are 80 people and 200 factions and it just depends on the issue.'

It is a minefield of interest groups, such that any views put down on paper are always hotly contested. But Western Mining's critics say that the company has done its best to keep the Marree community as divided as possible. Not surprisingly, perhaps, the most strident of these critics is Friends of the Earth (FOE), which has kept up a running campaign against Western Mining since the early 1990s.

Despite David Stokes' warm introduction to some of the Marree locals, it was not easy to speak candidly with them on a Western Mining tour! Such conversations followed several weeks later. In 1993, Reg Dodds from the Arabunna people lodged a Native title claim over land on Finniss Springs, a cattle station under a pastoral lease which in 1992 was resumed by the State Government to eventually create a national park. The Arabunna people were widely accepted as the guardians of the land and in fact, some Arabunna had been the lease-holders up to 1992. Unfortunately, this land also included an area which Western Mining had targeted for the development of Bore Field B and land over which their water pipes would have to run back to the mine.

Western Mining also had the ear of Government. In May 1993 the *Australian* reported that 'The South Australian Government promised yesterday to take whatever action was necessary to protect Western Mining's giant Olympic Dam copper, gold and uranium project from a new Mabo style land claim.'

Jan Whyte from FOE spends much of her time helping Reg Dodds in Marree. 'Western Mining set up their own Aboriginal group and all the agreements were signed with them, without anyone else involved. It's easy to be divided if there is money and influence.' The group in question was called the Dieri Mitha Council, formed in 1992 by people who had also been part of the Arabunna community. They too claimed rights to Finniss Springs Station. FOE believe that Western Mining provided money and vehicles to the Dieri Mitha Council. (The Dieri Mitha Council, it should be stressed, was not the same as the Dieri Land Council based at Birdsville, which supported the Arabunna claim.)

Of course one of the problems for mining companies in these circumstances is being caught out not negotiating with the right mob. Whyte says, 'There is a proper community voice, it's the Maree Arabunna Peoples Committee. It's been in place for a long time. The problem was that Western Mining didn't want to talk to people who would say no to them.'

According to Jan Whyte, Aboriginal people from as far away as Hermannsburg in the Northern Territory were brought down to hold a controversial ceremony in the area. That ceremony occurred in January 1995, where again, according to FOE, Western Mining appeared to have financed the Dieri Mitha Council to hold a traditional ceremony on Arabunna land, to cement their claim to the area. It was, said FOE, an act of sacrilege to the Arabunna people. Marree endured a week of violence. Twenty young men from outside the area were reported to have wreaked havoc, ending with the death of a man from Hermannsburg on 12 January.[16]

'There was a death, there were injuries, people went to jail,' explained Jan Whyte. 'Those people have less power now, because Western Mining has got what it wants, the pipeline across Finniss Springs.'

FOE stresses it does not hold Western Mining responsible for the violence, but it does say the miner worked to divide the town of Marree to get what it wanted. Kevin Buzzacott, one of the Aboriginal activists in the area, agreed. 'The breaking up of the Arabunna people to divide and conquer has been a tragedy for us. It's a reality, we've had people who've shot one another. Look, they can manipulate government, the banks . . . I'm born and bred in the area,' Buzzacott continued, 'I've seen the changes since the big company moved in.

A couple of springs are now just trickling out, all mucky. They used to pour out and you used to be able to drink the water like lemonade. The whole area is very sacred, Lake Eyre and the Basin, because it catches all the rivers from up north, the sacredness flows in. That's the spirit of the Lake Eyre basin. Water is the lifeline, the structure of our Dreamtime. So they're virtually sucking the life out of you. You know, we always had a policy, we never deprived people of water, people passing through, but this modern development means the water is taken for a use that is not a good use.'

Kevin Buzzacott is seen by Western Mining as one of the more extreme opponents to the company—indeed a militant, who set up a protest camp when the Stuart Creek Station, which sits over the GAB (next to Finniss Springs), was sold by the Kidman family to Western Mining.

Western Mining chief Hugh Morgan says that relations between the company and Marree are much improved from the old days, and shakes off any suggestion that Western Mining might have employed a divide-and-conquer strategy. 'Well . . . as if we were that smart! The reality is we're trying to do things for the community and groups like FOE, I think, want to maintain the rage and continue their campaign.'

Reg Dodds, well respected in the town and now running a tourism business, still believes water has continued to deteriorate in the Basin area and should not be taken out of it. 'It's hard to gauge the impact on the mound springs because there is so much fluctuation,' he said. Western Mining's pipes now run across Finniss Springs Station, which is finally in the process of being handed to the SA Aboriginal Lands Trust, a move approved by both Reg Dodds and Kevin Buzzacott. Naturally, Western Mining is determined this should not change the Bore Field B set-up. 'That's part of the game they are playing,' commented Kevin Buzzacott. 'They say, we'll give you the place, so long as our pipes can run through it. But we say that's not the point, the point is our water is being drained away.'

As for Native title problems, David Stokes is confident that Western Mining's bore fields qualify as a 'pre-existing use' under the *Native Title Act* and any moves to expand Bore Field B would also be deemed a past act, because the South Australian *State Indenture Act* for the project passed in 1982 anticipates this. That's not the view of the ACF's fellow in Adelaide, David Noonan. 'Remember, the *Indenture Act* is State law and Native title is Federal. There hasn't been a claim that has tested this in courts.'

Kevin Buzzacott is not so optimistic. 'Native title all looks good and rosy on paper, but it's watered down now. They're using it to negotiate. In South Australia, if they gave the land back, it would set such a precedent, I don't think it would ever happen.'

The spill

The uglier side of Olympic Dam is the tailings dams, the waste from uranium mining. The dams are massive and predicted at full mining capacity to cover 800 hectares. There is also groundwater pouring into the mine shaft which can be partly contaminated and is also pumped up to evaporation ponds.

Western Mining got a fright when in January 1994 monitoring bores on the mine site discovered what it calls 'seepage' and the Greens call 'a leak' into a local aquifer. (These were not bores to extract water, merely to test the content of sub-surface water.) The limestone floors of the dams, perfect for neutralising acidic tailings, were either thinning or cracked. Luckily for the company, the aquifer had no connection with GAB water and was about as salty as sea water. According to a Parliamentary inquiry in 1996, Western Mining had 'fixed' the problem. However, there are still serious concerns about Western Mining's water management. For a start, the company 'lost' the contaminated material, an estimated 5,000 megs (or 5 million cubic metres). Hydrology consultant Dr Dennis Matthews took a very close look at the leak for the ACF. 'They say it's fixed and that seepage into the aquifer doesn't matter because it's saline. But there's no such thing as useless water. In the future it will all depend on demand and we already have a desalination plant at Kangaroo Island. Western Mining are taking a very short-term view.' And at the rate the mine is expanding, Dennis Matthews reckons the mine life for minerals is much closer to 50 years than 200.

Under pressure

The GAB Council's bore capping and piping campaign is moving along, but slowly. The total number of flowing artesian bores is increasing as a result of improved water conservation practices.

What is certain is that the mining industry in South Australia will be looking for more water. 'South Australian Magnesium is about to start up,' David Stokes reminded me. 'That's over in the Willouran Ranges. There's the coal and iron-ore fields of Coober Pedy.' Federal Environment Minister Robert Hill also expects more action in the Basin and warned, 'I think in the decades ahead there will be greater reliance on groundwater, but it also needs to be more responsible. We need to adopt a precautionary approach to this, not the approach taken with our rivers.'

After a brainstorming session in Brisbane, Ross Blick and a few others at the Australian Conservation Foundation came up with a future vision for the GAB, a compromise that accepts settlement in arid and semi-arid areas of the Basin, but protects the land and biodiversity. According to the vision, 15 per cent of the GAB is national park. Land is returned to what it once was with bores fully capped and feral animals controlled. Bore water on grazing land is at least 5 kilometres from the national park borders on sheep properties and 10 kilometres on cattle stations, to prevent livestock invading the reserves, and stocking of herds is driven by seasons.

Interestingly, mining flourishes in the Blick future, accepted as a high-value use for GAB water, as is tourism, driven by Aboriginal heritage projects and a major roo industry for meat and skins. Overall, bore capping would restore pressure to 50 per cent of what it was.[17] 'If we fix things up it might take 300 years to recover,' said Ross Blick quite matter-of-factly.

13

Pipes and Pipedreams

We are such stuff
As dreams are made on, and our little life
Is rounded with a sleep

Shakespeare—The Tempest

The potential

It is no secret that on the north Queensland coast are some of the most exciting rivers in Australia: when in flood, the Burdekin, Herbert, Tully and Johnstone are a sight. Some of their water is used to irrigate the never-ending fields of sugarcane along the coast, but there is so much falling from the sky that irrigation is just a top-up in many areas. Indeed, a full day of sunshine in Tully, with the country's highest annual rainfall, is almost a drought. So it is that most of the water in these great rivers escapes to the sea.

These rivers and the ones that run north to the Gulf of Carpentaria, go nowhere near the Murray–Darling Basin. Given the political power that has been wielded to bring Queensland into line over the Murray–Darling Basin Cap, it is impossible to imagine major new diversions on the Condamine–Balonne system to water the dry inland. But the wild rivers further north, are undeveloped. For the pipe dreamers, this northern water has the potential to create a whole new food bowl for Australia, if only some of these rivers could be harnessed.

A water project that could green part of inland Australia is no pushover, however. It would need up-front investment, serious research, political wheeling and dealing, and a PR program capable of winning over communities in a social climate that is fast moving away from the damming and diverting of rivers. In this chapter, we'll meet the politically incorrect dreamers and their schemes—some brave, some foolhardy and some dead. The first of these was J.J.C. Bradfield.

J.J.C. Bradfield

Depending on how you see our water future, Dr John J.C. Bradfield was either ahead of his time or a dangerous planner whose ideas were fortunately nipped in the bud. Yet the Bradfield scheme is one that has never completely faded.

Born in 1867, John Bradfield was a highly respected engineer, involved in projects ranging from the Sydney Harbour Bridge and Brisbane's Story Bridge to dams such as Burrinjuck and Cataract in New South Wales. He was fascinated by irrigation. Yet it was a scheme never realised for which he is remembered. Bradfield was determined to solve the country's biggest natural resource challenge: to turn central Australia into a Ghirraween, or 'place of flowers', as he called it. His plan was to harness the torrid rivers of northern Queensland to bring enough water right the way down to fill Lake Eyre in South Australia, thereby creating evaporation that would bring rain to the interior.

First proposed in 1929, Bradfield's scheme involved damming and diverting the high flooding rivers: first the Tully into the Herbert, then the Herbert to the Burdekin, then to the Flinders River, and finally via a channel, water would flow to the Thomson in central Queensland and run across the vast inland to Lake Eyre.

Bradfield set off on horseback with basic equipment through the rainforests to map out the best points for dams and diversions. He worked with aneroid barometers, painstakingly measuring air pressures at different heights to map out the topography. When he presented his ideas to the Queensland Government in 1936 there was great excitement. The euphoria did not last, though. The critics took hold of Bradfield. The most damning report coming as late as 1974 from W.H.R. Nimmo, later head of the Queensland Water Resources Commission. Nimmo questioned both the storage capacity of the scheme and whether water would even be available, given the growing demands of northern expansion.

The Bradfield scheme was picked up again in 1981, when a Queensland National Party subcommittee headed by Bob Katter, then the Queensland Minister for Northern Development, proposed a variation, one closer to home—'using the upper reaches of the four coastal streams and diverting them onto the Mid-West and Central Western Plains for irrigation of crops, cattle fattening, timber farms and drought mitigation for sheep'.[1]

Pro-development Premier Joh Bjelke-Petersen liked the idea and secured $5 million from the Federal Government for the project. A further report commissioned by the Queensland Government confirmed that Bradfield's ideas were possible, but expensive. 'To justify

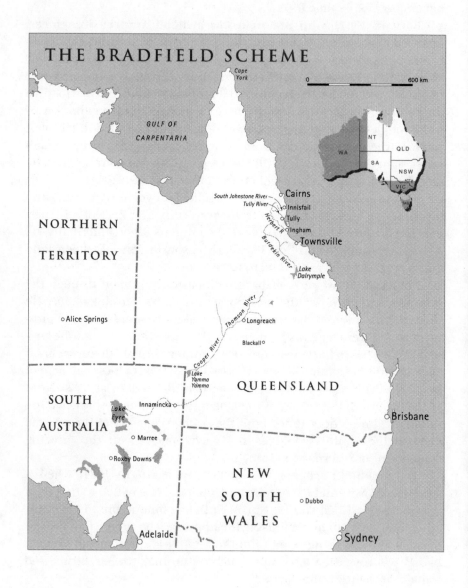

THE BRADFIELD SCHEME

the high cost of such a scheme, the water would have to be used for intensive cropping. Such a scheme would, however, generate substantial direct employment and even more through multiplier effects.'[2] Unfortunately for those involved, the incoming Hawke Government axed all spending programs on the scheme, and despite two further reports commissioned by the Queensland Government (and never made public) the project has lost favour.

Professor Eric Heidecker, who was the agricultural geologist working on the 1981 proposal with Bob Katter, has distant memories of meeting Dr Bradfield as a small child. 'I knew Bradfield. I grew up as a child in Hughenden where my father was a local engineer. Dr Bradfield came by as an old man in about 1939–40, just when the war was getting nasty.'

Bradfield had made a number of mistakes on his mounted trips, hardly surprising given the conditions. But to the delight of Eric Heidecker and Bob Katter, Bradfield's suggestion for the Herbert River diversion to the Burdekin was spot on, with a near perfect gradient and at just 30 kilometres between the two rivers. Eric Heidecker believes today a diversion from the Herbert to the Burdekin could be created at minimal cost, using a tunnel. The project would also prevent disastrous flood damage to cane-growing areas around Ingham, like the event in 1998 that wiped out close to one-third of the cane crop. 'The trouble is that this doesn't appeal to politicians, it's not big enough,' he said.

The Burdekin itself is a wonderful, long river, the catchment running north to south in such a way that it almost always captures some rainfall. Since Bradfield looked at it, however, a massive dam has been built at its lower end, near the coast. And as predicted, coastal development now makes upper diversion of the Burdekin highly sensitive. Townsville, sitting downstream of the Burdekin, is growing fast, particularly industrially. According to Eric Heidecker, despite the unwillingness of the city to become involved in the dam, it was industry that was quietly dropping pipes into the system during the six-year dry spell in the area during the 1990s.

As to the other rivers mentioned in the Bradfield scheme, there are certainly enough obstacles to stop the faint-hearted. The Tully River already has a small power-generating capacity that government might not want to forgo. And further north, the Johnstone would require enormous energy to pump water inland, as would any suggestion of diverting water from the Burdekin into the Flinders. Eric Heidecker believes nuclear power is about the only solution but community sentiment on things nuclear puts the kybosh on that option.

Not everyone has given up on Bradfield, however.

The politician

Thursday 23 November 2000 was a very important day for Ernie Bridge—the only indigenous Australian who has ever made it to a cabinet position in any government was making his valedictory speech in the Western Australian Parliament. Later that day, he still had time to talk about water.

Ernie Bridge's dreams for water have always been big. Very big. Born in the Kimberley seven years after Bradfield first proposed his grand plan, Ernie dusted off the concept and transposed it west, to open up the vast arid interior of Western Australia. It was as the State's minister for Water Resources that he called for action. 'I found that there were endless, dozens upon dozens, of very important communities that were existing in Australia, that had been built over 100 years ago, but in the year 1987 when I became a minister were still without a reliable water supply. It was then that I was prompted to think in terms of the big rivers of Australia.'

Ernie Bridge put all his energy into the Ord—he had quite different plans for the water now earmarked for the many hectares of sugarcane under the Ord Stage 2 scheme. Water from the Ord would be piped thousands of kilometres down to Perth and right into South Australia, opening up Western Australia's parched inland on the way down. For his supporters, it was a dream well worth fighting for, even inspiring a little poetry.

You've proven your worth, for this dry earth,
With your great Australian dream,
Bringing water overland 'from the Ord River Dam',
To me, that's a mighty good scheme,
And whatever it cost, it won't be lost
I'm behind you all the way,
Let your song be heard by the emu bird
From 'Argyle to Esperance Bay'.

This country needs more sons like you,
To make it stand up straight!
And when you bring that water down,
You'll open up this State.
Sydney has its Harbour Bridge,
And its Snowy River Scheme,
But we have got our 'Ernie Bridge',
And the great Australian dream.[3]

Ernie Bridge's big break came when Carmen Lawrence's cabinet agreed to part with $3 million to work up the Kimberley pipeline scheme. To this day, he believes the pipedream could have come true, as he made clear in his farewell speech to Parliament. 'The one regret I have is that I did not stitch up the Kimberley pipeline plan. Sadness surrounds the fact that, in my judgment, I was 12 months away from pulling it off when the people of this State, in their wisdom, decided to throw us out of government. They might have had good reasons to do so, but it was a tragedy for the State that I was not given the opportunity to at least start that project. That project would have meant more to this State in the future than all the other projects combined.'[4]

Today, Ernie Bridge is undeterred. And he has plans for many other rivers beyond the Ord. He marvels at what he calls the weak knee-jerk reaction of the water users in New South Wales towards water reform, where they are being hoodwinked by assurances from authorities and by Government policy. 'It is saying simply this. We shall never commit to any other water supply. What we will do for the future of New South Wales is to contain and continue to contain as much as we can out of the existing source, which means that if you can't accommodate that, you go off the land. Cut back on the usage of the existing source, but never contemplate putting an augmentation supply in to boost up the capacity to supply.'

Ernie Bridge is a true believer—in water. He eventually resigned from the Labor Party and ran successfully as an independent because of this belief. For several years he has been president of the Watering Australia Foundation, an independent privately funded body with the aim of harnessing 'idle water resources', in a national scheme which includes the Ord, Daly, Burdekin and Clarence river systems. After a pause, he spoke again with renewed intensity. 'Australia has got to realise that there is a bulging population north of us in the Asian region, it's going to expand by many millions of people in the next ten years. They will need to be fed and Australia sits pivotally in the most strategic place to respond to that human challenge. So get our backyards in order at a time when we can control the process, not at a time when we've got to react to demand and pressures upon us that won't give us too many opportunities of contemplating the future of Australia. Otherwise, I'll tell you now that if we don't do it, somebody else will decide to do it, that's a hard one but a truthful one.'

'That program there, Australia-wide,' he added, pointing to an ambitious map on the office wall of pipelines running in from the coast over the top half of the continent. 'We were looking at about

$4 billion all up for the whole nation, but it worked out that it would cost $470 per person on a pro rata basis.'

This costing is hotly disputed, but money is only one of the hurdles the river turners will have to clear. As we shall see, the other formidable issue is the growing concern over what such dramatic intervention would do to the environment.

The logician

Martin Albrecht has recently retired as chief executive of one of Australia's most successful construction and engineering companies, Thiess Contractors. He is a natural-born developer. He is also one of the real thinkers on the development side talking sustainable development.

Martin believes Australia is faced with a challenge. To the north, where water systems remain undeveloped, there are very different problems from the south. The rainfall patterns mean that almost all the rain falls within three months of the year. There is also what is known as the protein drought, whereby cattle start to lose condition from August and by October it's almost impossible for meatworks to find sufficient fat cattle to turn off until the following March. Yet much of the land is fertile. The solution? Just add water. Pasture could be irrigated to fatten cattle, but more importantly perhaps, higher value agriculture could be developed in a year-round growing season; cotton, horticulture, you name it.

In the south, we need to increase environmental releases into the stressed rivers, and recharge the old aquifers. But perhaps water users should be able to look to new sources for water. There is no reason to shackle production when we have plentiful supplies which can be redirected, this time in a more environmentally sensitive manner.

Martin Albrecht points to the national data. If the surface and groundwater resources that could be economically developed are taken together, then in the Murray–Darling Basin in New South Wales, 115 per cent of that water is being utilised (that is, overallocated). Compare that to around 2 per cent of the water utilised in Queensland's Gulf of Carpentaria and Cape York region, 1 per cent in Western Australia's Kimberley and 24 per cent in the Queensland part of the Lake Eyre drainage basin.[5]

Like Ernie Bridge, he is appalled by the blinkered thinking from the south which he believes could stymie development for years. While we all worry about the Murray–Darling Basin, he says, by ignoring resources to our north, we are making the existing infrastructure, already run-down, work twice as hard. 'Look how reliant Sydney is on

Warragamba Dam. Now if something were to go wrong with the water in it, what does that say about our risk management?'

COAG water reform and the push to claw back duckwater have virtually put an end to dam building, despite the latest ABS figures showing an increase of Australia's total net water consumption of nearly 20 per cent in the three years to 1996–97. That is going too far, Albrecht said. 'I challenge the findings of the AATSE [Australian Academy of Technological Sciences and Engineering] report, *Water and the Australian Economy*, which concludes that there will be no new Snowy-type schemes and then examines scenarios to take our economy forward constrained by limited water availability.'[6]

This is music to Ernie Bridge's ears. 'If the Kimberly pipeline had have been commenced by me, say nine odd years ago, what we would have seen by now is the fruits of that big project evolving, you know, and suddenly Australia as a nation would have grown to live with it and would have certainly said, 'Well, this is not pipedream, this is not pipe in the sky stuff, you can actually tap into these big rivers.' I would just like to get a slight window of opportunity of help from somewhere. The corporate sector, for example, in my view, has been as dead as a doornail on this project.'

Yes, it has in the past, but down the track Martin Albrecht and Ernie Bridge may find themselves some very powerful supporters.

The mogul, his ex-premier, their broadcaster and drought-proofing Australia

'You must capture on tape the fact that I am not an authority on water.' These are the first words of recycling king and philanthropist Richard Pratt. 'My closest link to understanding water was that I come from the Goulburn Valley and my father had an orchard and we used to use channel water to irrigate our tomatoes and our peaches and our pears.'

Dick Pratt's ambition for water in Australia is big. Part of his vision the Greens applaud, but some ideas they fear greatly, Dick Pratt wants to put Australia's water into pipes, recycled plastic pipes. 'Water in pipes as big as, if you like 10 metres in diameter, made of waste recycled plastic or a combination of waste recycled plastic and steel and concrete, and dams which stop seepage, have some sort of lining.' Like they do in Israel? 'Yes they do, but they do it here with landfills. It will minimise the evaporation and seepage.'

At a time when the country's leaking, creaking infrastructure is desperately in need of an expensive upgrade, one with environmentally friendly solutions, here is a man who can fix it. A recycler. A maker of plastic pipes.

Part of Dick Pratt's vision is that better areas of land in Australia could be more intensively and carefully farmed. 'I'm not going to tell farmers what to do. But I am a little bit disappointed when I drive through the Riverina, beautiful country, and I talk to one of the farmers who boasts that he and his son can run 10,000 acres and I say to myself this is not right. I know that properly farmed, that 10,000 acres could produce so much, and yet he's got to do that because he can't afford at the price of the fat lambs to employ labour. Now I'm saying that this is an unnecessary problem, if we have a plan.'

Cattle to carrots, lambs to peaches. Former Victorian premier Jeff Kennett couldn't agree more. He and Dick Pratt have spent quite a lot of time together talking about water. During 2000, they paid a visit to Mildura where moves to drip-filter irrigation and the new water trading regime have brought tremendous investment to the inland town and made skilled labour almost impossible to find. But even here, the main irrigation channels are open. 'I wanted Mildura to use recycled plastic,' said Jeff Kennett firmly, as if he had been in the business for as long as he was in politics. 'Put it into nodules, break it down to nodules and extrude it from the back of the truck as you run it along. All of those big channels that there are now could be reduced to something like that. You're going to save millions of litres of evaporation or seepage, and you're also going to get the benefit of having more land,' he added, referring to the space saved by getting rid of the channels.

Of the top 46 irrigation areas in the country, almost 19,000 kilometres of channel are uncovered.[7] Farmers wouldn't say no, as long as someone else forked out on the costs, and the Australian Conservation Foundation's Peter Garrett, too, is cautiously supportive. 'Any proposal that makes better use of water is welcome,' he said, 'but with the overriding caveat that ecological health has to become the prime issue that the policy makers are considering. Re-piping is great, but not so long as there is no environmental flow, not so long as there is degradation of our rivers and not so long as savings are re-used and there is this belief of an inexhaustible supply of goodies.'

Dick Pratt, however, wants to take his plastic much further than the existing grid. His plan is ultimately to drought-proof the whole of Australia. And he's calling for much more than Peter Garrett and Ian Donges' $65 billion. 'Quite frankly, I think that this is a 50-year project,' he explained calmly. 'First of all let me say there has to be an overall plan for Australia, and that plan should indicate that this whole project is going to take $10 billion for the next 50 years, which is $500 billion, to effect.' Five hundred billion dollars to drought-proof

Australia. And yes, this is about bringing northern (NSW) river water south. The recycling king continued, 'The nearest floodwater to the Murray–Darling is the Clarence River. We saw the TV pictures. Huge floods, 6 foot deep, rivers of water running through the main street in some of the towns in northern New South Wales. Now I'm sure that some of the prawn fishermen will be very disappointed if we take that floodwater away from them, but it's possible to do that.'

This is not the first time that locals along the Clarence, which reaches the sea between Grafton and Lismore on the northern NSW coast, have heard talk of their river being redirected. Leonie Blain is secretary of the Clarence Valley Conservation Coalition, a community group. 'Ernie Bridge came over here a number of years ago hoping to get some acceptance and support and the local people don't want a bar of it, quite frankly. We've seen what happens to rivers in the past when they're interfered with. The Snowy River is a testament to that. Mind you,' she added, 'there are plenty of people who've caused gross damage to the rivers out west who would be happy to see it happen and some very rich people, I'm talking about the cotton industry, who would be interested.'

This is also where Jeff Kennett parts company with Dick Pratt. He does not share Pratt's view on diverting northern river water. However, another highly influential figure does.

'It seems to me that on 105 fronts there is just appalling waste,' said broadcaster Alan Jones, moving into an easy editorial gait. 'Every time there is a drought you've got these headless chooks in Canberra run[ning] round on drought relief, which costs us millions and millions and millions of dollars but in no way can compensate for the damage that drought is doing and lost markets overseas, the lost employment opportunity, the lost revenue, therefore the lost taxation, the alienation of people as a result of drought and poverty, therefore the loss in family security, individuals willing to continue on the land, so all that is incalculable. Then when you have flood, you've got billions and billions of litres of water and no one knows what to do with it, and none of it is conserved. If we did that to the whale and the koala, it would be national tragedy!'

Dick Pratt, Jeff Kennett and Alan Jones all agree on one thing. Water is the key to opening up inland Australia and necessary to grow new thriving communities. 'You then have a total decentralisation of your population.' Alan Jones' pace picked up. 'There are many people who would love to live there rather than in the cities but it's not viable to live there, so once you then put a town along a water route, or build a new industry, your agriculture improves then there's another baker, another butcher and another dry cleaner, the multipliers are mammoth. You would make productive Australia more productive, you would make unproductive Australia productive.'

According to Jeff Kennett, the beauty of a big scheme is that it cuts across communities. 'Farmers understand it, environmentalists understand it. And most of it would be spent in rural Australia. It has political up side as well as all the economic and social. Look at all the crap that's going on at the moment in terms of country versus the city. The reality is this project, this national water agenda has the capacity of meeting all the concerns that are currently being expressed, because if you get the infrastructure right, development follows.'

John Seccombe from outback Longreach in Queensland takes a similar view. 'While governments maintain a focus on development in the southeast corner of Australia, which is less than a third of Australia's water, they're getting it wrong. They say people won't live up north, but they've never tried it with incentives.'

Dick Pratt is very careful and quite defensive about his plans. A long-time promoter of increasing Australia's population, he is quick to nip in the bud any ideas of self-interest. 'People will say, "Oh, that's your selfish way because you want to bring more migrants in and if we have water down here, you'll get cheap labour." I don't care about cheap labour. I don't care about any of those things. I'm neither a warmonger, nor am I an exploiter of Australia's resources. I don't exploit Australia's resources. I take the rubbish that the people who are exploiting Australian resources [produce] and turn it back into good product so you won't have to dig it out of the bloody ground!'

Dick Pratt is also sensitive to the criticism that he's the millionaire plastics recycler wanting to pipe up Australia. Now that would be good business.

He sits back in his chair. 'Let's a few of us who believe there is opportunity here, opportunity for Australia, not worry too much about Mr Pratt's going to make a buck out of this. If you like, I can undertake to put all that into a foundation, anything we make in that company will go into a foundation which will be distributed to charity. I'm prepared to make that undertaking. If it's a combination between government and private enterprise, maybe foreign government see this venture as being relevant, but can you tell me how much more profitable Melbourne is today as a result of Citylink? Nobody can, but I assure you it's going to put millions of dollars into the community, because it's infrastructure.'

Show me the money!

Dick Pratt is very serious about his plans to drought-proof Australia, but he also believes that government has a responsibility to take the lead and put up capital as well, even if it organises payback by charging the users.

Dick Pratt has the dosh. There's nothing surer. Would he be prepared to start with $100 million? He's recently spent about $850 million buying Southcorp's packaging division. 'Now $850 million or $100 million this money is identifiable, but normally public companies anyway need to see a return on it very quickly. This, however, is not something [for] which you're going to give somebody $100 million and expect to get a return at the end of three months or three years or 30 years. But it's good for Australia, it won't hurt me. I've given more than $100 million to charity in my lifetime.'

The whole funding issue makes Martin Albrecht very fed up. Why should water reform deal governments out of putting in some of the development kitty? On the contrary, Albrecht argues, there is a real role for government and private industry to work together to look at efficient schemes to replace the old ones.

The model project held up by promoters of irrigation development is the Emerald Irrigation Scheme in central Queensland, built in the 1970s around the Fairbairn Dam. The joint scheme between Federal and State governments has boosted the population of Emerald from 2,000 in 1966 to 11,000 in 1999. As a result, major coalfields, associated power stations that use water and other industries have developed, demonstrating the multiplier effect of such schemes referred to by Dick Pratt and Alan Jones and which now underpins the prosperity of the region.[8] Yet according to Martin Albrecht, the Emerald Irrigation Scheme simply would not have got off the ground without funding from government. The truth is that historically in Australia, almost all dams have been government funded. Looked at in isolation, new dams may not be economically viable. But Martin would argue the economy of surrounding communities will thrive and this justifies government funding.

In contrast to the Emerald Irrigation Scheme, the Nathan Dam proposal for the Dawson River, also in central Queensland, is on ice. A showpiece of private investment put up after the water reform process to provide irrigation for 30,000 hectares and support a coalmine and a new power station, it is not bankable without some government contribution, because the returns for private investors simply were not high enough, according to Martin Albrecht. It's a situation made all the more galling, he has said, by recent government decisions to spend $300 million of taxpayers' money on improved irrigation efficiency to return flows to the Snowy River, and $1.4 billion on fighting salinity on the Murray. 'Unless it is recovered directly from the water users through a levy, then this is nothing short of a straight-out subsidy—tilting the playing field firmly in favour of the existing environmentally inefficient schemes and making a folly of COAG's "economically viable" criteria!'[9] Albrecht's point here is that in terms of spending to be efficient and therefore environmentally sensitive, you get more return on your buck with new developments.

Martin Albecht is not arguing that we should stop funding a clean-up of the Murray–Darling, but he is raising the question about whether we should be throwing taxpayers' money to keep inefficient schemes going, rather than build new ones in less stressed areas. By doing this he believes we are knocking back good projects simply because they are unable to compete with the existing subsidised schemes.

The Greens are less than impressed with the logic. 'Well, what would you expect an engineering firm to say?' retorted the ACF's Tim Fisher of Martin Albrecht's argument. 'There is a huge risk of more Murray–Darlings, more Snowy Rivers. When you build a new dam the economics stack up best if you use as much water as possible.' Dam foe Jenifer Simpson has pointed to the Nathan project as a classic example of why people like Martin Albrecht should be stopped in their tracks. 'The potential dam is about twice the surface area of the Fairbairn Dam at Emerald, which is in a very similar climatic region. And it's about the same volume of water, but it's spread out over twice the area and it's half as deep as Fairbairn.'

Former National Party Senator Bill O'Chee is exasperated by the anti-development attitude. 'We used to go down to Canberra and put up proposals in the joint party room and the cat calls would follow, "Oh, here the Nats go again, another dam, another election".' Bill O'Chee's team did a little of their own analysis using postcodes and tax statistics comparing similar rural areas with and without irrigation. The result was that the population roughly doubled in regions with enhanced infrastructure like Emerald and the population's taxable income rose between $2,000 and $4,000, enough, according to Bill O'Chee, to justify government infrastructure spending.

The National Party is shy about northern water, despite what many of its constituents in the Murray–Darling think. When I spoke to her on her Riverina property in New South Wales, National President Helen Dickie dug out the Party's 1996 report entitled *Positive Progress to 2000 and Beyond*. This is what it had to say about dam policy:

The National Party should be leading policy research into and support for appropriate ways of harnessing the vast amounts of water that are currently lost into the sea in times of heavy rain and flood. Huge inland water storages and reservoirs strategically placed beside river systems could capture vast water resources in times of flood and help supplement town water supplies and drought-proof massive inland areas—without impacting in any way on natural river flows. Such concepts should be further investigated with the private sector encouraged to participate. Increasing the nation's water security and quality should be a national priority and one that is led by the National Party.

'What's the National Party doing?' Alan Jones flared up. 'They're talking about taking the votes away from Pauline Hanson. There's your way! We'll water Australia! That will be our project and we'll present it. And [the] number of people—you don't have to have a brain on your own. Dick Pratt'll help, Geoff White'll [civil engineer and founder of White Industries] help, Professor Endersbie [civil engineer at Monash University] will help, Ernie Bridge'll help.' Helen Dickie didn't bite on the issue of river turning. 'That's a subject we're not game to talk about in the National Party. The seats up north are important to us.'

'Why is it that our political leaders are so unwilling to take up the challenge?' asked Alan Jones, turning rhetorically to the other parties. 'Because they've got no guts, and this is the most astounding thing. I believe that this is the Snowy Mountains equivalent and a gigantic infrastructure project which would just mobilise the nation. Instead what are we doing? Alice Springs to Darwin railway line. The Alice Springs to Darwin railway line is useless if there is no water to furnish the route. It will be nothing but people looking out the windows at absolutely nothing at all. Now, it's not a matter of doing it, they won't even listen to people who say it should be done. The least they could do for us is to conduct an actuarial study to tell the nation how much it would cost to water Australia. They owe us that at least, but they won't even do that and I think political leaders since the war are condemned for their repeated and consistent failure to even embrace the notion.'

He drew breath.

Interestingly, the leader of the National Party is not closed to the idea of harnessing free rivers, but predictably, is measured. 'We're rapidly approaching the stage where we can have a long hard look at what can be done,' said John Anderson. 'The hitch is that while trade for agricultural product in the short term is looking at better times, in the long term, it continues to decline. So we're simply producing more for less and it's much harder to justify capital expenditure when this is happening, especially on a fragile continent like ours.'

Greens at the northern front

Parting with money is not the only reason a Federal Government is wary of the grand scheme. As can be imagined, environmentalists do not appreciate the vision of development up north. 'The hydrology and ecology of river systems have been developed over extensive periods,' explained Peter Garrett patiently. 'We need to firstly robustly address the river catchments we are using in the entirety. We should not be looking to take drawdowns in new rivers simply because we run short in others. It's a profound ecological question. It's the same argument, only bigger, of why should we let rivers run to the sea. The sea needs that water, fish stocks need that water. Of course we need human ingenuity to assist us, but that cannot be a replacement to us addressing the problems we have now.'

'I'd be happy to entertain that [northern rivers scheme], in 20 years' time, when we've demonstrated that we have in fact learned the lessons from the mistakes of the past,' said ecologist Peter Cullen. 'At the moment, that seems to be an excuse to walk away from the disasters that they've created and just create new disasters, they'll be slightly different disasters, but I don't see any evidence that they've been able to manage [the southern] river systems effectively.'

Most vulnerable are the coastal rivers, where diversions threaten to damage the delicate estuarine balance. Changes could damage the Great Barrier Reef. Arguments run late into the night on this subject, because the massive blanket of sugarcane covering over 500,000 hectares of coastal land running from Mackay up to Cairns has already changed the behaviour of our northern rivers. Like everywhere else in Australia, trees were pulled over to make way for this very successful monoculture. Little, if any, vegetation was left around the river edges to hold the banks in place. The result has been a massive shift in silt, much of which is dumped when it hits the sea. The subsequent rise in 'turbidity' levels has been blamed for some of the dramatic coral kills along the coast, although this is hotly disputed by the industry's 6,500 cane growers.

Diverting rivers to avoid damaging flood events could slow the passage of silt while riparian repair continues. Unfortunately, flood events also trigger breeding periods in marine life, and once again, our scientific knowledge in this area is too wobbly to be conclusive.

But any talk of diversions is enough to ring alarm bells in fishing communities along the coast.

Ernie Bridge sees few problems. The Argyle Dam, he has pointed out, was built to capture all the water flowing into the Ord. This is not the way of the future. The volumes involved in these northern rivers are so immense that, according to Ernie, we can easily divert a portion of the flow and still keep the fish happy. On his estimates, just 7 per cent of the annual flows of rivers earmarked could be used to drought-proof Australia. 'In the case of the Fitzroy alone, you are talking about an annual run-off of eight million megs. Now, bring it down to you understanding what that represents: Sydney, and the surrounding areas, the biggest and most populated city in Australia, has coming up to 5 million population. It consumes a little over a half a million megs per annum. We've got *eight* in a river running untouched into the sea! In the Clarence, you've got nearly five.'

Alan Jones believes the Greens are missing the point altogether. 'There is a lot of political correctness and I think this is what annoys me about the Peter Garretts of this world in that they want to focus on an agenda. Their agenda is the Green agenda, and therefore we've done these terrible things to the Murray–Darling.

'The real Green agenda is to recognise that we are a desert continent and we need water to survive, and anyone genuinely interested in the Greens would say, "Well, we want to put less reliance on the existing water courses and get water substitution from areas where water is aplenty and you'll balance the whole thing out." Now we've gone on for so long drawing on the Murray–Darling, how could you be surprised? I don't even read about it. I'm surprised the river even exists. They've built industries and agriculture on it, and now they're going to cripple that agriculture and say you don't have access, but we're giving you nothing to replace it. And we're not going to compensate you for it.'

There are many farmers along the Murray–Darling system who would see a lot of sense in what Alan Jones has to say. Dick Pratt has a more coaxing argument to allay the fears of those who say we risk creating another Murray–Darling crisis. 'There are so many people who say there is no way we are going to change the Australian countryside and not affect other areas. Now I have built a pulp mill in Tumut and it's taken me a long time, three or four years, to identify what the environmental people perceived as unacceptable and to find solutions to that problem. Finally we did and Garrett got up and he made a speech to say that this complies with all the environmental problems that we've had.'

Peter Garret did. 'We welcome genuinely the arrival of a project that satisfies the community's needs for sustained employment and for the environment. Well done Visy!' said Peter Garrett at the time.

Kerry Packer

Our richest man is famous for many things: transforming cricket, world series polo, gambling, outwitting boards of inquiry, cheating death, confounding the media and being rich. But one area is often overlooked. It is his connection with the land. Kerry Packer owns some of the more beautiful and more profitable cattle stations in Australia, mainly in the Top End. Newcastle Waters, Carlton Hill and Ivanhoe.

'You're doing a book on water. Kerry's interested in water,' were the words of Ken Warriner, who runs the big man's pastoral business. The fact that Kerry Packer has even thought about water is in itself interesting. But in what way is he interested? Most city suits will tell you KP only ever makes decisions with the bottom line in mind. That's why he is KP.

Another phone call from Ken Warriner came in the week after Christmas 2000. Post the 'op', apparently, Kerry Packer was bored. It was quite clear that he wanted to know more about the water game—about harnessing the powerful rivers up north that drive millions of megs out to sea every year. Which rivers were good candidates? Who were the best hydrologists with the answers?

Like Dick Pratt, however, Kerry Packer is waiting for the right climate. 'Kerry? Oh, well, he and I talk about it,' said Alan Jones. He just gets sick of it. The private sector will put the money up. But [Government should] give them a plan, they've never challenged the private sector.'

'The money is there,' insisted Dick Pratt. 'A number of people are prepared to put their shoulder behind it. I think they deserve to be congratulated, not to be abused by the larger community that they are doing something.'

So what should we make of these pipedreams, the dreams to drought-proof Australia and to use our abundant free-running rivers in the north to do it? Perhaps the first point is that they should no longer be pooh-poohed. They should be taken seriously. The second is that those who dream are as optimistic about what they may do to the environment as they are about the schemes themselves. To think that new projects would undergo the same sorts of rigour that Dick Pratt's paper mill underwent is also optimistic, particularly as some scientists warn of very serious long-term salinity problems that could be caused by moving water from one part of the country to another.

The answer must be the precautionary approach. Any river diversion proposal needs to demonstrate that the *long-term* benefits weigh in favour of development. To this end it would be no bad thing for Richard Pratt to direct some of his millions to the research that might help his cause.

High stakes

On the other side of Australia lies a region where pipedreams could well become pipes. In fact, a century ago this region was home to one of our most extraordinary pipeline achievements. In the heart of the goldfields of Western Australia, about 600 kilometres east of Perth, lies Kalgoorlie–Boulder. There are many remote towns in Australia, but there would be few of over 30,000 people that are without a local supply of water. In Kalgoorlie–Boulder there is no surface water, and deep underneath the town, natural groundwater is between five and six times as salty as the sea.

Kalgoorlie developed and exists today to service the vast goldfields of Western Australia. It is the gold capital of Australia where each year miners and industry folk gather and speculate at the annual 'Diggers and Dealers' conference. The old super pit in the goldfields is so big you can see it from the moon. Around 50 mines of mainly gold and nickel are spread over many kilometres of desert land around Kalgoorlie, but there is also silver, gypsum, sand and, it will be of no surprise, salt to be mined. The one substance not in plentiful supply, however, is water.

When the golden mile was discovered in 1893 the rush began. Fortune seekers from throughout the colony and even overseas swelled the population of Kalgoorlie to thousands, all without a proper water supply. The town experienced first-hand the tyranny of distance. Conditions were dreadful and inevitably many lives were lost through bad sanitation. The pressure was on to find a good source of water.

At the same time, engineer Charles Yelverton O'Connor had just completed the deepwater harbour at Fremantle to service Perth. What C.Y. O'Connor was to do next was one of the most extraordinary engineering feats ever accomplished in Australia. O'Connor convinced the then Premier at the time, Sir John Forrest, to fund the construction of a pipeline from Perth 557 kilometres east to Kalgoorlie and the goldfields.

The Premier had had his own obsession with water, as his great-great nephew, Andrew Forrest explained. 'John believed there was an inland sea and he tracked up most of Western Australia's major river courses looking for it. The accepted science at the time was that the river course ran to the central lake like Lake Victoria in Africa. Not many decades before, Lake Victoria was discovered as the source of many tributaries including the Nile, and they were punting that there was something like that over here, this high lake. It looks extremely stretched logic now, but in those days, over 100 years ago, it all made sense. And so those guys risked dying of thirst more times than you would ever want to imagine and lost camels and horses, not people. John never lost any people, but he lost practically everything else in his search for connecting up the east with the west of Australia, the search for the inland. The inland sea exists of course, just a couple of hundred feet below the horses hooves.'

Two years in the planning, the C.Y. O'Connor pipeline of steel pipes, tarred inside and out, was to pump water from Mundaring Reservoir in the hills east of Perth up an incline of almost 400 metres to the goldfields.

Ernie Bridge is envious. 'Let's face it, if the late C.Y. O'Connor had not been supported by the then Premier . . . the pipeline to Kalgoorlie probably would never have got up, and that was over 100 years ago. But he found an ally in the Premier you see, and the Premier's strength of commitment towards getting water into the goldfields enabled C.Y. O'Connor to have a powerful ally within government who said to the engineer, "You push on, regardless of all the attacks and the negatives." But you see, I never quite had that.'

Sadly, the Premier's support turned out to be not enough. In tragic circumstances C.Y. O'Connor never saw the completion of his most important work. Throughout the building, there had been a rash of vicious claims made against the project and calls for other solutions. According to Andrew Forrest, O'Connor 'used to have to wake up to read he was taking bribes and the only reason the pipeline was being built was because he was on the take from engineering firms. In the end he rode into the sea and shot himself, before the pipeline was complete.' That was in March 1902. Within a month, the project that drove him to his death began to pump a new life blood into the goldfields. Less than a year later, Kalgoorlie residents were drinking Perth water.[10]

A discovery

Until the dot.com crisis, the place where a punter could make or lose a fortune faster than anywhere else in Australia was in the west. It is not so much to do with the country being 'white shoe' in business, although the legacy of Alan Bond and WA Inc. is still hard to shake off. No. The main reason for the 'high beta' risk, as it is known, is mining.

One of most ambitious new start-ups in recent years is Anaconda Nickel. As its name suggests Anaconda has the potential to be one the biggest nickel producers worldwide. The company's big punt, a huge new nickel mine, Murrin Murrin, near Laverton, is now up and running. In May 2000, however, Anaconda made a new discovery, one that would send it off into a completely new area of development. It was water—water the company plans one day soon to pipe to Kalgoorlie and miners all over the goldfields.

By coincidence, the CEO of Anaconda, Andrew Forrest, was the same great-great nephew of the former Premier, and even more of a risk taker than his uncle. Just ask his competitors like Western Mining's Hugh Morgan about the controversial laterite nickel processing that the company is still pinning its hopes on. Or the investments in mining infrastructure that Anaconda has made to jump onto the world nickel stage. It was this sort of risk taking that almost cost Andrew Forrest his job. Teething problems with extraction technology sucked up funds at an alarming rate and the company's share price suffered. In May 2001, despite pressure from Anaconda's powerful shareholder, Anglo, Andrew Forrest held on to the Chief Executive position.

Late in 2000 in his Perth offices, talking water brought a smile to the face of this tough but clearly jovial player. 'It's a bit like the emperor with no clothes, no one actually wants to own up [to] how important water is because then they've got to admit they've got a water problem, and the whole industry is actually in private concern but public denial.'

Most people would remember headlines in the national papers claiming that a massive new body of water, the Officer Basin, had been discovered under Western Australia. Within days, a controversy ensued about the size of the discovery, the quality of the water and, indeed, whether it really was a new discovery at all. Then just as suddenly, the media went quiet and little has been heard of the Officer Basin since.

So what is the story? Andrew Forrest paused before answering. Quite apart from the Officer Basin find, he pointed out, Anaconda has plenty of water for its immediate projects. Ancient underground palaeochannel-water (not Officer Basin water) will provide water for the next 30 years. He would admit, however, that the pressure is on to try to mine sustainable water supplies and if the resource runs out after 30 years, this is hardly sustainable.

It was Anaconda's water man, Dr Richard Martin, who came up with the idea of turning to the desert for water. According to Andrew

Forrest, 'Dr Martin is paid to make sure we never run out of water, and he had achieved his long-term job security, with the palaeochannels.' (Richard Martin, who was in the office, looked up approvingly.) 'So you might say, "Why the hell go and do the Officer Basin?" With the Officer Basin, Richard had a theory. It was a gutsy theory and if it happened to come good it would be bigger than Ben Hur, and the consequences of that water is a multiple over anything Anaconda can use.'

Anaconda's find does not give it ownership of the water, but providing the company spends a few dollars developing bores, it does give the company first go at a water licence from the State Government.

Both men are indignant about criticisms of the discovery, particularly that the Officer Basin had not been discovered by Anaconda at all. Geologists had known about it for many years and the press even discovered that Aborigines at Wiluna on the edge of the Officer Basin had been using groundwater to grow fruit for many years.

It is true that water in the Officer Basin had been struck before by geologists drilling for oil and gas, but only in passing. 'They recorded that they'd hit H_2O,' said Andrew Forrest, 'but that was it. We ran into water hydrocarbon, full stop, no flow rates, no details, no salinity, nothing. Not one analysis, in at least 20 years. So when people say that, "Oh, we knew it was there," that's complete rubbish. Yes it had been drilled, but was water there of any significance whatsoever? Who the hell knew?'

The find was welcomed warily by the State's Water and Rivers Commission, which warned that the Officer Basin should not be seen as a 'newly discovered quick fix' to solve the long-term water problems for Perth or the rest of Western Australia.

What Anaconda is hoping for is much more of a long-term fix, one that folds perfectly with the company's other ambition driven by Andrew Forrest, the 'Three Nickel Provinces'. No one could ever accuse Twiggy, as his mates call him, of not thinking big. This project is nothing less than the development of the entire mineralised region of Western Australia, basically three massive nickel provinces in each of which the company happens to have dominant landholdings.

The aim is to have water, rail transport and energy all linked in a tight web. 'It's Anaconda's job over the next decade to just fill in the bits of the web, and it will create some 300 to 350,000 tonnes of new nickel production. The world will need around 450,000 tonnes of new nickel production, so we intend to fill most of that growth. And that's on pretty conservative parameters,' he added (just like a miner).

Critical to this development is a sustainable water supply. According to Richard Martin, Anaconda could need to increase the current 35 megs a day for Murrin Murrin to over 200 megs a day in the next four to five years for the Three Nickel Provinces. It is still early days for Anaconda. Another criticism at the time of the find was that the water is too saline to be useful. 'It would probably need a bit of treatment if it goes into a domestic,' conceded Richard Martin in a no-nonsense fashion, 'but at this stage we're looking mainly at mineral processing.'

The grand plan is to provide water not just for the future of Anaconda, but for all the mines in the Kalgoorlie goldfields region. With a significant shortage for the high-tech processing end of mining, Officer Basin water could provide the answer, and if the low salt levels prove true, purifying this water to drinking water would be far cheaper than purifying sea water, now the common desalination process.

There are many risks, not least that the pipeline would cross Aboriginal land. Andrew Forrest is no stranger to media criticism about some of his past dealings with Indigenous claims, but if he does see any issue with Native title, he's not giving much away. 'What have you got to argue about? What, you don't want water? Fine. It's no issue.' This 'non-issue' is negotiable, however.

No one is expecting the pipeline to be cheap, the biggest running cost being the pumping. It was the same back in the days of C.Y. O'Connor. 'If you look back to the Perth-to-Kalgoorlie pipeline, there we had, I think, nine pumping stations originally along the way and each of those were wood-fired power stations, so we cut the forest down to power the pumping stations, the old steam generators,' explained Richard Martin.

At least Andrew Forrest did not experience the same level of ridicule as C.Y. O'Connor. 'We haven't had any of that, we've had some gentlemanly competitor proposals—bit of narkiness in the press.' Coincidentally, the Officer Basin Kalgoorlie pipeline, or OBK as it is called, would run about the same length as the C.Y. O'Connor line. At its most promising, it might even alleviate the demand for Perth water (no bad thing as the city is already looking for alternative supplies to underwrite further expansion). The Anaconda team believes the OBK pipeline could bring forestry and even cotton to the arid region.

Some might write off the Officer Basin idea as fantasy, but several major international water companies are very interested in the Anaconda proposal. One of those potential partners is United Utilities, from Adelaide. Left stranded as the tide turned against privatisation in Australia over the last two years, United Utilities has turned its attention to the mining industry. The company's Don Richardson spread the plans out on the desk and ran through the details: a 350-kilometre pipeline from the Officer Basin to Murrin Murrin, a 100-kilometre branch up to Mount Margaret, down to Menzies and then into Kalgoorlie. According to Don, the price tag as of late 2000 for the scheme was $900 million, including a bore field of as many as 250 bores.

'We and two other parties, Lend Lease and Thames Water, [have been] talking with Anaconda and have been for some months now about developing that scheme. So is the WA Water Corporation, would you believe?' Here is a classic example, Don Richardson pointed out of government agencies getting into high-risk business (as discussed in chapter 2). 'Who takes the risk on that? What happens if they turn to the government at some stage in the future [and say], "I'm sorry guys, we've lost $900 million for you"?'

The whole scheme rests on what the Water and Rivers Commission will say about replenishment of the Officer Basin. The negotiating process has apparently involved some interesting conversations between Commission Chief Executive Roger Payne and Anaconda's water man, Richard Martin, on sustainability.

Demand for water is rising, but the OBK pipeline is not the only option that the State is considering for Kalgoorlie and goldfields water. Sensibly, United Utilities has a foot in two camps. The rival plan is a shorter pipeline carrying water from the coast at Esperance in the Bight. There is an extra cost, however, for desalination. A third option from a consortium of Thiess, Macquarie Bank and the WA Water Corporation to pipe raw sea water into Esperance was abandoned. With tailings dams on sterile land all around Kalgoorlie, the idea of piping salt in was always going to be hard to sell.

Don Richardson has said both the Esperance scheme and Anaconda's OBK pipeline rely on the WA Water Corporation buying water into Kalgoorlie. The good news for the developers is that delivering their water, at least on the back of an envelope, looks to cost around two-thirds of what the Corporation is paying. Even without the cost of debt, the Corporation's upkeep on the O'Connor pipeline is expensive, and then there's the cost of sourcing the water in Perth. For the Corporation to buy water for the developers at cheap rates, however, it would have to take on some of the development risk. Otherwise it would no doubt find itself paying closer to market rates.

Is Anaconda in this to make money? A big yes from Richard Martin. Water values are only going to go up. 'Very much so. We believe that this is a fundamental business for the whole area, in fact it stacks up on its own as an operating company.' Andrew Forrest was a little more circumspect. 'Yes, we get rights to transport the water and that puts us in a position of authority, but I'm not even fussy about that. Providing Anaconda gets a fair deal I don't care who owns the water or the water supply. I want to make sure the fair deal we get is as fair as everyone else gets and it causes a serious growth in long-term infrastructure, roads, railways towns, airports, so these big mining-based industries really turn into community-based industry.'

So there you have it. Andrew Forrest the altruist.

Outing the salt

Mention desalination to water users and moans usually follow. That old chestnut. The moans are not because the process doesn't work. It is because, even with 21st-century technology, and even though 97.5 per cent of the world's water is sea water, desalination is still expensive.

So when in November 2000, the WA Government announced that it was serious about sea water, Ernie Bridge could only draw breath in horror. 'We've had the Minister here, the current Minister, making a public statement that within the next ten years, Perth's water supplies are going to be drawn from the sea. When I went around the world and I went to Libya and went to UK and to America, they all said to me, we've thought hard and long about desalination in these countries, but we've come to the firm conclusion that it's too costly, it is too environmentally dangerous and there's no way we are going to turn to it, and yet here we've got this crazy notion of a relatively sparsely populated city in Perth that wants to talk about going into that process within the next ten years.'

It is a sign of the times that since Bridge's world trip in the 1980s, the negative thinking on desalination has mellowed. There are over 11,000 plants in the world today, 60 per cent of them in the Middle East. And while costs have held back desalination they are falling, albeit slowly.[11]

One of the most exciting desalination projects has commenced in the United States at Florida's Tampa Bay, where the local population has exploded to 2.5 million and existing freshwater resources are running out. The desalinator is contracted to produce water for as little as US$0.50 per kilolitre. This is remarkably cheap by world standards, partly assisted by the fact that the Florida water is not quite as salty as the sea. The pricing is expensive compared to the price of Australian river water (A$0.70), but in our remote areas comparisons on the cost of quality water puts desalination firmly onto the radar screen.

Australia is no stranger to desalination. 'My grandfather walked off to the goldfields in 1894,' said Don Rowe of Sunraysia Rural Water, 'and the first desalination plant was built at Coolgardie, just 40 kilometres from Kalgoorlie itself in 1896.' During the 1920s and 1930s, BHP desalinated water for its Whyalla factory by low-pressure vaporiser. Even more amazing, this industry water was topped up by fresh water brought in the ballasts of ships!

These days, Australia has a handful of desalination plants at Roxby Downs, Coober Pedy and on Kangaroo Island. Their costs range from around \$2.40 per kilolitre to \$5 per kilolitre.[12] 'Now it's a big investment, but there's nothing to say you couldn't put a plant down on the Great Bight and actually take a trunk up and actually spread arms off it and irrigate millions of acres,' enthused Jeff Kennett.

Technology is racing ahead for all desalination methods: distillation, electrolysis and the now popular 'reverse osmosis', whereby water is pushed through a membrane under pressure, leaving the salts behind. United Utilities' Don Richardson has pointed out that just four years ago, 3 litres of good water were produced from 10 litres in desalination. Today, it's more like 5 litres. And even the membranes which suffer from clogging are improving rapidly, some of the most novel being trialled on Kangaroo Island.

As with pipelines, the big ongoing cost for desalination plants is still the power to operate them, even where power is on hand and cheap. At Whyalla steelworks, with all the heat it generates, General Manager Leo Selleck is more than happy with the Morgan feed from the Murray River. 'Steelworks are a very large, complicated exercise and to capture some of that heat, you have to have water in specific areas, that in turn would require significant equipment. To be perfectly honest, that hasn't been the focus.'

For now, Ernie Bridge is still right. Desalination is a poor cousin to pipes and storages in Australia. 'I love Ernie Bridge, he's a hero,' said CSIRO's Tom Hatton when asked what he thinks of Bridge's criticism of desalination for Perth as totally irresponsible. 'It's expensive but it's in the realm of possibility, particularly once the State agencies apply a strict interpretation of COAG water reform and that is allocation of water to the environment.' According to Tom, if reform does reduce the water available it should make desalination more attractive because the price of water will inevitably go up.

Large projects also become sirens for their sponsors, however. While Ernie Bridge's dream pipeline would do many things, as

Anaconda's Richard Martin pointed out it is hardly small beer. 'A 2,500-kilometre transfer link to get water down there? That water is not going to flow on its own. The amount of energy required is absolutely huge. You'd get a better return from desalination here than you would do from bringing water down.'

The Silver Fox

After all these pipedreams, it might seem impossible that anyone has managed to get water money from government. One man seems to have it down to a fine art.

The oases of Loxton, Berri and Renmark in South Australia's Riverland are home of the Silver Fox, Jeff Parish. Between these oases and the Barossa Valley near Adelaide, the country is desolate mallee, occasionally cut through by a meandering River Murray which has dug its way through the ground to make startling cliffs in the flat landscape. 'Land between here and the Barossa,' explained Jeff Parish, 'crosses Goyder's Line. G.W. Goyder was a surveyor and he drew this line around Blanchetown [about 80 kilometres west of Berri] and recommended that farming didn't exceed beyond this. Of course what happened was that earlier this century, people had this philosophy that rain would follow the plough, so they moved way out beyond Goyder's Line because the seasons looked good but it wasn't long before drought caught up with them and it all went horribly wrong.'

Goyder has been long forgotten. South Australia's Riverland country, almost 300 kilometres from Adelaide, is prime vines and horticulture. A low 25 millimetres of annual rainfall makes the country almost disease free and gives divine weather control to irrigators. Less well known is that about 60 per cent of exported Aussie wine comes from the Riverland. 'Our biggest winery crushed over 100,000 tonnes,' said Jeff Parish. 'All those famous wineries you know about crush between 5 and 15 [thousand tonnes]. It's really hard for people to understand because they say, "How come we never hear of your wine?" Well, there's three companies running tankers. When you see those tankers moving around the country, they're not carrying water.'

Water in the Riverland is almost as precious as wine. At least that is the impression in Loxton, one of nine districts managed by Jeff Parish's Central Irrigation. Like so many of the other irrigation companies in Australia, it is now a private body delivering Murray water to farmers from Renmark to Murray Bridge. The days of water channels have almost gone. Instead, along the side of the road lie perhaps 20 massive sections of pipe, each weighing 9 tonnes with a 1.5 metre diameter, delivered one per semi-trailer and waiting to be buried. 'Thirteen metres long, they're half-inch steel-plate concrete lined on the inside. The whole thing is welded because this system will deliver high pressure from our main pumping station from the river. In terms of technology, it is the best in Australia and also very unusual in irrigation.' Jeff slapped a palm on the concrete. Dick Pratt would be impressed.

What is even more unusual in irrigation these days is that the Riverland irrigators are paying just 20 per cent of the cost of this brand-spanking $40 million irrigation system. The State Government and, yes, the Federal Government are each coughing up 40 per cent. The feat has irrigators in other parts of Australia burning up with jealousy. 'He's the Silver Fox,' said a salivating Don Rowe just across the border in Mildura.

To be fair, the 'subsidy' for Central Irrigation started long before the Silver Fox arrived. State Government-funded pipe-laying for eight of the nine State-owned irrigation districts began in the 1970s. It was stopped two-thirds of the way through in 1983, when low commodity prices and 17 per cent interest rates sent the irrigation business almost broke and left the State with huge interest payments on the capital. They were disastrous times. 'We had vine pulls and tree pulls for pears and tree pulls for citrus,' said Jeff Parish.

For a couple of years, the work lay unfinished. But the board of Central Irrigation was soon agitating government to finish the project. According to Jeff, the State Government came back to the irrigators with a proposal. 'You pay 50 per cent, we'll pay 50 per cent, and of course the growers said, "Get nicked!" So this board of seven people, who are now my company directors, approached the Federal Government. They put up the idea that the Feds should put in 40 per cent, the State 40 per cent and the farmers 20 per cent, and that in the end, it should all go private and the growers should get to own their 20 per cent, because it was a Government scheme and they should have fixed it.'

Jeff nodded. That was the first Riverland victory.

The second win was Jeff's doing, all the more remarkable because it happened right in the middle of water reforms to create full cost recovery. Central Irrigation was duly privatised in 1997, but the ninth district, Loxton, remained to be upgraded to piped irrigation, an anomaly because of its history as a Commonwealth-owned soldier settlement. A self-confessed 'anti-channel' man, he set out for Adelaide and Canberra to repeat the process.

At the top of Jeff Parish's obstruction list was a fellow from the South Australian Treasury, notorious for thrashing submissions before they went to cabinet. 'I said to my business manager who dealt with him, "Look, I don't know him, but I think I'll go and have lunch with him and see if he'll come on the committee," and he said, "You're mad, don't!" And I said, "I'm telling you, if he's that bad, let's get him on the committee to look at Loxton, so that when it goes up, it's his submission too." And he went on the committee, and look, he had plenty to say, but when it was ready to go to cabinet, he was a signatory. He can't crap on his own submission.'

Loxton is now transformed, a bonus for both irrigators and presumably, the environment. With the old leaky channels, farmers needed to put in their water orders a week ahead, and about eight people had to want water before the system could be cranked up and water moved into the channels. Today, the 1-kilometre backbone of pipe is chock-a-block with water. Irrigators can order individually on just a few hours' notice, over the Net rather than the phone. And new drainage systems have replaced the old drainage pipes that were put down in the 1940s, when water table problems first appeared.

There is one other irrigation region which is enjoying a little help converting channels to pipes. Running from north of the Grampians to the Murray in Victoria, is the Wimmera–Mallee region. Work commenced on a $52-million pipeline project in 1992 and is scheduled for completion in 2002. The hope is that the 2,300 kilometres of pipes, funded by State and Federal Government as well as farmers who were offered water by contributing to the 'sale of savings' scheme, will bring savings of up to 50,000 megs a year.

These are the sort of upgrades that water reformers and Greens would like to see all over the Murray–Darling Basin. It is also the type of government investment that Martin Albrecht dreams about.

Carrots and sticks

Just an hour's drive over the border from the Riverland lie Sunraysia Irrigation's vineyards and orchards with a slightly different set-up. Still highly productive with its grapes and stoned fruits, Sunraysia is one of the few irrigation companies that is government owned. Contrary to popular belief, not all grape growers are conservative with water. Don Rowe reckons many growers are still running at 50 per cent more water than necessary, water that does nothing more than create drainage problems.

Unfortunately for the Sunraysia growers, moves to greater efficiency look to be prompted more by stick than carrot. The wielder of the stick is wine giant Southcorp, which has aligned itself with none other than the Australian Conservation Foundation (ACF) to help get water users up to scratch. And with Lindemans now the biggest winemaker in the southern hemisphere, Southcorp has quite some pull as a buyer of product in the region. From 2005, Southcorp will not be buying fruit from properties unless they have moved towards complete ecological sustainability. 'We know that the infrastructure here on farm systems has got to be up to that standard by 2005, not necessarily pipe, but preferably,' explained Don Rowe. 'Southcorp is growing grapes at about 6 to 6.5 megs per hectare. The pumped irrigation districts in Sunraysia use every bit of 9.114 megs and there's still flood irrigation and still furrow irrigation'.

This is a startling admission. The grape industry prides itself on the most efficient irrigation techniques, either spraying at low levels (not over the crop) or underground pipes which drip filter water to the vines. Flood and furrow irrigation for grapes would surprise a lot of cotton growers.

Southcorp's tie-up with the ACF is a very clever one from both sides. For the company, it provides good PR, showing the company as an environmentally responsible corporate citizen, perhaps even a leader. And the ACF now has one of South Australia's most powerful companies on side in the water debate.

A minor embarrassment to both parties occurred in March 2000 when Southcorp was caught with its pants down, allowing 8,500 kilolitres of wine to spill into a Barossa Valley river. About 12 kilometres were affected by the infusion, which reportedly killed hundreds of fish.[13] Two months later, in May, Southcorp announced magnanimously that it would spend $70,000 to clean up the mess, but denied it had any responsibility. It was one of the rare occasions when the State Environmental Protection Authority (EPA) girded itself up and took the infringing company to court. In a major victory for the EPA, Southcorp pleaded guilty to failing to report the spill, causing serious environmental damage and breaching a condition of its licence.

The Barossa spill is a timely reminder that big business is not in the business of being 'environmental' for the environment's sake. Southcorp's green and friendly reputation may have taken a knock from the incident, but Don Rowe speculates that the company is planning to milk more than just good corporate PR from the ACF alliance. 'How did Banrock Station go with marketing their wine grown in that environmentally sustainable way? I think one of Great Britain's biggest chains was in Mildura talking about long-term contracts for their wine which will have "environmentally sustainable" on the label which puts them in front of everything from Europe.' If true, this is surely no bad thing. Australia has come a long way since Wallaby White and Roo Red.

Back at the Sunraysia vineyards, there is some belt tightening ahead. Unfortunately, converting from flood or furrow irrigation to drip filter is not a matter of flicking a switch. In a back issue of *The Irrigator*, Sunraysia's magazine for water users, there is a chirpy section titled 'Changing Over'—all a farmer needs to know about moving from furrow irrigation to drip irrigation but never dared to ask. The bottom line for a 20-hectare plot was a net outlay by the farmer of around $25,000, which would be gradually repaid by a $2,000 annual reduction in his water bill. There also appeared to be rather a lot of consulting and paperwork, something farmers love.[14]

Even around the beat of the Silver Fox, drip filter irrigation does not seem to be the norm. Most vineyards use sprayers, either overhead or under vine. Farmers prefer them to drip filter because you can see when the water is on or off. Drip filter irrigation might suit the Greens, but according to Jeff Parish, drip filter is a very sophisticated business. The fact that plants are watered only to the level they need leaves no margin for errors like clogging of the drip filter pores.

Despite the teething over drip filter, two scientists have taken micro-management a step further with so far remarkable success. About five years ago, Brian Loveys of the CSIRO and Peter Dry of Adelaide University came up with a technique called 'partial rootzone drying'. As the name suggests, at any one time only half the rootzone is dry. The scientists discovered that the root in a vine is the main organ which senses water stress. Their new method involves the adequately watered (healthy plant) producing a hormone which is sent up to the leaves and instructs them to slow down on their transpiration. In other words, the plant is tricked into believing it's stressed and the end result is a vine which uses the water very sparingly. What is really exciting is that the crops don't just use around half the water when slightly stressed; their yield actually improves. Other trials on fruit trees have similarly delivered better quality, colour and flavour in the fruit.

Despite the results, there is a big reason why the technique is not being taken up like Pokémon cards. It takes some skill and experience to get these great results and, particularly if the wet side is not kept wet enough, vines move very quickly from highly stressed and highly productive to overstressed and, effectively, rooted.

For most farmers in less intensive industries, like rice and dairying, the idea of turning channels into pipes is laughable. As George Warne is fond of pointing out, when the Mulwala Canal was built in the Murray Irrigation area, it was the longest canal this side of the Suez. The economics simply don't stack up. Indeed, the channels versus pipelines issue is hotly debated amongst farmers and between irrigators and Greens. Jeff Parish, for example, has no time for any form of channel system, even concrete, which he believes cannot be efficient. Other industry experts maintain that concrete-lined channels can often be less expensive to maintain and the losses are no worse than pipelines.[15] For the millions of dollars to be outlaid on a piped system, it is easy to see why irrigators are hanging back. The Silver Fox also has a rather large electricity bill to pay each month to power the pumps that drive the water through the pipes.

The frustration felt by some water users was well put by irrigator head Stephen Mills at a recent conference.

> The chairman of our Goulburn–Murray Water Board has a saying when referring to the general community. 'The great unwashed out there.' Well, we often hear the great unwashed giving the irrigation industry plenty of free advice such as 'concrete the channel' or better still, 'cover them with a roof to stop evaporation and all your problems will be fixed'. The general irrigation community does not understand irrigation and they do not understand the gains that we have been making by using water more efficiently.[16]

Dreams from the Promised Land

When it comes to water management, no one does it better than the Israelis. With two million megs a year for six million people, it's not hard to understand why. Since 1948, the harnessing of water has been the key to survival and security. Irrigators from Australia who visit the tiny nation—for instance, former United Dairy Farmers of Victoria President Max Fehring and cotton grower Paul McVeigh—come away in awe.

Two-thirds of Israel's water originates in the River Jordan in the Golan Heights, occupied by Israel since 1967, and from groundwater under the West Bank. Even now, the battle for the Golan Heights has as much to do with water as anything else. In 1990, King Hussein of Jordan made it clear that water was the sole reason for going to war with Israel.

Israelis have taken water management to a different level. The River Jordan flows into the main storage, the Sea of Galilee. Dams on the farm are the norm, using artificial liners to prevent seepage. And whereas almost all water in Australia is used just once, in Israel, necessity being the mother of invention, the average bucket of water is used between five and seven times! For example, water which may have been pumped from 800 metres underground, is used first for tourist spas (those visiting Israel will be relieved to learn), then to warm hothouses, and then on to different species of fish (eels, then catfish). This now enriched water is then taken for hydroponic tomatoes and herbs, with the rest going to drip irrigate field crops of olives, melons and alfalfa.[17] Since 1984, the use of fresh water on farms has halved, while the value of production is still rising. Brilliant! Even in Israel, however, Max Ferhing was surprised to find a lot of saline land, although with the sea only a few kilometres away in most parts of the country, easy disposal of highly saline water made him quite envious.

Israel is a great case study for Australian water managers. Inland aquaculture is a sunrise industry here—one way some rural communities might find new income. Whether it is red claw, eel bass, golden and silver perch, cod or local yabbies, fish farms are popping up around Australia, including on inland saltwater farms where salinity is rife. Those involved are huge enthusiasts, but one also gets the impression that, like so many new industries, it's a great deal harder than it first appears. And the biggest threat to aquaculture is not salt, but pesticides. As noted earlier, one of the most common chemicals, endosulphan, kills fish very effectively, making the aerial spraying of crops near dams a seriously precision business.

Waking up

In recent years, most politicians have closed the door on big schemes. Dams and diversions are destined for the politically incorrect basket, in the southern States at least. Yet there are still people talking grand plans. A former politician, a miner, the former head of a top engineering and construction company, and two moguls. You just need to know where to look.

Water entrepreneurs all over Australia are looking at opportunities. At one extreme, humble in-house and on-farm improvements. At the other, a massive solar-powered national water grid that collects and stores water and pipes it to wherever it is needed, anywhere in Australia. Proponents of the grid say it would end the dramatic flood–drought cycle. Water reserves could be stored in the many mine pits which still pockmark the countryside, and could support a new timber industry and a salt industry (from the salt interception work in rivers), through solar-powered desalination.[18]

The political landscape is also not as dire as some developers might think. At least two States and a Territory are open to the idea of waterway development. As luck would have it for entrepreneurs, all three are in Australia's north, where the big water runs.

Queensland is still an open question. With an announcement

in December 2000 of a major new dam for the water users of the Burnett, and other WAMPs (Water Allocation Management Plans) under review, the State is clearly going its own way. Western Australia's Government and particularly its water bureaucracy have been made slightly paranoid by events in the Murray–Darling. Beyond the Ord 2 Scheme, if that comes off, there is pressure for new developments on the Fitzroy.

For the Northern Territory, the clear focus is intensive agriculture in the Katherine–Daly Basin, an area of nearly half a million hectares between Katherine and Darwin with the Daly River, the biggest in the Territory in terms of flow, running through it. 'We're starting to subdivide pastoral leases and put in more improved pasture for the live cattle trade in Southeast Asia,' said head of the Resource Development Department, Howard Dengate. 'And the biggest challenge that we face is making sure that that's sustainable.'

The Territory is starting well, with plans for 30 per cent of land to remain uncleared and a 500-metre strip either side of the Daly to be fenced off and managed by Parks and Wildlife Commission. Howard Dengate said the environmental record of other States in Australia has been a good lesson. 'We're very jealous of our lifestyle—tourism is very important and a lot of people move to the Territory because of the lifestyle, which is the untouched bush. You can go bushwalking and nobody's ever been there before, well, no Europeans, and that's exciting.'

In the end, it is worth remembering that many of the pipes and pipedreams mentioned in this chapter are little different from the great Snowy Mountains Scheme. They might bring enormous social and economic benefits, but they also pose real threats to the environment. We *should* be in a more informed position than ever today to weigh up those goods and evils, and yet our knowledge is far from adequate. This then is the conundrum for developers.

14

Education, Leadership and Water

Give us the tools, and we will finish the job.
Winston Churchill, 9 February 1941

Education by osmosis

The character Watter Quandary, the posh drop that wants to be drinking water, and other playful books written by Jenifer Simpson put out mainly through the Australian Water Association, are a rare attempt to get the water message into classrooms and, indeed, mainstream thinking. 'It's got sex on page three so it's not a kid's book,' giggled Jenifer, 'but it's worked to [a] certain extent. There's a fellow who can make it into a kids' play and I think it could also be an animated video.' In fact the book also contains quite a complex biochemistry lesson worthy of university study.

On the whole, Jenifer Simpson's materials go to what she calls the 'motivated public', councillors and bureaucrats who are involved in decision making on water recycling. 'The unfortunate thing is that engineers don't have a clue about education, they don't think it's necessary. And the trainers are in the same position as everybody else in that they're ignorant. Even people with environmental qualifications don't know anything about waste water. It's crazy, isn't it?'

At the heart of the serious water quandary in this country is that water is not up there with health, education or the economy in the public or political imagination, although it is inextricably linked with all these areas. What has been created is a vicious circle where leadership is needed to raise awareness about water, yet no one wants the job because they believe it doesn't matter to the urban voter. If anything, water politics is a poisoned chalice.

The harsh reality is that the big enemy of sustainable water use is you and me, the amorphous public consumer block that makes the world go round and now has it spinning like a top. Market forces, and in particular consumer demand, drive water abuse in this country. And yet most consumers bother to think about water about as often as they wash the soles of their feet.

'Greed and fear are the two main drivers of life,' said a doleful CSIRO chief Dr Graham Harris. 'At the moment we are running off greed being good. Perhaps we need a good dose of fear to get some balance. Sometimes at 3 a.m., I think that there might be some ghastly cock-up that could bring everyone to their senses.' Graham Harris is absolutely right. In short, it is time for a reality check. 'Hello, earth to ugly consumer?' Trying to convince the politicians of this, however, is difficult, not least because John Howard believes that consumers aren't ugly at all. They are 'ordinary decent Australians'. At the moment, the task of whipping up concern is left to other people.

Love him or loathe him (few are indifferent), Alan Jones is one of the most influential voices in Australia. Advertisers think so, the Prime Minister thinks so, and his audience runs to around 20 per cent of Sydney's population. And as we've seen, one of Alan's favourite talking points is water. 'Every time I speak about it, you just get this universal endorsement of what's been said. I think it's almost the single biggest disgrace in the country, and our failure to do anything will undeniably and unchallengeably haunt us and damage us.'

At the most basic level, marketing the water issue is not helped by the appalling jargon which is an instant turn-off to the general public. The following words and phrases should be either removed completely or used as a last resort:

Economically sustainable development
Full cost recovery
Highest value use
Biodiversity
Integrated catchment management
Integrated pest management
Structural adjustment
Potable
Externalities

While useful to water professionals these words are frequently an irritant to journalists, which if the aim is to get the message of water out to the masses, is plain silly. We also have a multicultural society to woo. The instructions for the phone inquiry line for Sydney Water printed on its quarterly Consumer Confidence Report are in Arabic,

Chinese, Croatian, Greek, Italian, Macedonian, Serbian, Spanish and Vietnamese.

2002 is the 'Year of the Outback', a time for all Australians to take stock of our land, water and rural culture. Indeed, playing on the feel-good factor has merit. The thought that our wetlands might be relegated to the Discovery Channel is not one that the public relishes. In Britain, this feeling has been captured by a campaign to 'bring back the otter' to rural rivers. And it's working.

'Triage'

'I'm getting more militant about it as I get older,' said Graham Harris. 'If we continue to use market solutions, two things won't participate: other species and future generations.'

There are extreme campaigners, from those wanting to make farming 100 per cent organic to Dennis Avery, whose 1995 work *Saving the Planet Through Pesticides and Plastics* accused greenies and 'organic frenzies' of threatening the world with famine and loss of habitat for their sacred wildlife. Why? Because farming without synthetic pesticides, petrochemical fertilisers and biotechnology would require too much land. 'If you want pristine, everyone gets shot, that's easy. But if you come back from that, you have to decide on a balance,' commented former United Dairy Farmers of Victoria President Max Fehring.

During the First World War, the appalling casualties forced army medics to adopt a harsh but essential method of coping with the overwhelming numbers. It was known as 'triage' from the French 'trier' meaning to sort out. The walking wounded would look after themselves and anyone dying would be left unattended. The focus could be on those worth saving.

Triage is what agro-economist at the University of Western Australia, David Pannell, describes as the only real way forward for Australia in the war against salt. We need to be surgical. Some landscapes are fixable, some not, so efforts should be concentrated where pay-off is greatest. As we have seen in chapter 5, most scientists agree with him.

In many areas, salt will get a lot worse before it gets better, even if we are spending millions. CSIRO warns that 'in larger, intermediate and regional scale systems, the rise could extend to hundreds of years no matter what we do.'[1] Areas that are salted beyond human redemption, we leave to die.

We pick green winners, environmental treasures, and guard them obsessively. It's beginning to happen. Ramsar sites like the Moira Lake Wetlands in the NSW Barmah–Millewa Forests are now enjoying a rejuvenation thanks to a freeing up of Murray water. Others targeted could be the Cooper Creek and its plethora of lakes, or the troubled Coorong in South Australia.

Then there must be areas for intensive agriculture, but increasingly, they need to be land which can best stand the stress of irrigation. 'You've got to see irrigation as a factory farm,' explained the nation's plumber, Don Blackmore (of the Murray–Darling Basin Commission). 'It's on 2 per cent of the Basin that has huge input and potentially huge outputs, both in environmental pollutants but also in products. To use the analogy of the Ford Motor Company, if you drive up the highway at Broadmeadows, you have a look and you don't see too much biodiversity under the factory floor. It's bloody concrete for the whole lot . . . They have a pollution boundary around it. It's very much a managed outcome, because we're happy to drive the car.

'Irrigation isn't a low-impact industry. People have dressed it up as a low-impact industry and it's absolute crap. Now the moment you say it's a high-impact industry, you do the same as you do to the Ford Motor Company. What's your occupational health and safety requirements? What [are] the environmental management systems you work with? Can you afford to do it? If you can't, you don't do it. And that's the maturity we're now getting to.'

One of the ironies in 'triage' is that the graziers may well have the last laugh, as, in accepting the true costs, major irrigation areas slowly fall back to less intensive use and growing pressure from green groups forces the return of natural flows.

Solving for rural Australia

'I was challenged a little while ago to give a vision for the Murray–Darling Basin and I found it remarkably difficult,' said Peter Cullen, one of our top scientists. 'Everything I said seemed so waffly. I ended up saying that the driver for action was that we had to reduce the footprint of agriculture. When you've got agricultural activities in Queensland affecting the drinking water of Adelaide, most people think that's pretty much over the top, and it seems to me that we can't have no footprint, but my goodness, we should be doing a lot more within the economic framework and our technical framework to reduce the footprint of agriculture.'

Heresy. Put this to a farmer and you'll get a look as if you've just said something that goes against everything he stands for, and you have. Yet critics of irrigation say that from the Egyptians and the Venetians to the Aral Sea in Uzbekistan, history has proved that no irrigation culture can survive long term. Sandra Postel in her 1999 book on world irrigation, *A Pillar of Sand*, explained:

> The role of irrigation in the rise and demise of civilisations over the last 6,000 years is much more than a historical curiosity. On the cusp of a new millennium, human society is now as dependent on the ancient practice as ever. At the dawn of the modern irrigation age, in 1800, global irrigated area totalled just 8 million hectares, an area about the size of Austria; today, the irrigation base is 30 times larger, encompassing an area 2.5 times as large as Egypt. We now derive about 40 per cent of our food from irrigated land.[2]

According to Sandra Postel, one in five hectares of irrigated land worldwide is losing productivity because of salt problems.

Inevitably, it is the cities that are driving irrigation. Peter Cullen talks about reducing the agricultural footprint, yet that footprint belongs to every one of us, as described in chapter 1. The milk, rice, tomatoes and cotton are for us. And as Max Fehring pointed out, it is also the city footprint that is threatening. 'Here we are building cities and expanding on our best and most fertile areas. We say we'll just get the stuff from elsewhere, we couldn't give a stuff about how water is handled in another country.'

This is not to apologise or make allowances for irrigation. But there's no point in walking away from it. We are all part of the problem. 'Like it or not,' one irrigator explained, 'in rural Australia farmers are the real environmentalists, because vandals or caretakers, they are the ones managing the land on a day-to-day basis.'

Farmers understand the land better than any of us ever will, but it is the dark warnings about the big picture from chapter 6 that still go unheeded. Australia could not be more unlike England or the European continent, a fact that Mrs Macquarie looking wistfully towards the heads of Sydney Harbour is said to have felt only too deeply. It is not just bad management practice that is screwing up the land. The hard fact is that it is also best management practice. And it is dry land farming as well as irrigation. Moves by the sugar industry to harvest green cane rather than burn it, and water and pesticide cutbacks in crops like rice and cotton get a big tick, but they can't undo the damage of monocultures and lasered land.

Farmers cannot solve the environmental problems on their own, however. Anyone who believes the average cocky can pay for the sort of reform that is needed to fix or even arrest the damage done is misinformed. The average age of a farmer is now over 55 and the Murray–Darling Basin Council's work has found that 'even with off-farm employment, total household income is low, with up to 30 per cent of the population reliant on welfare and up to 48 per cent of children living in chronically poor families. These less favoured areas are often the worst affected by salinity and they impose significant costs on downstream water users.'[3]

Although it was understood rural communities would suffer when water was taken from them under the Murray–Darling Basin Cap, the $5 billion paid to the States by the Federal Government went straight into general coffers. The Federal Government is likely to pay for this in terms of its popularity in the bush. If farmers find themselves looking down a barrel in the next few years, we all have to take responsibility for that.

With a few notable exceptions there is one other inevitable trend happening in the bush which environmentalists like Peter Cullen find disturbing. 'All the kids have nicked off to the city, so we're going to see a real generational change, and it's not as though young people are going to be able to buy those properties. I see we're going to get more corporatisation of agriculture. And it will be big money and a lot of political power. I understand that when Queensland brought in their new *Water Act* just late last year that some of the irrigation interests just phoned around and within half a day had a half-a-million-dollar fighting fund to try and oppose that legislation, because it restricted floodplain harvesting.'

As water becomes more precious, corporations will get tougher and the regimes must toughen with them. The better news is that some companies, such as Southcorp, Visy and even Cotton Australia, are seeing the values (if not virtues) of demonstrating their good environmental citizenship. As Peter Garrett remarked, 'Salinity has emerged from the bush and into the boardroom.' Today, according to the Social Investment Forum, one in eight dollars of managed funds in the United States is invested for ethical reasons and one of the fastest growing areas in management consultancy is working on the 'triple bottom line' of social, economic and environmental inputs. The media can be cynical about the motives behind corporate Australia's new-found conscience, but if in absolute terms it helps the environment this needs to be acknowledged.

How much?

Communities need help. Triage needs cash too. And also overlooked is the tool to help make the tough decisions, research. 'We've reached a point where the investment in knowledge is probably a better investment for the future than spending on the military and defence

R&D,' Graham Harris said in his address to the year 2000 World Water Congress in Melbourne. 'If we are to be Y3K compliant—given the not so impressive record of the last thousand years—then we need to invest more in environmental science and water science in particular.'

Money on the ground is not enough, argued Don Blackmore. The Natural Heritage Trust proved that point. 'What the community is saying is we want three things. We want knowledge, we want institutional and community capacity built up. We want structures that work, government resources in natural resource agencies, all of that, and we want money for on-ground action. And we don't want money for on-ground action without these other two. The view that the simple solutions are on the shelf is just wrong, they're not there.'

Yet these asks all translate into still more funding. 'I would put to you that the seven-year stage-one strategy of $1.4 billion is a drop in the bucket,' said South Australian Premier John Olsen of the Prime Minister's National Action Plan (see chapter 5). The question remains, where is the money coming from? Telstra money won't last and there is little left in the way of public assets to flog off even if the community supported the idea.

One option is an environmental levy across the whole population, perhaps means-tested (quite proper as rich people certainly pollute more than poor ones). Calls for a national levy are coming from various quarters. The Nature Conservation Council wants a salinity tax and a similar idea was proposed to John Anderson in July 2000 by former head of the National Farmers Federation, Rick Farley, to raise up to $30 billion to fight salinity.

Would people be prepared to pay an environmental levy? It will be hard to convince John Howard. GST will have been quite enough for the PM and after all, once you're out of the lion's den, you don't go back for your hat. 'Well you see if I was the Prime Minister, I would be able to persuade them that this was a good thing,' said broadcaster Alan Jones without blinking. No doubt he probably could. 'It depends on how committed you are. I'd be happy to persuade the nation and I know they'd come with me. But I mean, you have to outline its virtue as a national project.' Alan Jones' pitch comes with the dream (described in chapter 13) of investment in water infrastructure that would create new towns. 'Everyone can understand the mental picture of what it means to be in a town. It means you're bitumenising the roads, it means someone has got to buy bulldozers and it means you're building houses and you're building schools. I mean, it is a phenomenal multiplying effect, simply as a result of committing yourself to the importance, urgency and necessity of water.'

The electorate has shown itself willing to commit to public spending for specific issues or levies. Recently, the guns buy-back cost $400 million to pay for the compensation of 660,000 firearms and the Timor tax levied $900 million. It is true that these are one-offs, but there are other annual levies. Perhaps the most interesting was the NSW 3x3 levy, introduced by the Government in September 1989. Promoted as three cents a litre for three years to fund new road infrastructure, the levy was quietly extended three times until a High Court ruling in 1997 found that only the Federal Government could raise levies and that all State levies were unconstitutional! Even then, it was agreed that the Feds would continue the levy and reimburse the State, which they did until the GST restructure of State and Federal revenue raising. Each time the levy came up for renewal, the community seemed happy that the money was being spent in the right places. By the end, the levy was bringing in over $200 million a year. This was masterly tax gathering.

People in Adelaide already pay a salinity tax, something which the electorate took on without fuss. The message on salt had already been received loud and clear. And despite the political worries of a backlash when Australians moved from fixed-rate volumetric charging for their water, the change hardly made a ripple.

Now, though, a Federal reality check. Even if we could raise the $65 billion that Peter Garrett and Ian Donges would like to see, Environment Minister Robert Hill thinks we're not ready for it. 'We don't really have the capability of spending it wisely at the moment. I don't think we could have spent more than the $1.5 billion NHT [National Heritage Trust] money wisely. And while we are getting better with research, I haven't seen anything to suggest a strong case for an environmental levy yet.' This is a sad state of affairs.

Ex-head of Thiess, Martin Albrecht, wants a consumption levy. Government can milk over $10 billion a year in petrol taxes at 47 cents a litre, so why not charge two cents a litre for water and raise over $4 billion? We could even have rebates for rural users, just like the diesel fuel rebate. 'I wouldn't like to distort the water market now by imposing a tax,' said a careful Don Blackmore. 'If you look at what's driving resource degradation it is really recharge to groundwater and if you put a resource tax on that, it would be incredibly unpopular, it would almost certainly lead to a change of government.'

Nevertheless, a green tax of some sort is certainly the norm in other parts of the world. 'We should have had it instead of the GST,' claimed Green senator Bob Brown. 'We are way behind Europe in eco taxes. It's inevitable.'

In Europe during 2000, a total of 2.2 billion euros (roughly $3.8 billion) was budgeted by the community to subsidise its rural and environmental projects. Farmers are required to set aside 10 per cent of their land to give it a rest. They get subsidies for doing so. They get more subsidies for looking after hedgerows and stone walls. Then there are subsidies to help village communities. If this is not a non-tariff barrier, what is? 'France subsidises for sociological reasons, the US does it for sociological reasons. In our case, it should be for sustainability,' said Ted Gardiner at Queensland's Department of Natural Resources. 'There's going to be a massive shift in [the] next 20 years where everyone pays their way or protects what they value.'

Direct management and stewardship payments are certainly what the National Farmers Federation is calling for. 'If we paid people, not for the water but for the ecosystems that they're managing and nurturing to provide quality water, you would get a much more direct control,' agreed CSIRO's John Williams, thinking about what a good way it would be to look after the precious Sydney Catchment. 'If the Catchment Authority paid an extraction fee of even five or ten cents a kilolitre, in some rural holdings that would be more money per hectare per year than [the holdings] get from the current enterprise. And if you then specified the fencing of streams, the maintenance of cover, the habitat quality, you'd buy the services, and I think the Sydney Catchment Authority is well placed within the budgetary situation that it currently has, rather than give grants. Look, even New York decided not to put $9 billion into a new treatment plant but manage the catchment through this sort of arrangement.'

One way or another, Australia will have to find more cash to deal with the growing demands of water management. The longer we leave it, the more that figure is going to be. Once again, though, a caution from Canberra. National Party leader John Anderson points out that Europe has 350 million people to raise levies from. How much would 20 million people be able to support? Now there's a loaded question.

How many?

Population is the one unknown that will have a massive impact on the way we use our primary resource. In a few decades, will we be 25 million or 50 million?

In the green corner sit those who decry 'growth for growth's sake', believing that a doubling of population will be a major threat to the environment. New South Wales Premier Bob Carr is one of them. 'A population of 50 million for this country—and there are people calling for it—would mean about ten million people living in the Sydney basin where you've got four million at the present time. Right down the east coast of Australia, you'd see the end of any farming. You'd see the end of any conservation, open space.' [4]

'I've said to my kids, there's only one environmental problem, and that's the number of people,' agreed CSIRO's urban water program director Andrew Speers.

In the red corner sits an equally passionate lot urging Australia to grow. A larger population base means a larger tax revenue base. It should be said, however, that anyone who believes that this would help ease the population bulges in the Third World should understand that India's annual increase is over 15 million a year, off a base of a billion people. But this is not the place for a treatise on population. What is interesting in the water resource context is how some of those pushing for development of the northern rivers believe that if we don't populate and use the resources, parts of Asia will. 'If you want the country to continue the way it is, and hopefully the Southeast Asians won't see it as an empty space for them to move into, fine. But I'm looking at this in a much broader and longer period'. Dick Pratt spoke quietly. The answer, he continued, is more people to drive growth. 'The point about the 50 years is that in 2050 we should have 50 million people in this country. That's all part of it. We shouldn't be precluded because of salt in our water.'

Alan Jones agrees. 'I don't believe in terms of population policy that we can go on with our natural resources as they are with the starving billions of people to our north and they'll keep looking and say, "Aren't they lucky down there? Isn't it wonderful that Australia's got all that space, they've only got a handful of people to share all those riches, and the water?" So as a corollary of defence policy, you'd try to build your population, as a corollary of trade, build your population, as a corollary of economic policy, fiscal policy, you'd build your population.'

As population pressure in Asia continues to increase, so no doubt will the fears of 'the hordes to our north'. The reality is that in a couple of decades, several parts of Asia will be short of drinking water. Yet long before international water wars affect Australia, we need to work out what we want our country to look like in 20 or 30 years' time. Political correctness has cloaked debate on population and the view seems to be that we'll muddle through without confronting this issue head on. But how else can we make a rational decision about what we do with the precious water resource in the northern part of the country?

We need more debate in the public arena, about population, about immigration and about the sensitivities surrounding Australia's pristine and 'empty' north. That takes leadership.

Treading water

During the now famous Rio Earth Summit in June 1992, Maurice Strong, the Canadian chair, made a telling observation.

> I went out to a little restaurant way out in the outskirts of Rio where I thought I wouldn't be known and I could maybe get a little peace and quiet. A lady came up to me and spoke English and said, 'You're Mr Strong.' I kind of admitted yes, and she said, 'Tell me how many world leaders are there actually here?' And I said, 'Well I count about 118 presidents, prime minister[s] and a few kings.' But then I added rather indiscreetly, 'But very few leaders.'[5]

While there are many different, sometimes violently opposed views about how we should rise to the challenge of sorting out water in Australia, almost all agree that what is missing is the leadership to do it.

'You're looking for someone to start a project that is going to be of immense value to Australia, socially and economically, not to mention environmentally,' explained Jeff Kennett. 'It's all about this issue of leadership, that's why it's got to be the prime minister because he or she over the next 100 years has got to drive it. It's got to be their pet project. Unless there's authority from the top, it will not go anywhere.'

'Kennet is right on that point,' said Deputy Prime Minister John Anderson. 'Water does need the prime minister to lead, and the reason is that the prime minister heads up COAG.'

ACF President Peter Garrett agrees. 'It should be driven by national leadership, highly purposeful, resourceful and collaborative, because of the States' involvement with water. We can't be held hostage to the dynamics of Federalism. This must be driven by supranational interests linked to the sensible custodianship of water, so we need to set aside the Federal bureaucracy.'

Resource management clearly hasn't inspired other world leaders to date, so why should Australia be any exception? 'We haven't got a politician at the moment that's really giving leadership anywhere in the water arena,' Peter Cullen, the Prime Minister's choice of Environmentalist of the Year, confirmed what everyone in the industry knows. 'I think probably Welford in Queensland has been the best of them, and he has the toughest battle getting the changes, and I think he's done a fabulous job. We need another Deakin, someone with a real vision for what water can be in this country.' You mean an Alfred Deakin who's walked the road to Damascus? 'Yes, a reconstructed Deakin!' Peter laughed.

David Suzuki, the great Green soothsayer, has some strong advice for any water leader. The crux of the problem lies with this buzz word, 'stakeholder', the use of which has taken off around the world. 'Get all the stakeholders together and work it out. Baloney it will get worked out!' he said in a recent interview. 'They're just going to argue and fight to get their little turf and protect it. You make damn sure the water will never be managed properly because water isn't the central piece—it's humans, human uses and human priorities—you've got it backwards. The water has got to determine what we do, not vested interests.'[6]

Taking people out of the main equation is a worthy idea. But unfortunately, as *Watershed* has found, people, the stakeholders, are not going anywhere in the near future. They are uninterested and will take some leading. 'We have to be bold, imaginative and resolute,' said Peter Garrett. 'Boldness [in] continuing with water reform and user pays; imaginative in devising a whole new set of ways of living with less water and understanding that we must stop trying to create a patch of Cornwall, or Devon or Sussex on the east or west coast of Australia. And we need to listen to our land managers, indigenous peoples, scientists and local communities who are trying to get past the weeds on salinity—understand their experience and culture and impart that into our psyche so that we can move beyond the cliches of the dry continent to realising the essential value of what we have.'

'It's not just about making sure you have water flows down rivers that have been changed,' rejoined Jeff Kennett. 'It's all-embracing. It's about rainfall, it's about collection, it's about storage, it's about distribution, it's about pricing, it's about recycling, it's about avoiding waste, evaporation, seepage.' It's about money, capital investment? 'It's about capital investment, but as much as it is about capital investment, it's about people's understanding to better utilise what we've already got.'

It is also about science. Graham Harris at CSIRO (just read a few of his speeches) is calling for a new era of ecological and social activism:

An era committed to repairing the damage so that we can truly claim to be Y3K compliant. In fact such an initiative would drive a whole new industry—the industry of ecological engineering. We have passed through or are passing through areas of civil, mechanical, electronic, software and genetic engineering—fortunes have been made in all areas. The next big challenge and opportunity is ecological engineering: and it will probably be the most important one. I continually meet people from governments and agencies that urgently require environmental systems advice because of the growing scale of wickedly complex environmental problems.[7]

Yes, a 21st-century Alfred Deakin. Is there such a person in Australia today? One of our most powerful businessmen, Dick Pratt, ponders the question. 'Let's say that in the next 12 years there will be. [Leadership's] not here today, it will be there in the next 12 years— four years, four years, four years—the next three governments, and it's not the leader of the party that's the leader of the country. I'm not sure whether he's a statesman or a politician, but I'm looking at a group of people beyond the politician. These are leaders of our country, maybe the hidden face of the establishment, they are the leaders and they have to be the leaders of the country.'

It cannot be lost on Jeff Kennett that were it not for a fight over water, Snowy water, he might still be premier. Few would deny his aggressive style of leading from the front or his ability to stir the community into excitement over big projects. 'We lifted the confidence levels and through confidence came a sense of pride. So the public were proud to be associated with change and development. Anything that's of value has got to come from people who are focussed on what they want to deliver.' The Liberal loss in Victoria was as much due to an exhaustion with Jeff Kennett's obsession with sometimes brutal reform and projects as anything else.

Leadership on the water issue is not something to be put off for a rainy day. The challenges are huge: finding the water that will secure a healthy River Murray and possibly paying for it; pulling all the States together, including Queensland; giving comfort to rural Australia on its water security or at the very least being serious about addressing the compensation issue; convincing all Australians that money is worth spending on pipes and water and that environmental obligations need to be shared. Neither the environment nor society can take any more damaging action—or inaction, such as over-allocation, shying away from sleepers and dozers, deferring WAMPs, avoiding compensation and bloating the bureaucracy.

Not surprisingly there are many motherhood statements from people calling for change. At a practical level, what can be done now is to raise public awareness and, in that way, force politicians to get wet. COAG should be acknowledged for what it has achieved, but much more must be done in the political arena to drive it further. And since Federal intervention, first on the Cap and then in pushing national reform, has goaded COAG into action, perhaps there should be more of it. But people have to understand why all this change is needed—the general public and the water users.

Leith Boully, the Chair of the Murray–Darling Basin's Community Advisory Committee, is forthright. 'Taxpayers will be faced with substantial costs. There will be difficult choices between different groups and individuals. The point that I am making is that to get a good result out of the pain which we are going to experience, we need a strong decision-making process that people will understand and trust.'

Watershed

The red tides in Asia, the salt of the Aral, the tightening of water resources in the great irrigation State of California. One only has to look around the world to realise this is still the Lucky Country. But Australia has a watershed decision confronting it.

A fundamental shift has occurred in Australia from water being plentiful and free to being scarce, threatened and valuable. But reforms can only go so far. 'Can we really put a price on biodiversity and ecosystem function?' asked Graham Harris. 'Economic policy demands that this be done and ecologists are rushing out to oblige. We do know that society values water resources for more than just money.'

It was Professor Ian Lowe, science policy analyst at Griffith University, who remarked that you can't run a First World economy in a Third World environment. There is a greater force than the market force. It's the natural force and there is not a water user in Australia who would disagree with this. We should attempt to work with it, not fight it.

'In the past, Australians found themselves by responding to external crises; exile, wars, depression. The crisis of today is one of meaning and of finding a way to live within the constraints of a finite natural world.' These words of Peter Garrett are a challenge for every Australian.

Epilogue

Nothing could be more tragic than the misuse of water, earth's primary resource, the beauty of which James Joyce captured in one of his euphoric eulogies in *Ulysses*.

What in water did Bloom, waterlover, drawer of water, watercarrier returning to the range, admire?

Its universality: its democratic equality and constancy to its nature in seeking its own level: its vastness in the ocean of Mercator's projection: its unplumbed profundity in the Sundam trench of the Pacific exceeding 8,000 fathoms: the restlessness of its waves and surface particles visiting in turn all points of its seaboard: the independence of its units: the variability of states of sea: its hydrostatic quiescence in calm: its hydrokinetic turgidity in neap and spring tides: its subsidence after devastation: its sterility in the circumpolar icecaps, arctic and antarctic: its climatic and commercial significance: its preponderance of 3 to 1 over the dry land of the globe: its indisputable hegemony extending in square leagues over all the region below the subequatorial tropic of Capricorn: the multisecular stability of its primeval basin: its luteofulvous: its capacity to dissolve and hold in solution all soluble substances including millions of tons of the most precious metals: its slow erosions of peninsulas and downwardtending promontories: its alluvial deposits: its weight and volume and density: its perturbability in lagoons and highland tarns: its gradation of colours in the torrid and temperate and frigid zones: its vehicular ramifications in continental lake contained streams and confluent oceanflowing rivers with their tributaries and transoceanic currents: gulfstream, north and south equatorial courses: its violence in seaquakes, waterspouts, artesian wells, eruptions, torrents, eddies, freshets, spates, groundswells, watersheds, waterpartings, geysers, cataracts, whirlpools, maelstroms, inundations, deluges, cloudbursts: its vast circumterrestrial a horizontal curve: its secrecy in springs, and latent humidity, revealed by rhabdomantic or hygrometric instruments and exemplified by the hole in the wall at Ashtown gate, saturation of air, distillation of dew: the simplicity of its composition, two constituent parts of hydrogen with one constituent part of oxygen: its healing virtues: its buoyancy in the waters of the Dead Sea: its persevering penetrativeness in runnels, inadequate dams, leaks on shipboard: its properties for cleansing, quenching thirst and fire, nourishing vegetation: its infallibility as paradigm and paragon: its metamorphosis as vapour, mist, cloud, rain, sleet, snow, hail: its strength in rigid hydrants: its variety of forms in loughs and bays and gulfs and bights and guts and lagoons and atolls and archipelagos and sounds and fjords and minches and tidal estuaries and arms of sea: its solidity in glaciers, icebergs, ice-floes: its docility in working hydrolic millwheels, turbines dynamos, electric power stations, bleachworks, tanneries, scutch-mills: its utility in canals, rivers, if navigable, floating and graving docks: its potentiality derivable from harnessed tides or water-courses falling from level to level: its submarine fauna and flora (anacoustic, photophobe) numerically, if not literally, the inhabitants of the globe: its ubiquity as constituting 90 per cent of the human body: the noxiousness of its effluvia in lacustrine marshes, pestilential fens, faded flowerwater, stagnant pools in the waning moon.

Endnotes

Chapter 1

1 Commonwealth Knowledge Network, Summary of the Discussion on Desalination, 14–30 October and 1–7 November 2000.
2 Asmal, Professor Kadar, chairman, World Commission on Dams, 'Water is a Catalyst for Peace', Stockholm Water Symposium Laureate Lecture, 14 August 2000.
3 Strong, Maurice, chairman, United Nations Earth Council, Inaugural Jack Beale Lecture on the Global Environment, delivered 11 February 1999 at the University of New South Wales.
4 Interview on *Landline*, ABC-TV, 27 February 2000.
5 White, Mary E., 'The Sad Brown Land', *Company Director*, May 2000.
6 Scanlon, John, Evidence to the South Australian House of Assembly Murray River Select Committee, 26 June 2000.

Chapter 2

1 Estimate from the Australasian Bottled Water Organisation, based on anecdotal evidence, personal communication to author.
2 Interviewed on 'Sydney's Cryptic Water Crisis', *Background Briefing*, ABC, 22 November 1998.
3 'Sydney's Cryptic Water Crisis', *Background Briefing*.
4 Sheil, Christopher, *Water's Fall*, Pluto Press, 2000.
5 McClellan, Peter, QC, *Sydney Water Inquiry*, Second Interim Report, NSW Premier's Department, September 1998.
6 Ibid.
7 Ibid.
8 Ibid.
9 Ibid.
10 Ibid.
11 Ibid.
12 Ibid.
13 Ibid.
14 Ibid.
15 *Sydney Water Legislation Amendment (Drinking Water and Corporate*

Structure) Act 1998 changed the Sydney Water Corporation and the Hunter Water Corporation into state-owned corporations giving the Minister powers to get information from and give directions to the boards in matters of public interest.

16   *Sydney Water Inquiry*, Second Interim Report, op. cit.
17   Ibid.
18   Ibid.
19   As quoted on 'Sydney's Cryptic Water Crisis'.
20   *Sydney Water Inquiry*, Fourth Report.
21   Ibid., Second Report.
22   Ibid., Third Report.
23   Ibid., Final Report.
24   Ibid., Third Report.
25   Ibid.
26   Clancy, J.L.J., 'Sydney's 1998 Water Quality Crisis', American Water Works Association, March 2000.
27   Interviewed on 'Sydney's Cryptic Water Crisis'.
28   *Sydney Water Inquiry*, Second Report.
29   Ibid., Final Report, vol 1.
30   *Audit of the Hydrological Catchments Managed by the Sydney Catchment Authority*, Final Report to the NSW Minister for the Environment, CSIRO Land and Water, December 1999.
31   Ibid.
32   Ibid.
33   Ibid.
34   Ibid.
35   Hogarth, Murray, 'Water at Risk After Peat Mine Spill', *Sydney Morning Herald*, 12 August 1998.
36   *Audit of the Hydrological Catchments*, op. cit.
37   *Sydney Water Legislation Amendment (Drinking Water and Corporate Structure) Act 1998*.
38   National Health and Medical Research Council, Commonwealth of Australia, September 2000.
39   *Sydney Water Inquiry*, Final Report, vol 2, ch 15.
40   Estimate from the Australasian Bottled Water Organisation, based on anecdotal evidence, personal communication to author.
41   Walker, Bob, and Walker, Betty C., *Privatisation, Sell Off or Sell Out? The Australian Experience*, ABC Books, 2001.
42   Quiggin, Professor John, Australian Research Council.
43   *Sydney Water Inquiry*, Final Report, vol 2, ch 15.
44   Sheil, Christopher, op. cit.
45   *Sydney Water Inquiry*, Final Report, vol 2, ch 15.
46   *The Hutton Report*, October 1992, confirmed levels of 143 oocysts per 10 litres of water in Sydney storages.
47   *Sydney Water Inquiry*, Fourth Report.

48 *Sydney Water Legislation Amendment (Drinking Water and Corporate Structure) Act 1998.*

Chapter 3

1   Water Services Association of Australia, 'The Australian Urban Water Industry, WSAA Facts 2000'.
2   Victorian Government, National Water Week, October 2000, www.nre.vic.gov.au/waterweek2000
3   Ibid.
4   Foss, John, chairman Surfrider Foundation, 'Ocean Outfalls—there is a solution to ocean pollution', issues paper, August 1998.
5   Water Services Association of Australia, op. cit.
6   Samual, Graeme, 'At Last a Looming Disaster is Being Addressed', *Australian Financial Review*, 23 March 2000.
7   Melbourne Water Corporation, 'Catchment to Coast, Long Term Strategic Directions', 2000.
8   Melbourne Water Corporation , 1999/2000 Public Health Report, 2000.
9   Water Services Association of Australia, op. cit.
10 Blackmore, Don, 'Reforming the Water Sector', speech to World Bank Conference, 16 December 1996.
11 Water Services Association of Australia, 'The Australian Urban Water Industry, WSAA Facts 2000', and *Water Reform and the Urban Sector*, Agriculture, Fisheries and Forestry Australia website, www.affa.gov.au
12 High Level Steering Group on Water, *Report to COAG on Progress in Implementation of the COAG Water Reform Framework*, Occasional Paper no. 1, 1999.
13 Productivity Commission, *GTE Performance Benchmarking Report 1991/2–1996/7.*
14 Fisher, Dr Peter, 'Dampened Expectations', *Australian Financial Review*, 11 August 2000.
15 Underwood, A.J., and Chapman, M.G., *Marine Pollution Bulletin*, vol 33, 1997.
16 Foss, John, op. cit.
17 Caldwell, Connell, *Sydney Submarine Outfall Studies,* 1976.
18 Simpson, Jenifer, and Oliver, Peter, 'Water Quality, from Wastewater to Drinking Water to Even Better, the Dilemma of Watter Quandary', Australian Water and Wastewater Association, Inc, 1996.
19 *The Environment Performance of Sydney's Deepwater Outfalls*, Findings of the Environmental Monitoring Programme, NSW Environment Protection Authority.
20 Water Services Association of Australia, 'The Australian Urban Water Industry, WSAA Facts 2000'.
21 Beder, Sharon, 'Getting into Deep Water: Sydney's Extended Ocean Sewage Outfalls', in Pam Scott, ed, *A Herd of White Elephants: Australia's Science and Technology Policy*, Hale & Iremonger, 1992.

22   Foss, John, op. cit.

23   SA Water 1998/1999 Annual Report.

24   Hartley, K.J., *Independent Audit of the Bolivar Wastewater Treatment Plant to Determine the Cause of a Major Odour Event*, 10 July 1997.

25   Survey undertaken by Australian Research Centre for Water in Society, CSIRO Land and Water.

26   Legge Wilkinson, Report for Surfrider Foundation, 1996; also NSW *State of the Environment Report*, 1997, states that in the Beachwatch program in New South Wales, enterocci pathogens for 1994–95 and 1995–96 all failed standards, mainly after rain.

27   Powers, Mary Buckner, 'Wastewater Treatment', *Engineering New Record*, 6 July 1999.

28   Manly Council media release, 7 September 2000.

29   'Carr Government to Clean Up Our Harbours, Rivers and Beaches', NSW Government media release, 1 May 1997.

30   Fisher, Dr Peter, op. cit.

31   Simpson, Jenifer, and Oliver, Peter, op. cit.

32   Productivity Commission, *Arrangements for Setting Drinking Standards*, International Benchmarking Report, April 2000.

33   Ibid.

34   Ibid.

35   *Recycling Strategy*, Price Waterhouse Coopers for Queensland Department of Natural Resources, April 1999.

Chapter 4

1   Tourism pamphlet, Murray–Darling Basin Commission.

2   White, Mary E., 'The Sad Brown Land', *Company Director*, May 2000.

3   With thanks to Native Fish Australia.

4   Fisher, Tim, 'Fish Out of Water', Special Habitat Supplement, Australian Conservation Foundation, October 1996.

5   Sturt, C., *Two Expeditions into the Interior of Southern Australia During the Years 1828, 1829, 1830 and 1831*, Smith & Elder, 1833.

6   Parsons, Ronald, *Ships of the Inland Rivers*, Gould Publishing, Gumeracha, SA, 1996.

7   Tolhurst, Michael, 'Song of Running Water', soon to be published.

8   Ibid.

9   Ibid.

10   Ibid.

11   Hill, Ernestine, *Water into Gold*, Robertson & Mullens, Melbourne, 1946.

12   Ibid.

13   Ibid.

14   Tolhurst, Michael, op. cit.

15   Op. cit.

16   Based on an estimate of farm-gate value of irrigation in 2000, by the

Murray–Darling Basin Commission, multiplied by a factor of four to give value of related processing activity.

17 Ricegrowers Cooperative Limited, 'Historical Development of the NSW Rice Industry', Rice Facts.

18 According to Ricegrowers Cooperative Limited, world average yield is 3.6 tonnes per hectare, Australia's is 8.8 tonnes per hectare.

19 National Registration Authority, *Cotton Resistance Management Strategy 1995/96.*

20 Bolt, Cathy, 'Beef Industry Plays Down Pesticide Scare', *Australian Financial Review*, 4 February 1999.

21 *Courier-Mail*, 18 March 1999.

22 Cotton Australia website, February 2001, www.cottonaustralia.com/au

Chapter 5

1 Murray–Darling Basin Ministerial Council, *Review of the Operation of the Cap*, Draft Overview Report of the Project Board, March 2000.

2 Scanlon, John, Evidence to the South Australian House of Assembly Murray River Select Committee, 26 June 2000.

3 CSIRO Land and Water, *A Revolution in Land Use: Emerging Land Use Systems for Managing Dryland Salinity*, CSIRO, 2000.

4 Murray Irrigation Ltd, Wakool Tullakool Sub Surface Drainage Scheme information flier, October 2000.

5 Murray–Darling Basin Council Salinity Audit, Murray–Darling Basin Commission, 1999.

6 Ibid.

7 Ibid.

8 Walker, Glen, Gilfedder, Mat, and Williams, John, *Effectiveness of Current Farming Systems in the Control of Dryland Salinity*, CSIRO, 1999.

9 Interviewed on *Insight*, SBS-TV, 13 July 2000.

10 Sturt, C., *Notes upon the History of Floods in the River Darling*, vol 1.

11 'Urban Salinity', Wagga Wagga community booklet, City of Wagga Wagga and Department of Land and Water Conservation, August 2000.

12 Personal comment from CSIRO representative in Wagga Wagga community booklet.

13 Department of Land and Water Conservation, *Impact of Urban Salinity and Waterlogging in Wagga Wagga*, 1998.

14 As reported on *Insight*, SBS-TV, 13 July 2000.

15 Midnight Oil, 'River Runs Red', *Blue Sky Mining*, Warrner/Chappell & Spirit Music, USA, 1990.

16 National Farmers Federation/Australian Conservation Foundation, 'Repairing the Country: A National Scenario for Strategic Investment', May 2000.

17 Boully, Leith, speech to the International Landcare Conference, Melbourne, March 2000.

18 Murray–Darling Basin Council Salinity Audit, op. cit.

19   Murray–Darling Basin Council Draft Basin Salinity Management Strategy 2001-2015.

20   Ibid.

21   Fisher, Tim, 'Water: Lessons from Australia's First Practical Experiment in Integrated Microeconomic and Environmental Reform', *Microeconomic Reform and the Environment*, part 2, Workshop proceedings, Productivity Commission, Melbourne, 8 September 2000.

22   Commonwealth Government, 'A National Action Plan for Salinity and Water Quality in Australia', 10 October 2000.

23   Scanlon, John, 'Saving the Murray. Problems and Prospects, Presentation Overview – Governance of the Basin' 16 February 2001.

24   Murray–Darling Basin Council Draft Basin Salinity Management Strategy, op. cit.

25   O'Kane, Bill, CEO Goulburn Broken Catchment, 'Water Reform, Outlook 2000', address to the Australian Bureau of Agriculture and Research Economics Outlook Conference Canberra, 2000, quoting Linke, Seker and Livingston, *Effect of Reafforestation on Stream Flows, Salinities and Groundwater Levels in the Pine Creek*, Sinclair Knight Mertz.

26   Murray–Darling Basin Council Draft Basin Salinity Management Strategy, op. cit.

27   Harris, Dr Graham, chief executive, CSIRO Land and Water, 'Water, Science and Ecology in an Age of Cynicism', keynote address to the 10th World Water Congress, Melbourne, 13 March 2000.

28   Cullen, Professor Peter, Fixing the Foundations Symposium, November 1999.

29   Murray–Darling Basin Council Draft Basin Salinity Management Strategy, op cit.

30   Stirzaker, Richard; Lefroy, Ted; Keating, Brian and Williams, John, CSIRO Land and Water, *A Revolution in Land Use: Emerging Land Use Systems for Managing Dryland Salinity,* CSIRO, 2000.

31   Ibid.

32   Murray–Darling Basin Council Draft Basin Salinity Management Strategy, op. cit.

33   Reilly, Peter J., 1999 pereilly@esc.net.au

Chapter 6

1   Carson, Rachael, *Silent Spring*, Houghton Mifflin Company, Boston, 1962.

2   White, Mary E., *The Greening of Gondwana,* Kangaroo Press, Sydney, 1993.

3   White, Mary E., *Running Down: Water in a Changing Land,* Kangaroo Press, Sydney, 2000.

4   Strong, Maurice, chairman, United Nations Earth Council, Inaugural Jack Beale Lecture on the Global Environment, delivered 11 February 1999 at the University of New South Wales.

5   Interviewed on *Timeframe*, ABC Science, 1 May 1997.

6   White, Mary E., *Running Down,* op. cit.

7   *Timeframe,* ABC Science, 1 May 1997.

8   Abbey, Edward, *The Monkey Wrench Gang,* Douglas Brinkley, 1975.

9   'Dams and Development: A New Framework for Decision Making', World Commission on Dams, 16 November 2000.

10   Ibid.

11   White, Mary E., *Running Down,* op. cit.

12   Simpson, Jenifer, 'Dinosaur Technology Syndrome', Australian Water and Wastewater Association, 1993.

13   Queensland Department of Natural Resources, *Overview of Water Resources and Related Issues, the Warrego/Paroo/Nebine Catchments,* February 2000.

14   Wildlife News, 'The Paroo, One of Australia's Last Free Flowing Rivers', *World Wildlife Fund for Nature,* no. 90, June 2000.

15   Kingsford, Richard, Boulton, Andrew, and Puckridge, Jim, *Viewpoint Challenges in Managing Dryland Rivers Crossing Political Boundaries: Lessons from Cooper Creek and the Paroo River, Central Australia,* John Wiley and Sons, 1998.

16   Ibid.

17   Interviewed on *Country Hour,* ABC Radio National, 9 September 1999.

18   Australian Museum website, www.amonline.net.au.

19   Ibid.

20   O'Rourke, Jim, 'Saving a Bug Makes 14 Mining Jobs Extinct', *Sun Herald,* 29 March 1998.

21   Rio Declaration on Environment and Development, Principle 15, 3–14 June 1992.

22   White, Mary E., *Running Down,* op. cit.

23   Department of Land and Water Conservation, *Macquarie Marshes Water Management Plan,* National Parks and Wildlife Service, 1996.

24   Stoneman Douglas, Marjory, *The Everglades: River of Grass,* Pineapple Press, Florida.

25   *A Water Secure World: Vision for Water, Life and the Environment,* March 2000.

Chapter 7

1   Mills, Stephen, speech to Australian National Committee for Irrigation and Drainage Annual Conference, Brisbane, 11 September 2000.

2   Fisher, Tim, 'Water: Lessons from Australia's First Practical Experiment in Integrated Microeconomic and Environmental Reform', *Microeconomic Reform and the Environment,* part 2, Workshop proceedings, Productivity Commission, Melbourne, 8 September 2000.

3   Tolhurst, Michael, 'Song of Running Water', soon to be published.

4   *Hanson v Grassy Gully Gold Mining Co.* 1900, 21 LR (NSW) 271, upheld as recently as 1995 in *Van Son v Forestry Commission of NSW 1996 86 LGERA 108.*

5   *Report of the Working Group on Water Resource Policy to COAG*, section 3.7, commissioned by COAG for COAG meeting, 25 February 1994.

6   A view supported by Marsden Jacob Associates, *Economic and Social Impacts*, companion paper 2, Murray–Darling Basin Ministerial Council Review of the Operation of the Cap, 2 March 2000.

7   Warne, George, 'Irrigation without Water', ABARE Outlook Conference, Canberra, 1999.

8   Mills, Stephen, speech to Agribusiness Australia Water Forum, Sydney, 4 November 1998.

9   Murray–Darling Basin Ministerial Council, *Review of the Operation of the Cap*, Draft Overview Report of the Project Board, 2 March 2000.

10   'The pursuit of river flow objectives to achieve environmental outcomes will ensure diversions below the Cap,' report by the Independent Audit Group, *Review of Cap Implementation 1996/97*. But 'while the environmental flow rules may influence diversions, they are not designed or targeted at achieving Cap compliance', Report by the Independent Audit Group, *Implementation and Compliance*, companion paper 4, Review of the Operation of the Cap, February 2000.

11   Donovan, Gary, former executive secretary, NSW Irrigators Council, 'Making Sense of Water Policy', September 1999.

12   *Report of the Working Group on Water Resource Policy to COAG,* op. cit., section 4.1.

13   *Sustainable Water Resource Management and Farm Dams*, discussion paper, Government of Victoria, 30 April 2000.

14   Donovan, Gary, op. cit.

15   'Victorian Water', *Landline,* ABC-TV, 15 October 2000.

16   Fisher, Dr Peter, 'Dampened Expectations', *Australian Financial Review,* 11 August 2000.

17   Victorian Farm Dams (Irrigation) Review Committee Draft Report, Government of Victoria, 14 December 2000.

18   *NSW Water Act 1912,* no. 44, section 17.

19   Murray–Darling Basin Ministerial Council estimate, from Australian National Committee on Irrigation and Drainage (ANCID) Annual Conference, September 2000.

20   Within sectors, water use can also vary dramatically depending on management practices.

21   Marsden Jacob Associates for Department of Land and Water Conservation, *Water Trading Development and Monitoring,* 20 May 1999.

22   Donovan, Gary, op. cit.

23   Postel, Sandra, 'Can the Irrigation Miracle Last?', W.W. Norton & Co Inc, NY, 1999.

24   Confirmed in Marsden Jacob Associates for Department of Land and Water Conservation, op. cit.

25   High Level Steering Group on Water, *Report to COAG on Progress in*

Implementation of the COAG Water Reform Framework, Occasional Paper no. 1, 1999.

26   Ibid.

27   *Impact of Competition Policy Reforms on Rural and Regional Australia,* Productivity Commission, 14 October 1999.

28   Marsden Jacob Associates for Department of Land and Water Conservation, op. cit.

29   Ibid. 'We see no valid reason why holdings of water entitlements should not be available on a public register, as are holdings of land through the Titles Office.'

30   National Competition Council, *Compendium of National Competition Policy Agreements,* 2nd ed, June 1998.

31   Tripartite Meeting on Water Reform, National Competition Council, Committee on Regulatory Reform and Representatives of the Australian and New Zealand Conservation Council, 14 January 1999.

32   Fisher, Tim, op. cit.

33   Fitzgerald, Peter, Central Goulburn Water, speech to ANCID Annual Conference, Brisbane, September 2000.

34   High Level Steering Group on Water, op. cit.

35   Boully, Leith, speech to the International Landcare Conference, Melbourne, 2–5 March 2000.

36   Marsden Jacob Associates for Department of Land and Water Conservation, op. cit.

37   Commonwealth Government, *A National Action Plan for Salinity and Water Quality in Australia,* 10 October 2000.

38   Ibid.

39   COAG Communiqué, 25 February 1994.

40   Fisher, Tim, op. cit.

41   Monteith, H.O., 'The Right to Water', *Australian Surveyor,* December 1947.

42   'Memo from Matho', *The Irrigator,* Sunraysia Magazine, vol 1, no. 6, Autumn 1999.

43   *National Program on Irrigation Research and Development,* Action Plan and Schedule, Phase 3, Land and Water Resources Research and Development Corporation, 1999–2002.

44   High Level Steering Group on Water, op. cit.

45   Meyer, Dr Wayne, keynote address to Irrigation Australia 2000 Conference, Melbourne, 23–25 May 2000.

Chapter 8

1   Hill, Ernestine, *Water into Gold*, Robertson & Mullens, Melbourne, 1946.

2   Tolhurst, Michael, 'Song of Running Water', soon to be published.

3   www.snowyhydro.com.au

4   *Every Little Drop: The Story of the Snowy*, ABC-TV, 14 November 2000.

5   Endersbee, Emeritus Professor L.A., 'The First Decade of Engineering for the Snowy Mountains Scheme', Australian Academy of

Technological Sciences and Engineering (AATSE) Academy
Symposium, November 1999.

6   Commonwealth and States Snowy River Committee, *Final Report*, May
1950.

7   Conference on the Snowy Mountains Agreement and the River Murray
Waters Agreement Conference, Canberra, 6 June 1958.

8   Scanlon, John, evidence to the South Australian House of Assembly
Murray River Select Committee, 26 June 2000.

9   Snowy Water Inquiry Final Report, sponsored by the New South Wales
and Victorian Governments, 1998.

10   Ibid.

11   Hill, Senator Robert, media conference speech, 20 January 2000.

12   Minchin, Senator Nick, media release, 6 October 2000.

Chapter 9

1   Queensland Conservation Council,
www.qccqld.org.au/Landclearing.htm

2   Commonwealth Government, *A National Action Plan for Salinity and
Water Quality in Australia*, 10 October 2000.

3   Hill, Senator Robert, speech to the 10th World Water Congress,
Melbourne, 13 March 2000.

4   McClellan, Peter, QC, 'Water Resources: Judicial Dispute Resolution',
address to the Rosenberg Forum, University of California, San
Francisco, 1 November 1997.

5   Scanlon, John, evidence to the South Australian House of Assembly
Murray River Select Committee, 26 June 2000.

6   Scanlon, John, 'Saving the Murray: Problems and Prospects,
Presentation Overview—Governance of the Basin', 16 February 2001.

7   Hill, Senator Robert, speech to the 10th World Water Congress,
op. cit.

8   Scanlon, John, evidence to the Murray River Select Committee,
op. cit.

9   Scanlon, John, 'Saving the Murray', op. cit.

10   Boully, Leith, speech to the International Landcare Conference,
Melbourne, 3 March 2000.

11   Samuel, Graeme, 'Political Panic Mustn't Derail Economic Reform',
The Australian, 22 February 2001.

12   Morris, Graham, *Lateline*, ABC-TV, 9 September 1998.

13   *Mabo vs the State of Queensland* (No 2) (1992), 175 CLR 1.

14   *Wik Peoples vs the State of Queensland and Others* (1996), 121 ALR 129.

15   *Members of the Yorta Yorta Aboriginal Community v State of Victoria*,
8 February 2001, FCA 45.

16   Morgan, Monica, 'A Place for Indigenous People, a Place for Us All',
special supplement, *Habitat*, Australian Conservation Foundation,
October 1996.

17 *Members of the Yorta Yorta Aboriginal Community v State of Victoria 1998*, AILR 33, 3 AILR 91.

18 *Members of the Yorta Yorta Aboriginal Community v State of Victoria*, 8 February 2001, FCA 45.

19 Yu, Sarah, *Report on the Aboriginal Cultural Values of Groundwater in the La Grange Sub-Basin,* Centre for Anthropological Research, University of Western Australia, for the Waters and Rivers Commission of Western Australia, 2nd ed, 2000.

20 Ibid.

Chapter 10

1 Punch, Gary, speech to Agribusiness Water Forum, 4 November 1998.

2 McClellan, Peter, QC, 'Water Resources—Judicial Dispute Resolution', address to the Rosenberg Forum, University of California, San Francisco, 1 November 1997.

3 Murray, Wal, submission to Sir Laurence Street, commissioner of the Whalan Creek Inquiry, 8 January 1994.

4 *Coulton & ors v Holcombe & ors,* Supreme Court (no. 630/1984).

5 McClellan, Peter, QC, op.cit.

6 *Coulton vs Holcombe* (1986), 162 CLR 1 F.C. 86/032.

7 *Coulton & ors v Holcombe & ors* and *Holcombe v Water Administration Ministerial Corporation,* 1990, 20 NSWLR 138.

8 Street, Sir Laurence, *Whalan Creek Inquiry*, New South Wales Premier's Department, 13 March 1996.

9 McClellan, Peter, QC, op. cit.

10 Syme, G.J., and Nancarrow, B.E., *Fairness and its Implementation in the Allocation of Water*, Australian Research Centre for Water in Society, CSIRO Land and Water, 12 March 2000.

11 Murray, Wal, op. cit.

12 Dickie, Phil, 'Sold Down the River', www.onlineopinion.com.au, 13 September 2000.

13 Ibid.

14 *Socio Economic Impact Assessment, Condamine–Balonne WAMP*, a report prepared for the Balonne Community Advancement Committee, Price Waterhouse Coopers, November 2000.

15 www.smartrivers.com.au

16 *Lower Balonne Environmental Condition Report*, Sinclair Knight Merz, October 2000.

17 Queensland Department of Natural Resources licence conditions, sections 1.001, 1.002.

Chapter 11

1 *Ord River Irrigation Area*, Kununurra, Western Australia, Agriculture Western Australia, 6th ed, 1999.

2 *Background to Developing a Revised Interim Allocation Strategy*, draft briefing of the Kununurra reference panel, Water and Rivers Commission, 22 October 2000.

3 Community of Kununurra, *Ord Land and Water Management Plan 2000*, H.G. Gardiner & Associates, Local Action Group, Natural Heritage Trust, AGWEST.

4 *Report of the Scientific Panel on Interim Ecological Water Requirements*, Water and Rivers Commission, June 2000.

5 *Ord River Irrigation Area*, op. cit.

6 Environment Protection Authority, *Report on the Draft Interim Allocation Plan*, December 1999.

7 *Background to Developing a Revised Interim Allocation Strategy*, op. cit.

8 Ward, Ben, on behalf of the Miriuwung and Gajerrong People vs State of Western Australia and ors, 1998, 1478 FCA.

Chapter 12

1 *Landline*, ABC-TV, 13 February 2000.

2 Cook, Alex, and McKenzie, Don, *The Great Artesian Basin*, M'Choinneach Publishers, Ilfracombe, Queensland, 1997.

3 Ibid.

4 Paterson, A.B., 'Song of the Artesian Water'.

5 Land and Water Resources Research Development Corporation, *Dependence of Ecosystems on Groundwater and its Significance to Australia*, Canberra, December 1998.

6 Heathcote, R.L., *The Back of Bourke: A Study of Land Appraisal and Settlement in Semi-arid Australia*, Melbourne University Press, 1865.

7 Blick, Ross, 'Managing Rangelands Better by Managing Artesian Water', in *Saving Our Natural Heritage? The Role of Science in Managing Australia's Ecosystems*, Halstead Press, 1997.

8 Sturt, C., *Two Expeditions into the Interior of Southern Australia During the Years 1828, 1829, 1830 and 1831*, Smith & Elder, 1833.

9 Blick, Ross, op. cit.

10 Smith, Bob, Department of Land and Water Conservation, Agribusiness-KPMG Water Issues Forum presentation, Sydney, 4 November 1998.

11 *Great Australian Basin, Strategic Management Plan*, Agriculture Fisheries and Forestry Australia, September 2000.

12 High Level Steering Group on Water, *Report to COAG on progress in Implementation of the COAG Water Reform Framework*, occasional paper, no 1, 1999.

13 Great Australian Basin Consultative Council, *Great Australian Basin, Strategic Management Plan, Summary*, September 2000.

14 'Uranium, the Facts', Western Mining Corporation pamphlet.

15 Mudd, Dr Gavin M., 'Cases and Solutions: Mound Springs of the Great

Artesian Basin in South Australia: A Case Study from Olympic Dam',
Environmental Geology, Victoria University of Technology, Abstract,
vol 39, issue 5, 2000.

16   Friends of the Earth, 'Native Title Claim Puts Roxby in Fluid Situation',
no. 69, and 'Undermining Aboriginal Interests', nos 73–74, *Chain
Reaction*.

17   Blick, Ross, op. cit.

Chapter 13

1   Queensland National Party of Australia Water Resources
Subcommittee, *Tully, Herbert, Burdekin Rivers, On to the Inland Plains of
North and Central Queensland,* November 1991.

2   Cameron McNamara Pty, *Report for the Premier's Department,
Queensland Government*, 1981.

3   Edwards, Warwick, 'Ode to Ernie Bridge', unpublished.

4   Bridge, Ernie, valedictory speech to WA Parliament, Hansard,
23 November 2000.

5   Australian Academy of Technological Sciences and Engineering and
the Institution of Engineers, Australia, *Water and the Australian
Economy*, Parkville, Victoria, April 1999.

6   Albrecht, Martin, 'Stewardship of Our Water for Responsible Growth',
presented at Australian Academy of Technological Sciences and
Engineering Symposium, 22 November 2000.

7   Alexander, Peter, consultant with Hydro Environmental Pty,
'Benchmarking of Australian Irrigation Water Providers', Australian
National Committee for Irrigation and Drainage, 1999.

8   Albrecht, Martin, op. cit.

9   Ibid.

10   WA Water Corporation, *History of C.Y. O'Connor*,
www.watercorporation.com.au

11   Commonwealth Knowledge Network, *Summary of the Discussion on
Desalination,* 14–30 October and 1–7 November 2000,
www.commonwealthknowledge.net

12   Schonfeldt, Claus, 'Future of Water Resources for South Australia',
address to the Australian Academy of Technological Sciences and
Engineering, Melbourne, 24 September 1999.

13   Hockley, Catherine, 'Wine Giant Comes Clean on Spill', *Adelaide
Advertiser,* 15 December 2000.

14   'Changing Over', *The Irrigator*, Sunraysia magazine, vol 1, no. 6,
Autumn 1999.

15   Clemmens, Bert J., US Department of Agriculture Research Laboratory,
Phoenix, Arizona, commenting in *The Irrigator*, Sunraysia magazine
vol 1, no. 7, Winter 2000.

16   Mills, Stephen, Australian National Committee for Irrigation and
Drainage Annual Conference, 11 September 2000.

17   McVeigh, Paul, 'Multiple Water Use', speech to Australian Bureau of
     Agriculture and Resource Economics (ABARE) Outlook Conference,
     Canberra, 2000.
18   Hogan, L.J., and Dunn, B.W., Proposal for a National Water Grid,
     Water for Australia Pty Ltd, Sydney.

Chapter 14

1   CSIRO Land and Water, *A Revolution in Land Use: Emerging Land Use
    Systems for Managing Dryland Salinity*, CSIRO, 2000.
2   Postel, Sandra, 'Can the Irrigation Miracle Last?', W.W. Norton & Co
    Inc, New York, 1999.
3   Murray–Darling Basin Ministerial Council Draft Basin Salinity
    Management Strategy 2001 to 2015, September 2000.
4   Fisher, Dr Peter, 'Dampened Expectations', *Australian Financial Review*,
    11 August 2000.
5   Strong, Maurice, chairman, United Nations Earth Council, Inaugural
    Jack Beale Lecture on the Global Environment, University of New
    South Wales, delivered 11 February 1999.
6   Interviewed on *Landline*, ABC-TV, 27 February 2000.
7   Harris, Graham, chief executive, CSIRO Land and Water, 'Water,
    Science and Ecology in an Age of Cynicism', keynote address to the
    10th World Water Congress, Melbourne, 13 March 2000.

INDEX